A Primer for Computational Biology

A Primer for Computational Biology

Shawn T. O'Neil

Open Textbook Initiative
Oregon State University Libraries and Press
in partnership with Open Oregon State
Corvallis, Oregon

Oregon State University Libraries and Press
121 The Valley Library
Corvallis, Oregon 97331-4501

ISBN (print) 978-0-87071-926-4
ISBN (ebook) 978-0-87071-936-3

Published in the United States of America

Oregon State University Libraries and Press
Open Oregon State
Center for Genome Research and Biocomputing

Dedication

To the amazing instructors I've been fortunate enough to learn from, and to the amazing students I've been fortunate enough to learn with.

Contents

Preface

It has become almost cliché to state that contemporary life scientists work with a staggering amount and variety of data. This fact makes it no less true: the advent of high-throughput sequencing alone has forced biologists to routinely aggregate multi-gigabyte data sets and compare the results against multi-terabyte databases. The good news is that work of this kind is within the reach of anyone possessing the right computational skills.

The purpose of this book is perhaps best illustrated by a fictional, but not unreasonable, scenario. Suppose I am a life scientist (undergraduate or graduate research assistant, postdoc, or faculty) with limited but basic computational skills. I've identified a recently developed data collection method—perhaps a new sequencing technology—that promises to provide unique insight into my system of study. After considerable field and lab work, the data are returned as a dozen files a few gigabytes in total size.

Knowing this data set is too large for any web-based tool like those hosted by the National Center for Biotechnology Information (NCBI), I head to my local sequencing center, which conveniently hosts a copy of the latest graphical suite for bioinformatics analysis. After clicking through the menus and panes, searching the toolbox window, and looking at the help manual, I come to realize this software suite cannot process this newly generated data. Because the software is governed by an expensive license agreement, I send an email to the company and receive a prompt reply. It seems the development team is working on a feature for the type of analysis I want, but they don't expect it to be ready until next year's release.

After a quick online search, I find that no other commercial software supports my data, either. But I stumble upon a recent paper in a major bioinformatics journal describing not only a novel statistical methodology appropriate for the data, but also software available for download! Sadly, the software is designed for use on the Linux command line, with which I'm not familiar.

Realizing my quandary, I head to the local computing guru in the next lab over and explain the situation. Enthusiastically, she invites me to sit with her and take a look at the data. After uploading the data to the remote machine she regularly works on, she opens a hacker's-style terminal interface, a black background with light gray text occasionally dotted with color. Without even installing the bioinformatics software, she begins giving me an overview of the data in seconds. "Very nice! Looks like you've got about 600 million sequences here . . . pretty good-quality scores, too." After a few more keystrokes, she says, "And it looks like the library prep worked well; about 94% of these begin with the expected sequence bases. The others are probably error, but that's normal."

Still working in the terminal, she proceeds to download and install the software mentioned in the bioinformatics paper. Typing commands and reading outputs that look to be some sort of hybrid language of English and whatever the computer's native language is, she appears to be communicating directly with the machine, having a conversation even. Things like `./configure --prefix=$HOME/local` and `make install` flash upon the screen. Within a few more minutes, the software is ready to use, and she sets it to work on the data files after a quick check of its documentation.

"I'm guessing this will take at least a half hour or so to run. Want to go get some coffee? I could use a break anyway." As we walk to the cafe, I tell her about the commercial software that couldn't process the data. "Oh yeah, those packages are usually behind the times because they have so many features to cover and the technology advances so quickly. I do use them for routine things, but even then they don't always publish their methods, so it's difficult to understand exactly what's going on."

"But aren't the graphical packages easier to use?" I ask. "Sure," she replies, "sometimes. They're not as flexible as they look. I've written graphical versions of my own software before, but it's time consuming and more difficult to update later. Besides, it's easier for me to write down what commands I ran to get an answer in my lab notebook, which is digital anyway these days, rather than grabbing endless screenshots of a graphical interface."

When we get back to her office, she opens the results file, which shows up in the same gray-on-black typewriter font in nicely formatted rows and columns of numbers and identifiers. I could easily imagine importing the results into a spreadsheet, though she mentions there are about 6.1 million rows of output data.

"Well, here it is! The *p* values in this last column will tell you which results are the most important," she says as she sorts the file on that column (in mere seconds) to reveal the top few records with the lowest *p* values. Recalling that the significant results should theoretically correlate to the GC content of the sequences in certain positions, I ask if it's possible to test for that. "Yes, it's definitely possible," she answers. "Well, extracting just the most significant sequences will be easy given this table. But then I'll have to write a short program, probably in Python, which I just started learning, to compute the aggregate GC content of those sequences on a position-by-position basis. From there it won't be hard to feed the results into an R script to test for differences in each group compared to all the others. It should only take a few hours, but I'm a bit busy this week. I'll see what I can do by next Friday, but you'll owe me more than just coffee!"

A Few Goals

Bioinformatics and computational biology sit at the intersection of a variety of disciplines, including biology, computer science, mathematics, and statistics. Whereas bioinformatics is usually viewed as the development of novel analysis methods and software, computational biology focuses on applying those methods to data of scientific interest. Proficiency in both requires an understanding of the language of computing. This language is more like a collection of languages or dialects—of basic commands, analysis tools, Python, R, and so on.

It may seem odd that so much of modern computational research is carried out on the comparatively ancient platform of a text-based interface. Graphical utilities have their place, particularly for data visualization, though even graphics are often best described in code. If we are to most efficiently communicate with computational machinery, we need to share with the machinery a language, or languages. We can share powerful dialects, complete with grammar, syntax, and even analogs for things like nouns (data) and verbs (commands and functions).

This book aims to teach these basics of scientific computing: skills that even in fields such as computer science are often gained informally over a long period of time. This book is intended for readers who have passing familiarity with computing (for example, I assume the reader is familiar with concepts such as files and folders). While these concepts will likely be useful to researchers in many fields, I frame most of the discussion and examples in the analysis of biological data, and thus assume some basic biological knowledge, including concepts such as genes, genomes, and proteins. This book covers topics such as the usage of the command-line interface, installing and running bioinformatics software (without access to administrator privileges on the machine), basic analysis of data using built-in system tools, visualization of results, and introductory programming techniques in languages commonly used for bioinformatics.

There are two related topics that are not covered in this book. First, I avoid topics related to "system administration," which involves installing and managing operating systems and computer hardware, except where necessary. Second, I focus on computing for bioinformatics or computational biology, rather than bioinformatics itself. Thus this book largely avoids discussing the detailed mathematical models and algorithms underlying the software we will install and use. This is not to say that a good scientist can avoid mathematics, statistics, and algorithms—these are simply not the primary focus here.

Bioinformatics and computational biology are quickly growing and highly interdisciplinary fields, bringing computational experts and biologists into close and frequent contact. To be successful, these collaborations require a shared vocabulary and understanding of diverse skill sets; some of this understanding and vocabulary are discussed here. Although most of this book focuses on the nuts and bolts of data analysis, some chapters focus more on topics specifically related to computer science and programming, giving newcomers a chance to understand and communicate with their computational colleagues as well as forming a basis for more advanced study in bioinformatics.

Organization

This book is divided into three parts, the first covering the Unix/Linux command-line environment, the second introducing programming with Python, and the third introducing programming in R. Though there are some dependencies between parts (for example, chapter 21, "Bioinformatics Knick-knacks and Regular Expressions," forgoes duplication of topics from chapter 11, "Patterns (Regular Expressions)"), readers sufficiently interested in only Python or only R should be able to start at those points. Nevertheless, the parts are given their order for a reason: command-line efficiency introduces "computational" thinking and problem solving, Python is a general-purpose language that emphasizes good coding practice, and R specializes in certain types of analyses but is trickier to work with. Understanding these three general topics constitutes a solid basis for computational biologists in the coming decade or more.

The text within each part follows, more or less, a Shakespearean plot, with the apex occurring somewhere in the middle (thus it is best to follow chapters within parts in order). For Part I, this apex occurs in chapter 6, "Installing (Bioinformatics) Software," wherein we learn to both install and use some bioinformatics software, as well as collect the procedure into a reusable pipeline. In Part II, chapter 23, "Objects and Classes," describes an involved custom analysis of a file in variant call format (VCF) using some principles in object-oriented design. Finally, the apex in Part III occurs in chapter 33, "Split, Apply, Combine," which describes some powerful data processing techniques applied to a multifactor gene expression analysis. Following each apex are additional, more advanced topics that shouldn't be overlooked. The second half of Part I covers the powerful paradigm of data pipelines and a variety of important command-line analysis tools such as awk and sed. Part II covers some topics related to software packages and concludes with an introduction to algorithms and data structures. Finally, Part III follows its apex with handy functions for manipulating and plotting complex data sets.

Finally, the text includes an extensive number of examples. To get a proper feel for the concepts, it is highly recommended that you execute the commands and write the code for yourself, experimenting and trying variations as you feel necessary. It is difficult for even the sharpest of minds to absorb material of this nature by reading alone.

Availability

This book is available both as an open-access online resource as well as in print. The open-access license used for the online version is the Creative Commons CC BY-NC-SA, or "Attribution-NonCommercial-ShareAlike" license. According to https://creativecommons.org/licenses/, "This license lets others remix, tweak, and build upon [the] work non-commercially, as long as they credit [the author] and license their new creations under the identical terms."

The data files and many of the completed scripts mentioned within the text are available for direct download here: http://library.open.oregonstate.edu/computationalbiology/back-matter/files/.

For comments or to report errors, please feel free to contact oneilsh@gmail.com. Should any errata be needed post-publication, they will appear in this preface of the online version.

FAQ

Your example for [topic] isn't the standard way to do it. Why not?

Many of the technologies covered in this book (and the communities that use them) are highly "opinionated." Examples for the command-line and R include the "useless use of `cat` award" on the command-line, folk wisdom surrounding loops in R, and conventions for `ggplot2` syntax. This is particularly true of Python, where certain syntax is said to be more "Pythonic" than other syntax, and some syntax is simply not shared between different versions of the language.

There are a few reasons for breaks with convention. The first is that I'm always learning myself, and I'm not aware of all aspects of these technologies. More deliberately, I've attempted to create a resource that is consistent and provides a base for future learning. Ideally a student familiar with the strongly object-oriented nature of Python could map concepts to similar languages like Java or C++, and a student familiar with R's functional aspects could explore other functional languages such as JavaScript. Given the complex histories of the command-line, R, and Python, as well as their predilections for language-specific syntactic sugar, this goal is sometimes at odds with standard practice.

The code blocks appear to be rasterized images. Why?

At the outset of writing this book, I knew we would be targeting multiple formats: print on demand, web, ebook, and so on. Additionally, during the editing process it spent some time as a Microsoft Word document. As code is very sensitive to formatting changes, this ensured that every version would be identical and correct, and it allowed me to easily overlay some blocks with visual annotations. Although I have scripts that automate the production of these images, it does have the downside of making adjustments more cumbersome.

An additional feature–which some readers might consider a bug–is that it is not possible to copy-and-paste these code blocks from electronic versions of the book. In my experience copying and pasting code actively hinders the learning process, as it encourages skipping details of concepts and prevents learning from mistakes.

Acknowledgements

I'd like to thank the Oregon State University Press, the OSU Library, OSU Ecampus, and the Center for Genome Research and Biocomputing at OSU for their support in the production of this work–in particular, Brett Tyler, Tom Booth, Faye Chadwell, Dianna Fisher, and Mark Kindred from those departments. I am grateful to all those who contributed feedback including Matthew Peterson, Kevin Weitemier, Andi Stephens, Kelly Stratton, Jason Williams, John Gamble, Kasim Alomari, Joshua Petitmermet, and Katie Carter. Finally, I especially thank Gabriel Higginbotham for his heroic work in typesetting, and to Stacey Wagner for her comprehensive and insightful comments.

Part I: Introduction to Unix/Linux

Chapter 1

Context

Command Lines and Operating Systems

Many operating systems, including Microsoft Windows and Mac OS X, include a command line interface (CLI) as well as the standard graphical user interface (GUI). In this book, we are interested mostly in command line interfaces included as part of an operating system derived from the historically natural environment for scientific computing, Unix, including the various Linux distributions (e.g., Ubuntu Linux and Red Hat Linux), BSD Unix, and Mac OS X.

Even so, an understanding of modern computer operating systems and how they interact with the hardware and other software is useful. An operating system is loosely taken to be the set of software that manages and allocates the underlying hardware—divvying up the amount of time each user or program may use on the central processing unit (CPU), for example, or saving one user's secret files on the hard drive and protecting them from access by other users. When a user starts a program, that program is "owned" by the user in question. If a program wishes to interact with the hardware in any way (e.g., to read a file or display an image to the screen), it must funnel that request through the operating system, which will usually handle those requests such that no one program may monopolize the operating system's attention or the hardware.

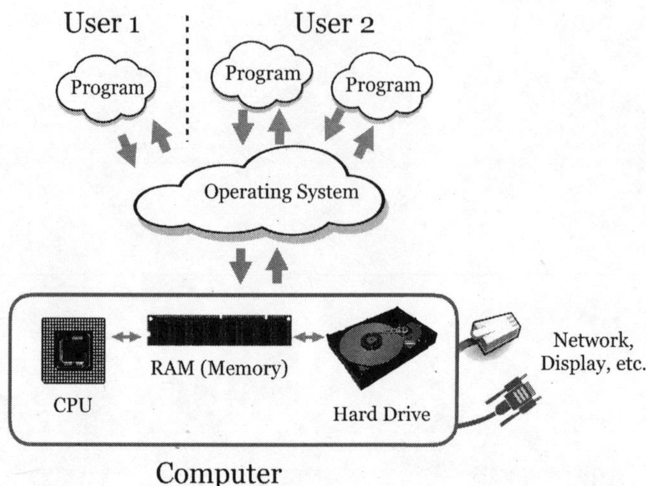

The figure above illustrates the four main “consumable” resources available to modern computers:

1. The CPU. Some computers have multiple CPUs, and some CPUs have multiple processing “cores.” Generally, if there are *n* total cores and *k* programs running, then each program may access up to *n*/*k* processing power per unit time. The exception is when there are many processes (say, a few thousand); in this case, the operating system must spend a considerable amount of time just switching between the various programs, effectively reducing the amount of processing power available to all processes.
2. Hard drives or other “persistent storage.” Such drives can store ample amounts of data, but access is quite slow compared to the speed at which the CPU runs. Persistent storage is commonly made available through remote drives “mapped in” over the network, making access even slower (but perhaps providing much more space).
3. RAM, or random access memory. Because hard drives are so slow, all data must be copied into the “working memory” RAM to be accessed by the CPU. RAM is much faster but also much more expensive (and hence usually provides less total storage). When RAM is filled up, many operating systems will resort to trying to use the hard drive as though it were RAM (known as “swapping” because data are constantly being swapped into and out of RAM). Because of the difference in speed, it may appear to the user as though the computer has crashed, when in reality it is merely working at a glacial pace.
4. The network connection, which provides access to the outside world. If multiple programs

wish to access the network, they must share time on the connection, much like for the CPU.

Because the software interfaces we use every day—those that show us our desktop icons and allow us to start other programs—are so omnipresent, we often think of them as part of the operating system. Technically, however, these are programs that are run by the user (usually automatically at login or startup) and must make requests of the operating system, just like any other program. Operating systems such as Microsoft Windows and Mac OS X are in reality operating systems bundled with extensive suites of user software.

A Brief History

The complete history of the operating systems used by computational researchers is long and complex, but a brief summary and explanation of several commonly used terms and acronyms such as BSD, "open source," and GNU may be of interest. (Impatient readers may at this point skip ahead, though some concepts in this subsection may aid in understanding the relationship between computer hardware and software.)

Foundational research into how the physical components that make up computing machinery should interact with users through software was performed as early as the 1950s and 1960s. In these decades, computers were rare, room-sized machines and were shared by large numbers of people. In the mid-1960s, researchers at Bell Labs (then owned by AT&T), the Massachusetts Institute of Technology, and General Electric developed a novel operating system known as Multics, short for Multiplexed Information and Computing Service. Multics introduced a number of important concepts, including advances in how files are organized and how resources are allocated to multiple users.

In the early 1970s, several engineers at Bell Labs were unhappy with the size and complexity of Multics, and they decided to reproduce most of the functionality in a slimmed-down version they called UNICS—this time short for Uniplexed Information and Computing Service—a play on the Multics name but not denoting a major difference in structure. As work progressed, the operating system was renamed Unix. Further developments allowed the software to be easily translated (or ported) for use on computer hardware of different types. These early versions of Multics and Unix also pioneered the automatic and simultaneous sharing of hardware resources (such as CPU time) between users, as well as protected files belonging to one user from others—important features when many researchers must share a single machine. (These same features allow us to multitask on modern desktop computers.)

During this time, AT&T and its subsidiary Bell Labs were prohibited by antitrust legislation from commercializing any projects not directly related to telephony. As such, the researchers licensed, free of cost, copies of the Unix software to any interested parties. The combination of a robust technology, easy portability, and free cost ensured that there were a large number of interested users, particularly in academia. Before long, many applications were written to operate on top of the Unix framework (many of which we'll use in this book), representing a powerful computing environment even before the 1980s.

In the early 1980s, the antitrust lawsuit against AT&T was settled, and AT&T was free to commercialize Unix, which they did with what we can only presume was enthusiasm. Unsurprisingly, the new terms and costs were not favorable for the largely academic and research-focused user base of Unix, causing great concern for many so heavily invested in the technology.

Fortunately, a group of researchers at the University of California (UC), Berkeley, had been working on their own research with Unix for some time, slowly reengineering it from the inside out. By the end of AT&T's antitrust suit, they had produced a project that looked and worked like AT&T's Unix: BSD (for Berkeley Systems Distribution) Unix. BSD Unix was released under a new software license known as the BSD license: anyone was free to copy the software free of charge, use it, modify it, and redistribute it, so long as anything redistributed was also released under the same BSD license and credit was given to UC Berkeley (this last clause was later dropped). Modern versions of BSD Unix, while not used heavily in academia, are regarded as robust and secure operating systems, though they consequently often lack cutting-edge or experimental features.

In the same year that AT&T sought to commercialize Unix, computer scientist Richard Stallmann responded by founding the nonprofit Free Software Foundation (FSF), which was dedicated to the idea that software should be free of ownership, and that users should be free to use, copy, modify, and redistribute it. He also initiated the GNU operating system project, with the goal of re-creating the Unix environment under a license similar to that of BSD Unix. (GNU stands for GNU's Not Unix: a recursive, self-referencing acronym exemplifying the peculiar humor of computer scientists.)

The GNU project implemented a licensing scheme that differed somewhat from the BSD license. GNU software was to be licensed under terms created specifically for the project, called the GPL, or GNU Public License. The GPL allows anyone to use the software in any way they see fit (including distributing for free or selling any program built using it), provided they also make available the human-readable code that they've created and license it under the GPL as well (the essence of "open source"[1]). It's as if the Ford Motor Company gave away the blueprints for a new car, with the requirement that any car designed using those blueprints also come with its own blueprints and similar rules. For this reason, software written under the GPL has a natural tendency to spread and grow. Ironically and ingeniously, Richard Stallmann and the BSD group used the licensing system, generally intended to protect the spread of intellectual property and causing the Unix crisis of the 1980s, to ensure the perpetual freedom of their work (and with it, the Unix legacy).

While Stallmann and the FSF managed to re-create most of the software that made up the standard Unix environment (the bundled software), they did not immediately re-create the core of the operating system (also called the kernel). In 1991, computer science student Linus Torvalds began work on this core GPL-licensed component, which he named Linux (pronounced "lin-ucks," as prescribed by the author himself). Many other developers quickly contributed to the project, and now Linux is available in a variety of "distributions," such as Ubuntu Linux and Red Hat Linux, including both the Linux kernel and a collection of Unix-compatible GPL (and occasionally non-GPL) software. Linux distributions differ primarily in what software packages come bundled with the kernel and how these packages are installed and managed.

Today, a significant number of software projects are issued under the GPL, BSD, or similar "open" licenses. These include both the Python and R projects, as well as most of the other pieces of software covered in this book. In fact, the idea has caught on for noncode projects as well, with many documents (including this one) published under open licenses like Creative Commons, which allow others to use materials free of charge, provided certain provisions are followed.

1. Modern software is initially written using human-readable "source code," then compiled into machine-readable software. Given source code, it is easy to produce software, but the reverse is not necessarily true. The distinctions between the BSD and GPL licenses are thus significant.

Chapter 2

Logging In

This book assumes that you have access to an account on a Unix-like operating system (such as Linux) that you can use directly or log in to remotely via the SSH (Secure-SHell) login protocol. Accounts of this type are frequently available at universities and research institutions, though you may also consider using or installing Linux on your own hardware. Additionally, the CyVerse Collaborative provides free command-line access to biological researchers through their Atmosphere system; see the end of this chapter for information on how to access this system using your web browser.

Before explaining anything in detail, let's< cover the actual process of logging in to a remote computer via SSH. To do this, you will need four things:

1. Client software on your own computer.
2. The address of the remote computer, called its "host name," or, alternatively, its IP (Internet protocol) address.
3. A username on the remote computer with which to log in.
4. A corresponding password on the remote machine with which to log in.

If you are using a computer running Mac OS X, the client software will be command line oriented and is accessible from the Terminal utility. The Terminal is located in the Utilities folder, inside of the Applications folder.[2]

2. The style of your terminal session likely won't look like what we show in this book, but you can customize the look of your terminal using preferences. We'll be spending a lot of time in this environment!

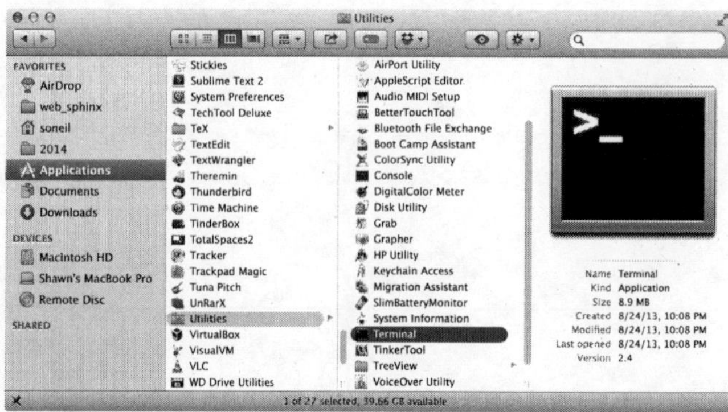

In the window that opens up, you will see a prompt for entering commands. On my computer it looks like this:

```
[soneil@mbp ~]$
```

At this prompt, enter the following: `ssh <username>@<hostname or ip>`. Note that you don't actually type the angle brackets; angle brackets are just some commonly used nomenclature to indicate a field you need to specify. To log in to my account with username `oneils` at the Oregon State University's main Linux computer (`shell.onid.oregonstate.edu`), for example, I would type:

```
[soneil@mbp ~]$ ssh oneils@shell.onid.oregonstate.edu
```

To log in to a CyVerse instance with IP address `128.196.64.193` (and username `oneils`), however, I would use:

```
[soneil@mbp ~]$ ssh oneils@128.196.64.193
```

After pressing Enter to run the command, you may be asked to "verify the key fingerprint" of the remote computer. Each computer running the SSH login system uses a unique pair of "keys" for cryptographic purposes: the public key is accessible to the world, and only the remote computer knows the private key. Messages encrypted with the public key can only be decrypted with the private key—if you cared to verify with the owner of the remote computer that the public key "fingerprint" was correct, then you could be assured that no one between you and the correct remote computer could see your login session. Unless you have a reason to suspect espionage of some sort, it's usually safe to enter `yes` at this prompt. Normally, you will be prompted once for each fingerprint, unless your local computer forgets the fingerprint or the system administrator changes it (or there is indeed an espionage attempt!).

In any case, you will next be asked to enter your password. Note that *as you type the password, you won't see any characters being shown on the screen.* This is another security feature, so that no passersby can see the contents or even the length of your password. It does make password entry more difficult, however, so take care when entering it. After logging in to a remote computer, the command prompt will usually change to reflect the login:

```
oneils@atmosphere ~$
```

If you are running Microsoft Windows, you will need to download the client SSH software from the web, install it, and run it. One option is PuTTy.exe, which is available at http://www.chiark.greenend.org.uk/~sgtatham/putty/download.html.

Once downloaded, it will need to be installed and run. In the `Host Name (or IP address)` field, enter the host name or IP address, as discussed above. Leave the port number to the default of `22`, and click `Open`.

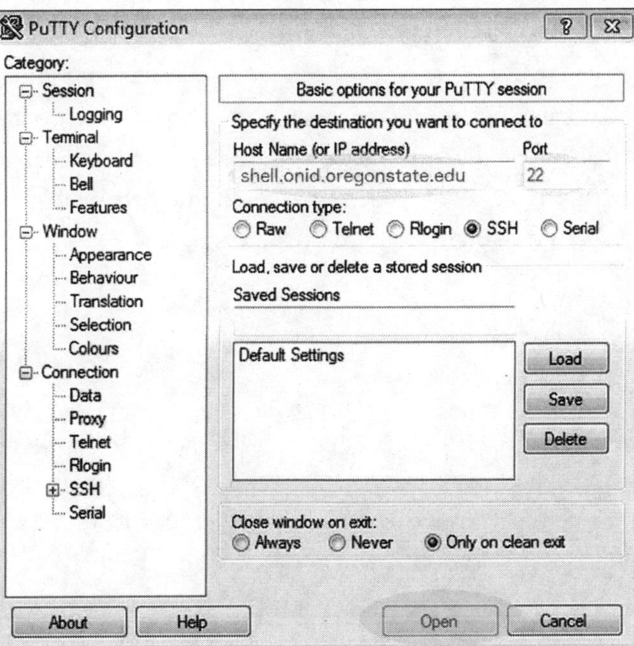

If you have successfully entered the host name or IP, you will be prompted for your username and password (and potentially to verify the key fingerprint).

Finally, if you are already running a Linux or other Unix-like operating system, the steps for logging in remotely via SSH will be similar to those for Mac OS X, though you'll need to find the Terminal utility on your own. If you are using CyVerse Atmosphere, then you can utilize a terminal window right in the web browser by clicking on the Access By Shell tab or Open Web Shell link. In all cases, the text-based interface will be identical, because the remote computer, rather than your local desktop, determines the display.

Logging in Further, Changing Your Password

Depending on where you are logging in to, you may not be done with the login process. At many universities and research centers, the administrator would prefer that you not do any work on the computer you initially logged in to, because that computer may be reserved just for initial logins by dozens or even hundreds of researchers. As a result, you may need to "check out" a secondary, internal computer for your actual computational needs. Sometimes this can be done by SSHing on the command line to the secondary computer, but at some institutions, you will be asked to run other commands for this checkout process. Check with your local system administrator.

In addition, you may want to change your password after your initial login. Running the `passwd` command usually suffices, and you will be prompted to enter both your old and new passwords. As usual, for security reasons, no characters will appear when entered. Further, good system administrators never ask for your password, and they are unable to recover it if it gets lost. The best administrators can do is reset it to a temporary one.

SSH: Secure Shell

One might rightfully ask: what did we just accomplish with all of this logging in? On our desktop computer, we used a program called a client to connect to another program, called a server. A *server* is a program that waits in the background for another program (a client) to connect to it.[3] This connection often happens over a network, but connections can occur between programs on the

3. Sometimes server programs are called "daemons," terminology that evokes Maxwell's infamous "demon," an impossible theoretical entity working in the background to sort gaseous molecules.

same computer as well. A *client* is a program that is generally run by a user on an as-needed basis, and it connects to a server. While it's more correct to define a server as a program that waits for a connection from a client, colloquially, computers that primarily run server programs are also referred to as servers.

The SSH server and the SSH client communicate using what is known as the SSH "protocol," simply an agreed-upon format for data transfer. The SSH protocol is quite lightweight: because all of the actual computation happens on the remote machine, the only information that needs to be transferred to the server are the keystrokes typed by the user, and the only information that needs to be transferred to the client are the characters to be displayed to the user. As its name implies, the SSH protocol is very secure owing to its reliance on advanced public-key cryptography.[4]

The SSH server may not be the only server program running on the remote computer. For example, web servers allow remote computers to serve web pages to clients (like Mozilla Firefox and OS X's Safari) using HTTP (hypertext transfer protocol). But because there is only one host name or IP address associated with the remote computer, an extra bit (byte, actually) of information is required, known as the "port number." By way of analogy, if the remote computer were an apartment building, port numbers would be apartment numbers. By convention, SSH connects on port 22 and HTTP connects on port 80, although these ports can be changed by administrators who wish to run their services in nonstandard ways. This explains why port 22 is specified when connecting via Putty (and is the default when using command-line `ssh`, but it can be adjusted with a parameter).

4. The public-key infrastructure currently in use by SSH is only secure as far as anyone in the academic sphere suspects: the mathematics underlying the key exchange protocol haven't yet been proven unbreakable. Most mathematicians, however, suspect that they are unbreakable. On the other hand, bugs have been known to occur in the software itself, though they are usually fixed promptly when found.

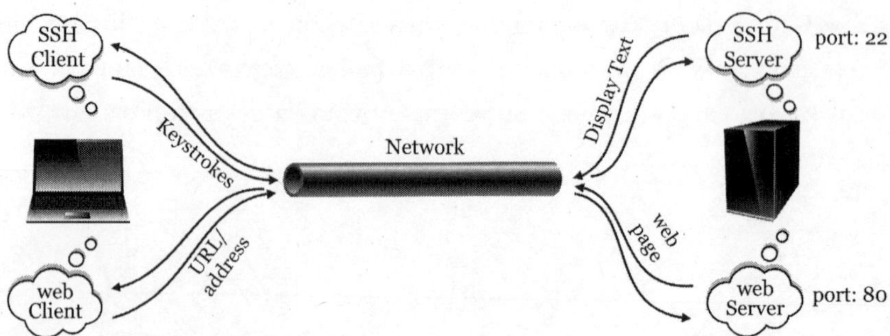

Other protocols of note include FTP (file transfer protocol) and its secure version, SFTP (secure file transfer protocol), designed specifically for transferring files.

Command-Line Access with CyVerse Atmosphere

Readers of this book will ideally have access to a Unix-based operating system (e.g., Linux) with command-line access. This is often the case for individuals at universities and other research or educational institutions. Many users have access to such systems but don't even realize it, as the service is not often widely advertised.

For those without institutional access, there are a few alternatives. First, Mac OS X machines are themselves Unix-based and come with a command-line interface (via the Terminal application), though the command-line tools differ vary slightly from the GNU-based tools found on most Linux distributions. A web search for "OS-X GNU core utils" will turn up some instructions for remedying this discrepancy. Second, it is possible to install a Linux distribution (like Ubuntu Linux) on most desktops and laptops by configuring the computer to "dual boot" Linux and the primary operating system. Alternatively, Linux distributions can be installed within "virtual machine" software, like VirtualBox (http://virtualbox.org).

A rather exciting, relatively recent addition to these options is the Atmosphere system, run by the CyVerse (previously iPlant) collaborative, a cyber-infrastructure project founded in 2008 to support the computational needs of researchers in the life sciences. CyVerse as a whole hosts a variety of projects, from educational resources to guided bioinformatics analyses. The Atmosphere system is the most relevant for us here, as it provides cloud-based access to systems running Linux. To get started using one of these systems, navigate to http://cyverse.org/atmosphere and click on the link for "Create an Account" or "Launch Atmosphere." If you need to create a new account, do so—CyVerse requests a variety of information for account creation to help gauge user interest and garner funding support. You may also need to request special access to the Atmosphere system through your account on the Atmosphere homepage, as Atmosphere is more resource intensive than other tools provided by CyVerse.

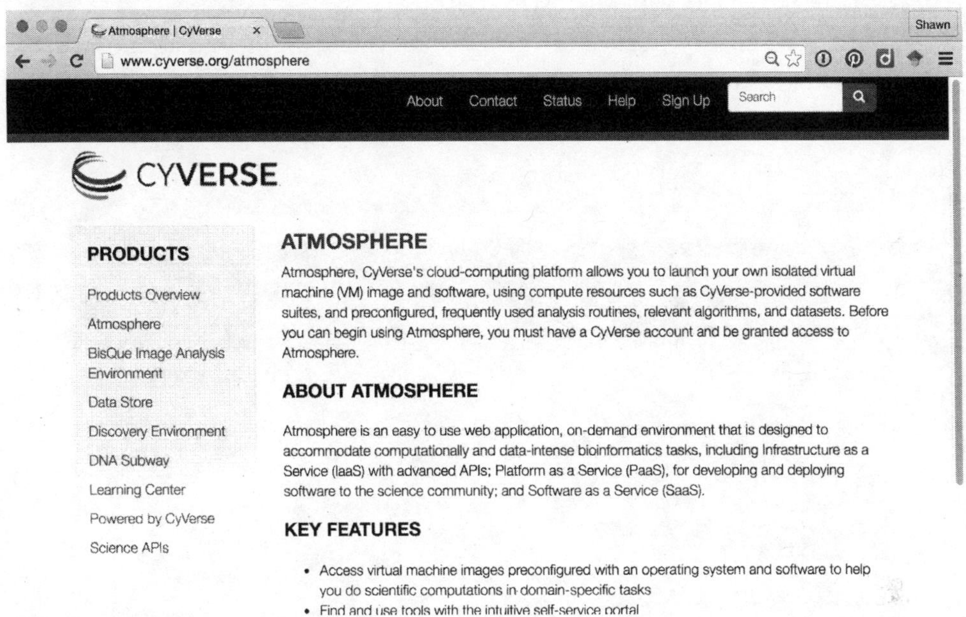

After clicking on the "Launch" link, you will have the opportunity to enter your username and password (if you have been granted one).

The Atmosphere system works by divvying up clusters of physical computers (located at one of various providers around the country) into user-accessible virtual machines of various sizes. When performing a computation that requires many CPU cores, for example, one might wish to access a new "instance" with 16 CPUs, 64 gigabytes (GB) of RAM, and 800 GB of hard disk space. On the other hand, for learning purposes, you will likely only need a small instance with 1 CPU and 4 GB of RAM. This is an important consideration, as CyVerse limits users to a certain quota of resources. Users are limited by the number of "atmosphere units" (AUs) they can use per month, defined roughly as using a single CPU for an hour. Users are also limited in the total number of CPUs and total amount of RAM they can use simultaneously.

After determining the instance size needed, one needs to determine which operating system "image" should be loaded onto the virtual machine. All users can create such images—some users create images with software preinstalled to analyze RNA sequencing data, perform de novo genome assembly, and so on. We've created an image specifically to accompany this book: it is fairly simple and includes NCBI Blast+ (the most modern version of BLAST produced by the National Center for Biotechnology Information), R, Python, `git`, and a few other tools. It is called "APCB Image."

To activate a new instance with this image, click on the "New -> Instance" button in the Atmosphere interface. You may first need to create a "Project" for the instance to live in. You can search for "APCB Image" in the search box of instance types. Here's the view of my APCB project after creating and starting the instance:

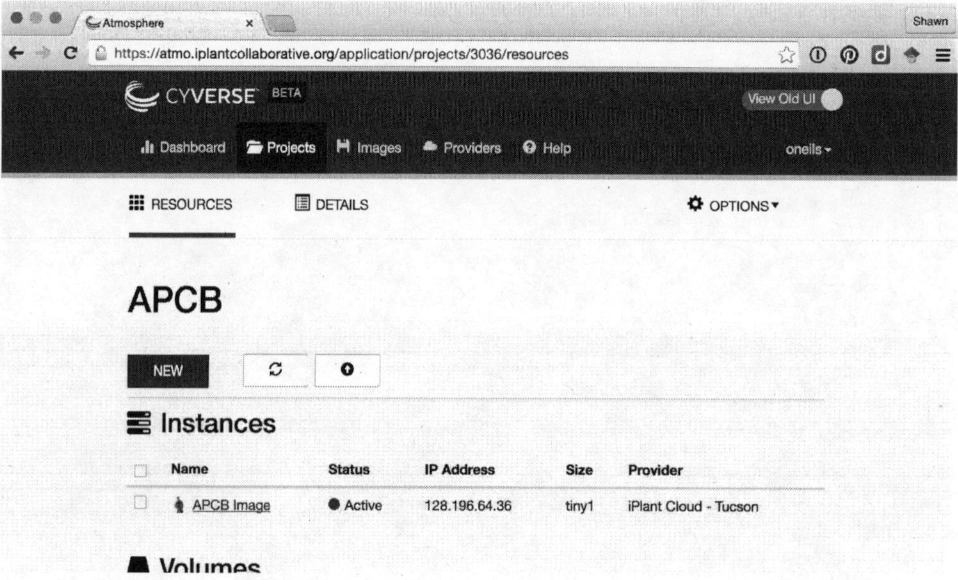

After creating the instance, it may be in one of several states; usually it will be either "running" (i.e., available for login and consuming resources) or "suspended" (effectively paused and not consuming resources). The interface for a given instance has buttons for suspending or resuming a suspended instance, as well as buttons for "Stop" (shutting an instance down), "Reboot" (rebooting the instance), and "Delete" (removing the instance and all data stored in it).

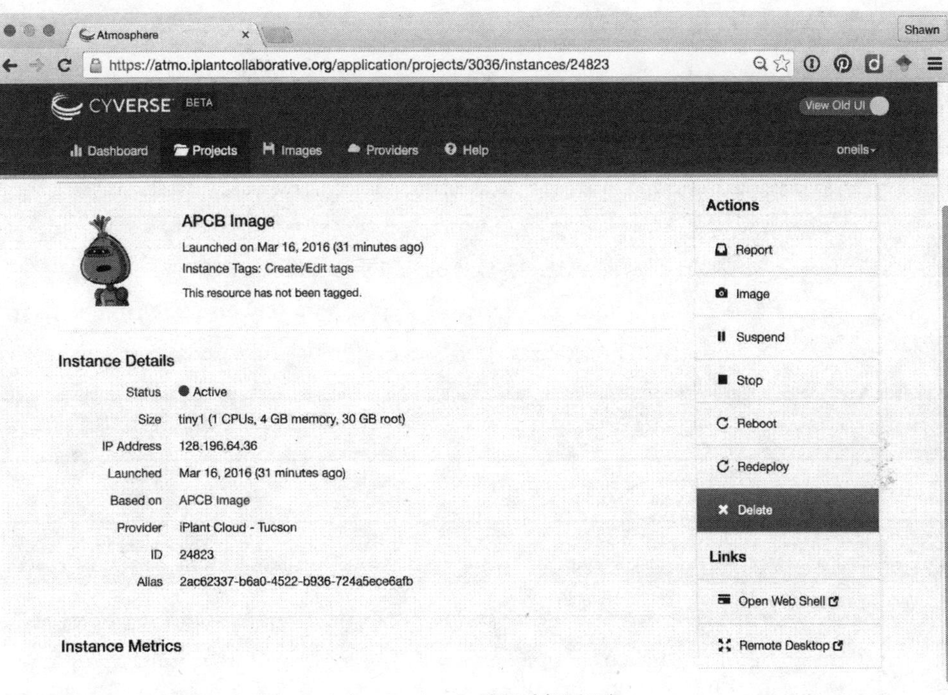

Once an instance is up and running, there are several ways to access it. First, it is accessible via SSH at the IP address provided. Note that this IP address is likely to change each time the instance is resumed. Above, the IP address is shown as `128.196.64.36`, so we could access it from the OS X Terminal application:

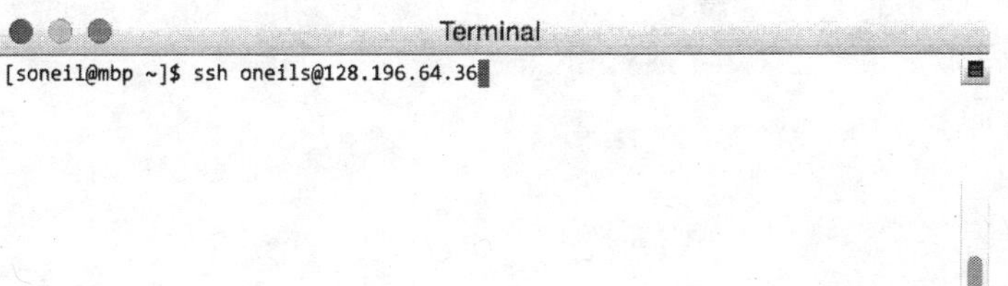

The Atmosphere Web interface also provides an "Open Web Shell" button, providing command-line access right in your browser. When you are done working with your Atmosphere instance, it's important to suspend the instance, otherwise you'll be wasting computational resources that others could be using (as well as your own quota).

Logging Out

Once finished working with the remote computer, we should log out of the SSH session. Logging out is accomplished by running the command `exit` on the command line until returned to the local desktop or the SSH client program closes the connection. Alternatively, it suffices to close the SSH client program or window—SSH will close the connection, and no harm will be done. Note, however, than any currently executing program will be killed on logout.

If you are working on an Atmosphere instance or similar remotely hosted virtual machine, it's a good idea to also suspend the instance so that time spent not working isn't counted against your usage limit.

Exercises

1. Practice logging in to and back out of the remote machine to which you have access.
2. Change your password to something secure but also easy to remember. Most Linux/Unix systems do not limit the length of a password, and longer passwords made up of collections of simple words are more secure than short strings of random letters. For example, a password like `correcthorsebatterystaple` is much more secure than `Tr0ub4dor&3`.[5]
3. A program called `telnet` allows us to connect to any server on any port and attempt to communicate with it (which requires that we know the correct messages to send to the server for the protocol). Try connecting with `telnet` to port `80` of `google.com` by using the "Telnet" radio button in PuTTY if on Windows, or by running `telnet google.com 80` in the Terminal on OS X. Issue the command `GET http://www.google.com/` to see the raw data returned by the server.

5. These example passwords were drawn from a webcomic on the topic, located at http://xkcd.com/936/.

Chapter 3

The Command Line and Filesystem

Computer users are used to interacting with a "user interface." On many computers, this interface displays the desktop or task bar, icons, file previews, and so on. It takes input from the user in the form of keystrokes, mouse movements, and in some cases voice commands, and presents the results of the user's actions. Perhaps most importantly, the user interface is itself a *program* (it is software running on a computer, after all) we interact with to *execute other programs.*

The same thing happens when we use SSH to log in to a remote machine, or open up the Terminal application on a Linux or OS X desktop. In this case, however, instead of interacting with a GUI (Graphical User Interface), we interact with a CLI (Command-Line Interface), or *shell*, which does the job of displaying the command prompt. The shell is the software we interact with on the command line. In some sense it *is* the command line, as it displays the command prompt, accepts input via typed text, runs other programs on our behalf, and displays the results textually. A *command prompt* is a line of status information provided in a text-based interface, indicating that commands are to be entered and run by pressing Enter. Command prompts often include information about what computer or network one is logged in to, the username one is logged in with, and an indication of the "present working directory" (discussed below).

The first command that we'll learn for the Linux command line is `echo`, which prints the parameters we give it.

```
oneils@atmosphere ~$ echo hello there
hello there
```

Let's break down the command prompt and the program that we ran, which consisted of a program name and several parameters, separated by spaces. In the figure below, the command prompt consists of `oneils@atmosphere ~$`.

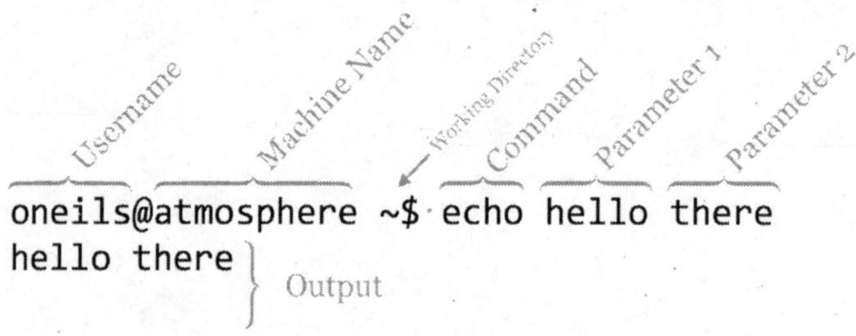

The `echo` program might seem absurdly simple, as it just prints its parameters. But it is quite useful in practice and as a learning tool. For example, we can use `echo` to print not only simple strings, but also the contents of an *environment variable*, which is a variable bit of information (usually holding strings of text) that is accessible by the shell and other programs the user runs. Accessing the contents of an environment variable requires prefixing it with a `$`.

The shell (and other programs) commonly uses environment variables to store information about your login session, much like how, in a GUI interface, a "variable" remembers the wallpaper picture for the desktop. Environment variables control many aspects of the command-line environment, and so they are quite important. Many of these are set automatically when we log in. For example, `$USER`.

```
oneils@atmosphere ~$ echo $USER
oneils
```

Setting Environment Variables, Dealing with Spaces

Setting environment variables is something we'll have to know how to do eventually, but learning now will give us an opportunity to discuss some of the finer points of how the shell interprets our commands. In `bash`, the most commonly used shell, setting environment variables is done with the `export` command, taking as the first parameter what the variable name should be (without the `$`) and what it should be set to.

```
oneils@atmosphere ~$ export GREETING=hello
oneils@atmosphere ~$ echo $GREETING
hello
```

Because `export` expects as its first parameter the variable description, we'll get an odd result if we include a space in our greeting, as the shell uses spaces to separate parameters.

```
oneils@atmosphere ~$ export GREETING=hello everyone
oneils@atmosphere ~$ echo $GREETING
hello
```

In the above, `GREETING=hello` was taken to be first parameter, and the second, `everyone`, was ignored by the `export` command. There are at least two ways to get around this problem. The first is to prefix any character that the shell deems special (like spaces) with a backslash, or \, thereby "escaping" it so that it is ignored by the shell as a special character. Alternatively, we can wrap a string in quotation marks to have it treated literally by the shell.

```
oneils@atmosphere ~$ export GREETING=hello\ everyone
oneils@atmosphere ~$ echo $GREETING
hello everyone
oneils@atmosphere ~$ export GREETING='hello everyone'
oneils@atmosphere ~$ echo $GREETING
hello everyone
```

The primary difference between using single and double quotation marks is whether variables inside the string are expanded to their contents or not.

```
oneils@atmosphere ~$ export GREETING='hello $USER'
oneils@atmosphere ~$ echo $GREETING
hello $USER
oneils@atmosphere ~$ export GREETING="hello $USER"
oneils@atmosphere ~$ echo $GREETING
hello oneils
```

Note that when setting an environment variable, we do not use the `$`. By convention, environment variable names contain only capital letters.[6] Further, this expansion (from environment variables to their contents) is done by the shell; the command itself is changed from `export GREETING="hello $USER"` to `export GREETING="hello oneils"`.

Alternative Shells

There is a special environment variable, `$0`, that generally holds the name of the currently running program. In the case of our interaction with the command line, this would be the name of the interface program itself, or shell.

```
oneils@atmosphere ~$ echo $0
-bash
```

The above command illustrates that we are running `bash`, the most commonly used shell.[7]

Depending on the system you are logged in to, running `echo $0` may not report `bash`. The reason is (although it may seem odd) that there are a variety of shells available, owing to the long history of Unix and Linux. In the beginning, the interfaces were quite simple, but over time better interfaces/shells were developed that included new features (consider how the "Start" menu has changed over the years on Microsoft Windows versions). We can run a different shell, if it is installed, by simply running it like any other program. The `tcsh` shell, itself an outgrowth of the `csh` shell, is sometimes the default instead of `bash`. (Both `csh` and `tcsh` are older than `bash`.)

```
oneils@atmosphere ~$ tcsh
172:~> echo $0
tcsh
```

When running `tcsh`, the `setenv` command takes the place of `export`, and the syntax is slightly different.

6. There is another type of variable known as a "shell variable," which operates similar to environment variables. There are some differences: (1) by convention, these have lowercase names; (2) they are set differently, in `bash` by using `declare` instead of `export`; and (3) they are available only to the shell, not other programs that you might run. The distinction between these two types can cause headaches on occasion, but it isn't crucial enough to worry about now.

7. Because `$0` holds the name of the currently running program, one might expect `echo $0` to result in `echo` being reported, but this isn't the case. As mentioned previously, the shell replaces the environment variables with their contents *before* the command is executed.

```
172:~> setenv GREETING "hello $USER"
172:~> echo $GREETING
hello oneils
```

Although `bash` and similar shells like `dash` and `zsh` are most commonly found (and recommended), you might find yourself needing to use a shell like `csh` or its successor, `tcsh`. In this book, the assumption is that you are using `bash`, but when different commands would be needed to accomplish the same thing in the older `tcsh` or `csh`, a footnote will explain.

To get back to `bash` from `tcsh`, a simple `exit` will suffice:

```
172:~> exit
exit
oneils@atmosphere ~$ echo $0
-bash
```

In general, it can be difficult to determine which shell is running on the basis of the look of the command prompt; using `echo $0` right on the command line is the most reliable way.

Files, Directories, and Paths

With some of the more difficult concepts of the shell out of the way, let's turn to something a bit more practical: understanding how directories (also known as folders) and files are organized.

Most filesystems are hierarchical, with files and directories stored inside other directories. In Unix-like operating systems, the "top level" directory in which everything can be found is known as `/` (a forward slash). This top-level directory is sometimes called the *root* of the filesystem, as in the root of the filesystem tree. Within the root directory, there are commonly directories with names like `bin`, `etc`, `media`, and `home`; the last of these is often where users will store their own individual data.[8]

8. Computer scientists and mathematicians usually draw trees upside down. One theory is that trees are easier to draw upside down when working on a blackboard, the traditional (and still considered by many the best) medium of exposition for those fields. Our language reflects this thinking: when we move a file, we move it "down" into a subdirectory, or "up" into a directory "above" the current one.

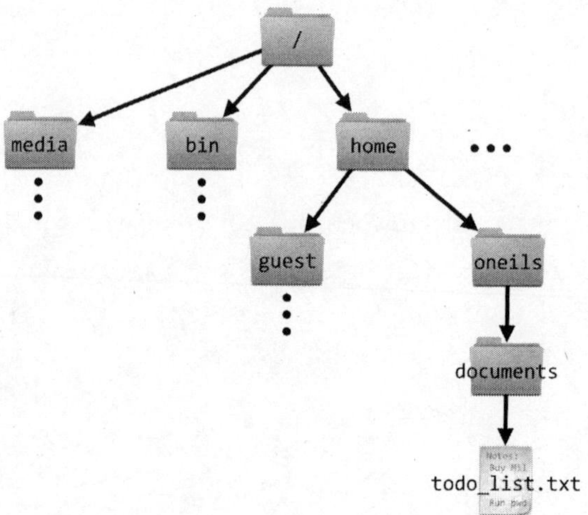

Each file and directory in the filesystem can be uniquely identified by its *absolute path*, a unique locator for a file or directory in the filesystem, starting with the root folder `/` and listing each directory on the way to the file. In the figure above, the absolute path to the `todo_list.txt` file is `/home/oneils/documents/todo_list.txt`.

Note that an absolute path *must* start with the leading forward slash, indicating that the path starts at the root folder `/`, and contain a valid path of folder names from there. (If you prefer to consider `/` as a folder name itself, an absolute path can also be specified like `//home/oneils/documents/todo_list.txt`, though using two forward slashes is considered redundant.)

Every user normally has a *home directory*, serving as a personal storage locker for files and directories. (Often the amount of space available this location is not as much as users would like.) The shell and other programs can find out the absolute path to your home directory via the environment variable `$HOME`; try running `echo $HOME` to see the absolute path to your own home directory.

What about special devices like CD-ROM drives and network drives? On a Windows operating system, these would get their own "root" directory with names like `D:` and `E:` (with the main hard drive usually having the name `C:`). On Unix-based operating systems, there is only ever one filesystem hierarchy, and the top level is always `/`. Special devices like CD-ROM drives and network drives are *mounted* somewhere in the filesystem. It may be that the directory `/media` remains empty, for example, but when a CD-ROM is inserted, a new directory may appear inside that directory, perhaps with the absolute path `/media/cdrom0`, and the files on the CD-ROM will appear in that location.

Determining how and where such devices are mounted is the job of the system administrator. On OS X machines, inserted devices appear in `/Volumes`. If you are logged in to a large computational infrastructure, your home directory likely isn't located on the internal hard drive of the remote computer, but rather mounted from a network drive present on yet another remote computer. This configuration allows users to "see" the same filesystem hierarchy and files no matter which remote computer they happen to log in to, if more than one is available. (For example, even `/home` might be a network mount, and so all users' home directories might be available on a number of machines.)

Getting around the Filesystem

It is vitally important to understand that, as we are working in the command-line environment, we always have a "place," a directory (or folder) in which we are working called the *present working directory*, or PWD. The shell keeps track of the present working directory in an environment variable, `$PWD`.

When you first log in, your present working directory is set to your home directory; `echo $PWD` and `echo $HOME` will likely display the same result. There is also a dedicated program for displaying the present working directory, called `pwd`.

```
oneils@atmosphere ~$ echo $HOME
/home/oneils
oneils@atmosphere ~$ echo $PWD
/home/oneils
oneils@atmosphere ~$ pwd
/home/oneils
```

We can list the files and directories that are stored in the present working directory by using the `ls` command.

```
oneils@atmosphere ~$ ls
apcb     Documents  Music     Public     todo_list.txt
Desktop  Downloads  Pictures  Templates  Videos
```

This command reveals that I have a number of directories in my home directory (`/home/oneils`) with names like `Music` and `Pictures` (helpfully colored blue) and a file called `todo_list.txt`.

We can change the present working directory—that is, move to another directory—by using the `cd` command, giving it the path that we'd like to move to.

```
oneils@atmosphere ~$ cd /home
oneils@atmosphere /home$ echo $PWD
/home
oneils@atmosphere /home$ ls
lost+found  oneils
```

Notice that the command prompt has changed to illustrate the present working directory: now it shows `oneils@atmosphere /home$`, indicating that I am in `/home`. This is a helpful reminder of where I am in the filesystem as I work. Previously, it showed only ~, which is actually a shortcut for `$HOME`, itself a shortcut for the absolute path to my home directory. Consequently, there are a number of ways to go back to my home directory: `cd /home/oneils`, or `cd $HOME`, or `cd ~`, or even just `cd` with no arguments, which defaults to moving to `$HOME`.

```
oneils@atmosphere /home$ cd $HOME
oneils@atmosphere ~$ pwd
/home/oneils
```

Relative Paths, Hidden Files, and Directories

It isn't always necessary to specify the full, absolute path to locate a file or directory; rather, we can use a *relative path*. A relative path locates a file or directory *relative* to the present working directory. If the present working directory is `/home/oneils`, for example, a file located at the absolute path `/home/oneils/Pictures/profile.jpg` would have relative path `Pictures/profile.jpg`. For the file `/home/oneils/todo_list.txt`, the relative path is just `todo_list.txt`. If the present working directory was `/home`, on the other hand, the relative paths would be `oneils/Pictures/profile.jpg` and `oneils/todo_list.txt`.

Because relative paths are always relative to the present working directory (rather than the root directory `/`), they *cannot* start with a forward slash, whereas absolute paths *must* start with a leading forward slash. This distinction occasionally confuses new users.

We can use the `cd` command with relative paths as well, and this is most commonly done for directories present in the current working directory, as reported by `ls`. In fact, both relative and absolute paths are allowed wherever a file or directory needs to be specified.

```
oneils@atmosphere ~$ ls
apcb     Documents  Music     Public     todo_list.txt
Desktop  Downloads  Pictures  Templates  Videos
oneils@atmosphere ~$ cd Pictures
oneils@atmosphere ~/Pictures$ ls
profile.jpg
```

A few notes about moving around directories: (1) Unix-like operating systems are nearly always case sensitive, for commands, parameters, and file and directory names. (2) Because of the way spaces are treated on the command line, it is uncommon to see spaces in file or directory names. The command `cd mydocuments` is much easier to type than `cd my\ documents` or `cd 'my documents'`. (3) If you forget to specify `cd` before specifying a path, a cryptic `permission denied` error will occur, rather than an error specific to the lack of `cd` command.

By default, when running the `ls` program to list the contents of the present working directory, all files and directories that start with a `.` are hidden. To see everything including the hidden files, we can add the `-a` flag to `ls`.

```
oneils@atmosphere ~/Pictures$ cd $HOME
oneils@atmosphere ~$ ls -a
.              .config    .gstreamer-0.10  .profile       .vim
..             .dbus      .gvfs            Public         .viminfo
apcb           Desktop    .ICEauthority    .pulse         .vimrc
.bash_history  Documents  .local           .pulse-cookie  .vnc
.bash_login    Downloads  Music            .ssh           .Xauthority
.bash_logout   .gconf     .netrc           Templates      .Xdefaults
.bashrc        .gnome2    Pictures         todo_list.txt  .xscreensaver
.cache         .gnupg     .pip             Videos         .xsession-errors
```

It turns out there are quite a few hidden files here! Many of those starting with a `.`, like `.bash_login`, are actually configuration files used by various programs. We'll spend some time with those in later chapters.

On the topic of `ls`, there are many additional parameters we can give to `ls`: include `-l` to show "long" file information including file sizes, and `-h` to make those file sizes "human readable" (e.g.,

4K versus 4,196 bytes). For `ls`, we can specify this combination of options as `ls -l -a -h` (where the parameters may be given in any order), or with a single parameter as `ls -lah`, though not all utilities provide this flexibility.

```
oneils@atmosphere ~$ ls -lah
total 168K
drwxr-xr-x 25 oneils iplant-everyone 4.0K Sep 23 22:40 .
drwxr-xr-x  4 root   root            4.0K Sep 15 09:48 ..
drwxr-xr-x  4 oneils iplant-everyone 4.0K Sep 15 11:19 apcb
-rw-------  1 root   root            2.2K Sep 15 10:49 .bash_history
-rw-r--r--  1 oneils iplant-everyone   61 Sep 16 19:46 .bash_login
-rw-r--r--  1 oneils iplant-everyone  220 Apr  3  2012 .bash_logout
-rw-r--r--  1 oneils iplant-everyone 3.6K Sep 15 09:48 .bashrc
drwx------  7 oneils iplant-everyone 4.0K Sep 15 09:52 .cache
...
```

Some of these columns of information are describing the permissions associated with the various files and directories; we'll look at the permissions scheme in detail in later chapters.

When using `ls -a`, we see two "special" directories: `.` and `..` are their names. These directories don't really exist per se but are instead shortcuts that refer to special locations relative to the directory they are contained in. The `.` directory is a shortcut for the same directory it is contained in. Colloquially, it means "here." The `..` directory is a shortcut for the directory *above* the directory containing it. Colloquially, it means "up."

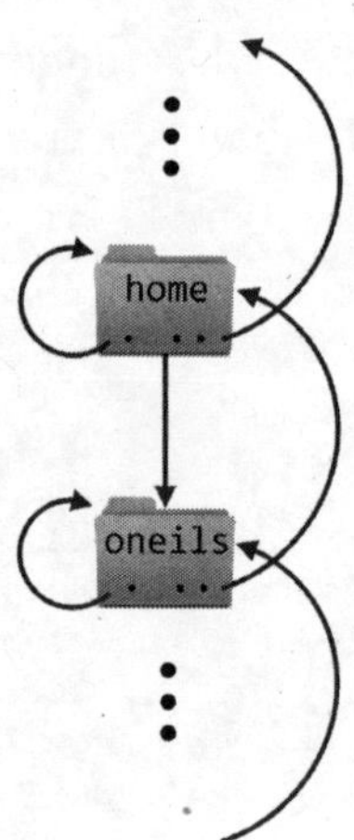

Every directory has these "virtual" links, and they can be used as relative paths themselves or in relative or absolute paths. Here, for example, we can `cd .` to stay where we are, or `cd ..` to go up one directory:

```
oneils@atmosphere ~$ echo $PWD
/home/oneils
oneils@atmosphere ~$ cd .
oneils@atmosphere ~$ echo $PWD
/home/oneils
oneils@atmosphere ~$ cd ..
oneils@atmosphere /home$ echo $PWD
/home
```

If we like, we can go up two directories with `cd ../..`:

```
oneils@atmosphere /$ cd $HOME
oneils@atmosphere ~$ echo $PWD
/home/oneils
oneils@atmosphere ~$ cd ../..
oneils@atmosphere /$ echo $PWD
/
```

We can even use `.` and `..` in longer relative or absolute paths (number 2 in the figure below illustrates the relatively odd path of `/media/./cdrom0`, which is identical to `/media/cdrom0`).

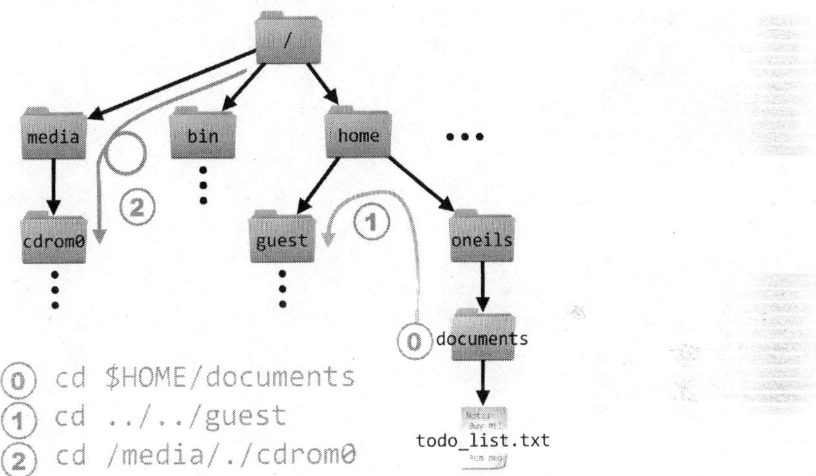

Exercises

1. Practice moving about the filesystem with `cd` and listing directory contents with `ls`, including navigating by relative path, absolute path, `.` and `..`. Use `echo $PWD` and `pwd` frequently to list your present working directory. Practice navigating home with `cd ~`, `cd $HOME`, and just `cd` with no arguments. You might find it useful to draw a "map" of the filesystem, or at least the portions relevant to you, on a piece of paper.

2. Without explicitly typing your username, create another environment variable called `$CHECKCASH`, such that when `echo $CHECKCASH` is run, a string like `oneils has $20` is printed

(except that `oneils` would be your username and should be set by reading from `$USER`).

3. The directory `/etc` is where many configuration files are stored. Try navigating there (using `cd`) using an absolute path. Next, go back home with `cd $HOME` and try navigating there using a relative path.
4. What happens if you are in the top-level directory `/` and you run `cd ..`?
5. Users' home directories are normally located in `/home`. Is this the case for you? Are there any other home directories belonging to other users that you can find, and if so, can you `cd` to them and run `ls`?

Chapter 4

Working with Files and Directories

Now that we know how to locate files and directories in the filesystem, let's learn a handful of important tools for working with them and the system in general.

Viewing the Contents of a (Text) File

Although there are many tools to view and edit text files, one of the most efficient for viewing them is called `less`, which takes as a parameter a path to the file to view, which of course may just be a file name in the present working directory (which is a type of relative path).[9]

```
oneils@atmosphere ~$ ls
apcb     Documents  Music        Pictures  Templates      Videos
Desktop  Downloads  p450s.fasta  Public    todo_list.txt
oneils@atmosphere ~$ less p450s.fasta
```

The invocation of `less` on the file **`p450s.fasta`** opens an "interactive window" within the terminal window, wherein we can scroll up and down (and left and right) in the file with the arrow keys. (As usual, the mouse is not very useful on the command line.) We can also search for a pattern by typing `/` and then typing the pattern before pressing Enter.

9. There is a similar program called `more`, originally designed to show "more" of a file. The `less` program was developed as a more full-featured alternative to `more`, and was so named because "less is more."

```
>sp|Q3LFU0|CP1A1_BALAC Cytochrome P450 1A1 OS=Balaenoptera acutorostrata GN=CYP1
A1 PE=2 SV=1
MFSVFGLSIPISATELLLASATFCLVFWVVRAWQPRVPKGLKSPPGPWSWPLIGHVLTLG
KSPHLALSRLSQRYGDVLQIRIGCTPVLVLSGLDTIRQALVRQGDDFKGRPDLYSFTLVA
DGQSMTFNPDSGPVWAARRRLAQNALKSFSIASDPASSSSCYLEEHVSKESEYLIGKFQE
p450s.fasta
```

When finished with `less`, pressing `q` will exit and return control to the shell or command line. Many of the text formats used in computational biology include long lines; by default, `less` will wrap these lines around the terminal so they can be viewed in their entirety. Using `less -S` will turn off this line wrapping, allowing us to view the file without any reformatting. Here's what the file above looks like when viewed with `less -S p450s.fasta`:

```
>sp|Q3LFU0|CP1A1_BALAC Cytochrome P450 1A1 OS=Balaenoptera acutorostrata GN=CYP1
MFSVFGLSIPISATELLLASATFCLVFWVVRAWQPRVPKGLKSPPGPWSWPLIGHVLTLG
KSPHLALSRLSQRYGDVLQIRIGCTPVLVLSGLDTIRQALVRQGDDFKGRPDLYSFTLVA
DGQSMTFNPDSGPVWAARRRLAQNALKSFSIASDPASSSSCYLEEHVSKESEYLIGKFQE
p450s.fasta
```

Notice that the first long line has not been wrapped, though we can still use the arrow keys to scroll left or right to see the remainder of this line.

Creating New Directories

The `mkdir` command creates a new directory (unless a file or directory of the same name already exists), and takes as a parameter the path to the directory to create. This is usually a simple file name as a relative path inside the present working directory.

```
oneils@atmosphere ~$ ls
apcb     Documents  Music        Pictures  Templates     Videos
Desktop  Downloads  p450s.fasta  Public    todo_list.txt
oneils@atmosphere ~$ mkdir projects
oneils@atmosphere ~$ ls
apcb     Documents  Music        Pictures  Public     todo_list.txt
Desktop  Downloads  p450s.fasta  projects  Templates  Videos
```

Move or Rename a File or Directory

The `mv` utility serves to both move and rename files and directories. The simplest usage works like `mv <source_path> <destination_path>`, where `<source_path>` is the path (absolute or relative) of the file/directory to rename, and `<destination_path>` is the new name or location to give it.

In this example, we'll rename `p450s.fasta` to `p450s.fa`, move it into the `projects` folder, and then rename the `projects` folder to `projects_dir`.

```
oneils@atmosphere ~$ mv p450s.fasta p450s.fa
oneils@atmosphere ~$ mv p450s.fa projects
oneils@atmosphere ~$ mv projects projects_dir
oneils@atmosphere ~$ ls
apcb     Documents  Music     projects_dir  Templates      Videos
Desktop  Downloads  Pictures  Public        todo_list.txt
oneils@atmosphere ~$
```

Because `mv` serves a dual role, the semantics are important to remember:

- If `<destination_path>` doesn't exist, it is created (so long as all of the containing folders exist).
- If `<destination_path>` does exist:
    - If `<destination_path>` is a directory, the source is moved inside of that location.
    - If `<destination_path>` is a file, that file is overwritten with the source.

Said another way, `mv` attempts to guess what it should do, on the basis of whether the destination already exists. Let's quickly undo the moves above:

```
oneils@atmosphere ~$ mv projects_dir/p450s.fa p450s.fasta
oneils@atmosphere ~$ mv projects_dir/ projects
```

A few other notes: First, when specifying a path that is a directory, the trailing `/` is optional: `mv projects_dir/ projects` is the same as `mv projects_dir projects` if `projects_dir` is a directory (similarly, `projects` could have been specified as `projects/`). Second, it is possible to move multiple files into the same directory, for example, with `mv p450s.fasta todo_list.txt projects`. Third, it is quite common to see `.` referring to the present working directory as the destination, as in `mv ../file.txt .` for example, which would move `file.txt` from the directory *above* the present working directory (`..`) *into* the present working directory (`.`, or "here").

Copy a File or Directory

Copying files and directories is similar to moving them, except that the original is not removed as part of the operation. The command for copying is `cp`, and the syntax is `cp <source_path> <destination_path>`. There is one caveat, however: `cp` will not copy an entire directory and all of its contents unless you add the `-r` flag to the command to indicate the operation should be recursive.

```
oneils@atmosphere ~$ cp todo_list.txt todo_copy.txt
oneils@atmosphere ~$ cp -r projects projects_dir_copy
```

Forgetting the `-r` when attempting to copy a directory results in an `omitting directory` warning.

It is possible to simultaneously copy and move (and remove, etc.) many files by specifying multiple sources. For example, instead of `cp ../todo_list.txt .`, we could have copied both the to-do list and the `p450s.fasta` file with the same command:

```
oneils@atmosphere ~/projects$ cp ../todo_list.txt ../p450s.fasta .
```

Remove (Delete) a File or Directory

Files may be deleted with the `rm` command, as in `rm <target_file>`. If you wish to remove an entire directory and everything inside, you need to specify the `-r` flag for recursive, as in `rm -r <target_dir>`. Depending on the configuration of your system, you may be asked "are you sure?" for each file, to which you can reply with a `y`. To avoid this checking, you can also specify the `-f` (force) flag, as in `rm -r -f <target_dir>` or `rm -rf <target_dir>`. Let's create a temporary directory alongside the file copies from above, inside the projects folder, and then remove the `p450s.fasta` file and the `todo_list.txt` file as well as the temporary folder.

```
oneils@atmosphere ~/projects$ mkdir tempdir
oneils@atmosphere ~/projects$ ls
p450s.fasta  tempdir  todo_list.txt
oneils@atmosphere ~/projects$ rm todo_list.txt
oneils@atmosphere ~/projects$ rm -rf tempdir/
oneils@atmosphere ~/projects$ ls
p450s.fasta
```

Beware! *Deleted files are gone forever.* There is no undo, and there is no recycle bin. Whenever you use the `rm` command, double-check your syntax. There's a world of difference between `rm -rf project_copy` (which deletes the folder `project_copy`) and `rm -rf project _copy` (which removes the folders `project` and `_copy`, if they exist).

Checking the Size of a File or Directory

Although `ls -lh` can show the sizes of files, this command will not summarize how much disk space a directory and all of its contents take up. To find out this information, there is the `du` (disk usage) command, which is almost always combined with the `-s` (summarize) and `-h` (show sizes in human-readable format) options.

```
oneils@atmosphere ~/projects$ cd $HOME
oneils@atmosphere ~$ ls
apcb     Documents  Music        Pictures  Public     todo_list.txt
Desktop  Downloads  p450s.fasta  projects  Templates  Videos
oneils@atmosphere ~$ du -sh p450s.fasta
16K p450s.fasta
oneils@atmosphere ~$ du -sh projects/
4.0K projects/
oneils@atmosphere ~$ du -sh .
11M .
```

As always, `.` is a handy target, here helping to determine the file space used by the present working directory.

Editing a (Text) File

There is no shortage of command-line text editors, and while some of them—like `vi` and `emacs`—are powerful and can enhance productivity in the long run, they also take a reasonable amount of time to become familiar with. (Entire books have been written about each of these editors.)

In the meantime, a simple text editor available on most systems is `nano`; to run it, we simply specify a file name to edit:

```
oneils@atmosphere ~$ nano todo_list.txt
```

If the file doesn't exist already, it will be created when it is first saved, or "written out." The `nano` editor opens up an interactive window much like `less`, but the file contents can be changed. When done, the key sequence `Control-o` will save the current edits to the file specified (you'll have to press Enter to confirm), and then `Control-x` will exit and return control to the command prompt. This information is even presented in a small help menu at the bottom.

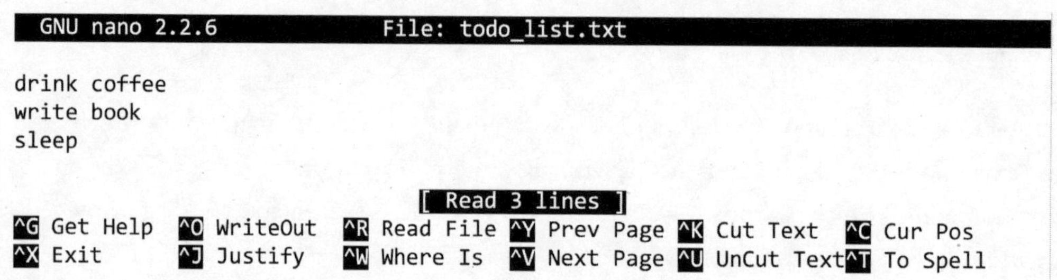

Although `nano` is not as sophisticated as `vi` or `emacs`, it does support a number of features, including editing multiple files, cut/copy/paste, find and replace by pattern, and syntax highlighting of code files.

Code files are the types of files that we will usually want to edit with `nano`, rather than essays or short stories. By default, on most systems, `nano` automatically "wraps" long lines (i.e., automatically presses Enter) if they would be longer than the screen width. Unfortunately, this feature would cause an error for most lines of code! To disable it, nano can be started with the `-w` flag, as in `nano -w todo_list.txt`.

Command-Line Efficiency

While the shell provides a powerful interface for computing, it is certainly true that the heavy reliance on typing can be tedious and prone to errors. Fortunately, most shells provide a number of features that dramatically reduce the amount of typing needed.

First, wildcard characters like `*` (which matches any number of arbitrary characters) and `?` (which matches any single arbitrary character) allow us to refer to a group of files. Suppose we want to move three files ending in `.temp` into a `temp` directory. We could run `mv` listing the files individually:

```
oneils@atmosphere ~/apcb/intro$ ls
fileAA.temp  fileA.temp  fileB.temp  temp
oneils@atmosphere ~/apcb/intro$ mv fileA.temp fileAA.temp fileB.temp temp/
```

Alternatively, we could use `mv file*.temp temp`; the shell will expand `file*.temp` into the list of files specified above *before* passing the expanded list to `mv`.[10]

```
oneils@atmosphere ~/apcb/intro$ ls
fileAA.temp  fileA.temp  fileB.temp  temp
oneils@atmosphere ~/apcb/intro$ mv file*.temp temp/
```

Similarly, we could move only `fileA.temp` and `fileB.temp` (but not `fileAA.temp`) using `mv file?.tmp temp`, because the `?` wildcard will only match one of any character. These wildcards may be used anywhere in an absolute or relative path, and more than one may be used in a single path. For example, `ls /home/*/*.txt` will inspect all files ending in `.txt` in all users' home directories (if they are accessible for reading).

Second, if you want to rerun a command, or run a command similar to a previously run command, you can access the command history by pressing the up arrow. Once you've identified which command you want to run or modify, you can modify it using the left and right arrows, backspace or delete, and then typing and pressing Enter again when you are ready to execute the modified command. (You don't even need to have the cursor at the end of the line to press Enter and execute the command.) For a given login session, you can see part of your command history by running the `history` command.

Finally, one of the best ways to navigate the shell and the filesystem is by using *tab completion*. When typing a path (either absolute or relative), file or directory name, or even a program name, you can press Tab, and the shell will autocomplete the portion of the path or command, up until the autocompletion becomes ambiguous. When the options are ambiguous, the shell will present you with the various matching options so that you can inspect them and keep typing. (If you want to see all options even if you haven't started typing the next part of a path, you can quickly hit Tab twice.) You can hit Tab as many times as you like while entering a command. Expert command-line users use the Tab key many times a minute!

10. This is important to consider when combining `rm` with wildcards; the commands `rm -rf *.temp` and `rm -rf * .temp` are very different! The latter will remove all files in the present directory, while the former will only remove those ending in `.temp`.

Getting Help on a Command or Program

Although we've discussed a few of the options(also known as arguments, or flags) for programs like `ls`, `cp`, `nano`, and others, there are many more you might wish to learn about. Most of these basic commands come with "man pages," short for "manual pages," that can be accessed with the `man` command.

```
oneils@atmosphere ~$ man ls
```

This command opens up a help page for the command in question (usually in `less` or a program similar to it), showing the various parameters and flags and what they do, as well as a variety of other information such as related commands and examples. For some commands, there are also "info" pages; try running `info ls` to read a more complete overview of `ls`. Either way, as in `less`, pressing `q` will exit the help page and return you to the command prompt.

Viewing the Top Running Programs

The `top` utility is invaluable for checking what programs are consuming resources on a machine; it shows in an interactive window the various processes (running programs) sorted by the percentage of CPU time they are consuming, as well as which user is running them and how much RAM they are consuming. Running `top` produces a window like this:

```
top - 02:23:20 up 15 days, 16:34,  2 users,  load average: 0.03, 0.02, 0.05
Tasks: 127 total,   1 running, 126 sleeping,   0 stopped,   0 zombie
Cpu(s):  0.0%us,  0.0%sy,  0.0%ni, 99.7%id,  0.3%wa,  0.0%hi,  0.0%si,  0.0%st
Mem:   4050124k total,  1801608k used,  2248516k free,   142652k buffers
Swap:        0k total,        0k used,        0k free,  1256080k cached

  PID USER      PR  NI  VIRT  RES  SHR S %CPU %MEM    TIME+  COMMAND
    1 root      20   0 90540 4264 2696 S  0.0  0.1   0:07.85 init
    2 root      20   0     0    0    0 S  0.0  0.0   0:00.00 kthreadd
    3 root      20   0     0    0    0 S  0.0  0.0   0:43.50 ksoftirqd/0
    5 root      20   0     0    0    0 S  0.0  0.0   0:00.35 kworker/u:0
    6 root      RT   0     0    0    0 S  0.0  0.0   0:00.00 migration/0
```

From a users' perspective, the list of processes below the dark line is most useful. In this example, no processes are currently using a significant amount of CPU or memory (and those processes that are running are owned by the administrator `root`). But if any user were running processes that required more than a tiny bit of CPU, they would likely be shown. To instead sort by RAM usage, use the key sequence `Control-M`. When finished, `q` will quit `top` and return you to the command prompt.

Of particular importance are the `%CPU` and `%MEM` columns. The first may vary from 0 up to 100 (percent) *times the number of CPU cores on the system*; thus a value of `3200` would indicate a program using 100% of 32 CPU cores (or perhaps 50% of 64 cores). The `%MEM` column ranges from 0 to 100 (percent). It is generally a bad thing for the system when the total memory used by all process is near or over 100%—this indicates that the system doesn't have enough "working memory" and it may be attempting to use the much slower hard drive as working memory. This situation is known as swapping, and the computer may run so slowly as to have effectively crashed.

Killing Rogue Programs

It sometimes happens that programs that should run quickly, don't. Perhaps they are in an internal error state, looping forever, or perhaps the data analysis task you had estimated to take a minute or two is taking much longer. Until the program ends, the command prompt will be inaccessible.

There are two ways to stop such running programs: the "soft" way and the "hard" way. The soft way consists of attempting to run the key combination `Control-c`, which sends a stop signal to the running process so that it should end.

But if the rogue program is in a particularly bad error state, it won't stop even with a `Control-c`, and a "hard" kill is necessary. To do this requires logging in to the same machine with another terminal window to regain some command-line access. Run `top`, and note the `PID` (process ID) of the offending process. If you don't see it in the `top` window, you can also try running `ps augx`, which prints a table of all running processes. Suppose the PID for the process to kill is `24516`; killing this process can be done by running `kill -9 24156`. The `-9` option specifies that the operating system should stop the process in its tracks and immediately clean up any resources used by it. Processes that don't stop via a `kill -9` are rare (though you can't `kill` a process being run by another user), and likely require either a machine reboot or administrator intervention.

Exercises

1. Create the following directories inside your home directory, if you don't already have them:

`downloads`, `local`, and `projects`. Inside of `local`, create a directory called `bin`. These folders are a common and useful set to have in your home directory—we'll be using them in future chapters to work in, download files to, and install software to.

2. Open not one but *two* login windows, and log in to a remote machine in each. This gives you two present working directories, one in each window. You can use one to work, and another to make notes, edit files, or watch the output of `top`.

   Create a hidden directory inside of your home directory called `.hidden`. Inside of this directory, create a file called `notes`. Edit the file to contain tricks and information you fear you might forget.[11]

3. Spend a few minutes just practicing tab completion while moving around the filesystem using absolute and relative paths. Getting around efficiently via tab-completion is a surprisingly necessary skill.

4. Skim the `man` page for `ls`, and try out a few of the options listed there. Read a bit of the `info` page for `nano`.

11. You might also consider keeping a paper notebook, or something in a Wiki or other text document. If you prefer to keep digital notes, try to use a simple text editor like `nano`, TextEdit on OS X, or Notepad on Windows. Rich-text editors like Microsoft Word will often automatically replace things like simple quote characters with serif quotes, which don't work on the command line, leading to headaches.

Chapter 5

Permissions and Executables

As mentioned previously, the administrator `root` configures most things on the system, like where network drives are mounted and how access to the system is granted.[12] Most of the files that are important for these configurations are "owned" by root, and for security purposes, other users can't tamper with them. In fact, all users of the system usually own the files in their own home directory, and these files can't be tampered with by other users (except for `root`, who has access to everything). These security settings are controlled via *permissions* associated with every file and directory.

Root can also put users together into *groups*, allowing users in the same group to share files with each other but not outsiders, if they so choose. Groups may have more than one user, and a single user may be part of more than one group, as in the following example illustrating three groups and eight users.

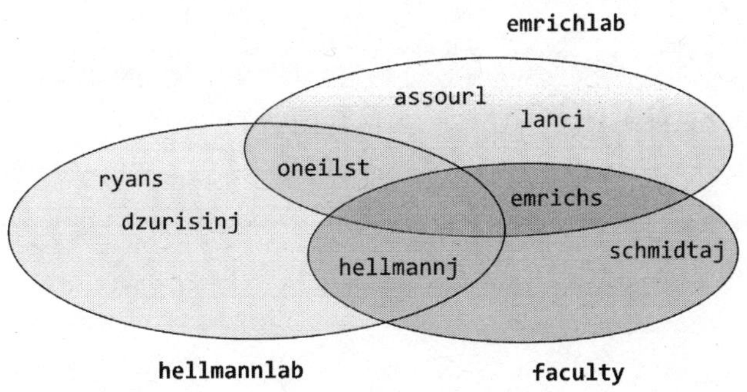

12. The administrator, or `root` user, is sometimes also called the "superuser." This has been known to go to some administrators' heads.

The `groups` command shows the groups to which a user belongs; in the above example, `groups oneilst` would report `emrichlab` and `hellmannlab`. To see your own groups, you can always use something like `groups $USER` (relying on the shell to replace `$USER` with your username).

Unfortunately, there's no surefire or easy way to list all the members of a particular group—at least not without some deep knowledge of how the system is configured or some programming expertise. On some systems, a command like `getent group <groupname>` will provide the answer; `getent group faculty` would report `emrichs`, `schmidtj`, and `hellmannj` for the example above.

If you are unsure of a person's username, the `finger` command may come to the rescue. You can supply `finger` with either a first name or last name to search for (or even the username, if known), and it will return information—if entered by the system administrator—about that user.

```
oneils@atmosphere ~$ finger Shawn
Login: oneils                Name: Shawn O'neil
Directory: /home/oneils                 Shell: /bin/bash
On since Mon Oct 20 16:52 (MST) on pts/0 from 8-169.ptpg.oregonstate.edu
No mail.
No Plan.
```

Each file and directory is associated with one user (the owner) and one group; unfortunately, in normal Unix-like permissions, one and only one group may be associated with a file or directory. Each file and directory also has associated with it permissions describing:

1. what the owner can do,
2. what members of the group can do, and
3. what everyone else (others) can do.

This information is displayed when running `ls -l`, and is represented by a combination of `r` (read), `w` (write), and `x` (execute). Where one of these three is absent, it is replaced by a `-`. Here's an example, showing two entries owned by `oneils` and in the `iplant-everyone` group; one has permissions `rwxrwxrwx` (an insecure permission set, allowing anyone to do anything with the file), and the other has `rwxr-xr-x` (a much more reasonable permission set).

```
oneils@atmosphere ~/apcb/intro$ ls -l
total 20
-rwxrwxrwx 1 oneils iplant-everyone 15891 Oct 20 17:42 p450s.fasta
drwxr-xr-x 2 oneils iplant-everyone  4096 Oct 20 17:40 temp
```

There is an extra entry in the first column; the first character describes the type of the entry, `-` for a regular file and `d` for directory. Let's break down these first few columns for one of the entries:

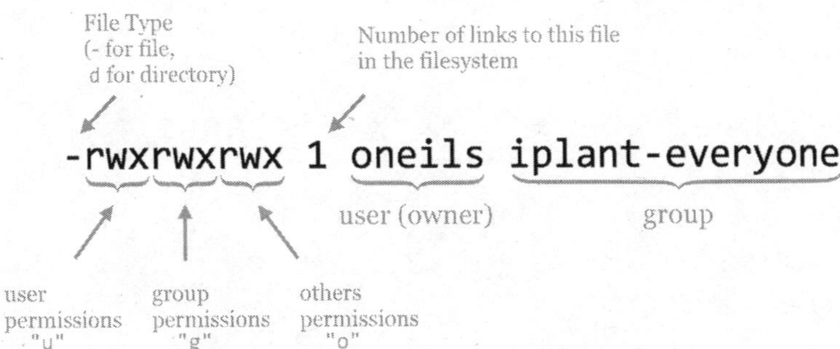

Each file or directory may have some combination of `r`, `w`, and `x` permissions, applied to either the user, the group, or others on the system. For files, the meanings of these permissions are fairly straightforward.

Code	Meaning for Files
`r`	Can read file contents
`w`	Can write to (edit) the file
`x`	Can (potentially) "execute" the file

We'll cover what it means for a file to be executable in a bit. For directories, these permissions take on different meanings.

Code	Meaning for Directories
`r`	Can see contents of the directory (e.g., run `ls`)
`w`	Can modify contents of the directory (create or remove files/directories)
`x`	Can `cd` to the directory, and potentially access subdirectories

The `temp` directory above gives the user all permissions (`rwx`), but members of the group and others can only `cd` to the directory and view the files there (`r-x`); they can't add or remove files or directories. (They may be able to *edit* files in `temp`, however, depending on those files' permissions.)

The `chmod` (change mode) utility allows us to add or remove permissions. There are two types of syntax, the simpler "character" syntax and the numeric "octal" syntax. We'll describe the simpler syntax and leave discussion of the octal syntax for those brave enough to read the manual page (`man chmod`).

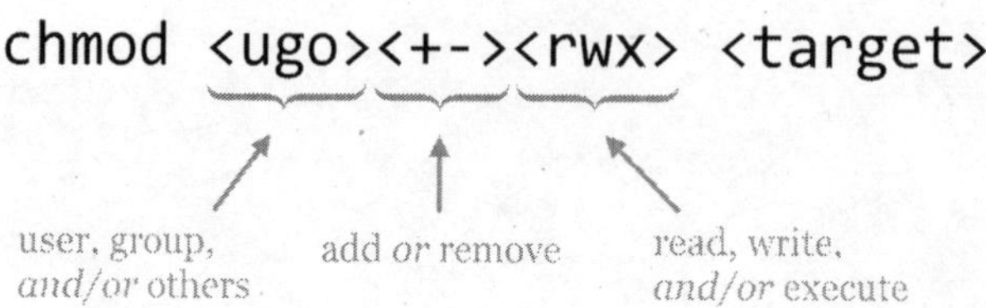

To clarify, here are some examples of modifying permissions for the `p450s.fasta` file.

Command	Effect
`chmod go-w p450s.fasta`	Remove write for group and others
`chmod ugo+r p450s.fasta`	Add read for user, group, and others
`chmod go-rwx p450s.fasta`	Remove read, write, and execute for group and others
`chmod ugo+x p450s.fasta`	Add execute for user, group, and others
`chmod +x p450s.fasta`	Same as `chmod ugo+x p450s.fasta`

If you wish to modify a directory and everything inside, you can add the `-R` flag (capital R this time for recursive) to `chmod`. To share a `projects` directory and everything inside for read access with group members, for example, you can use `chmod -R g+r projects`.

There are a few small things to note about file and directory permissions. The first is that while it is possible to change the group of a file or directory, you can only do so with the `chgrp` command if you are a member of that group.

```
oneils@atmosphere ~/apcb/intro$ groups $USER
oneils : iplant-everyone users community de-preview-access atmo-user dnasubway-
users myplant-users
oneils@atmosphere ~/apcb/intro$ chgrp community p450s.fasta
```

Second, you own the files that you create, but generally only the root user has access to the `chown` utility that changes the owner of an existing file (it wouldn't be very nice to "gift" another user a nefarious program).

Third, while it is convenient to be able to open up a directory for reading by group members, doing so is only useful if all of the directories above it are also minimally accessible. In particular, all the directories in the path to a shared directory need to have at least `x` for the group if they are to be accessed in any way by group members.

Executables and $PATH

What is a "program?" On a Unix-like system, it's a file that has executable permissions for some user or users. (It also helps if that file contains some instructions that make sense to execute!) Often these are encoded in a "binary" format—simply long strings of `0`'s and `1`'s representing machine code that the CPU can interpret—but they may in some contexts also be human-readable text files. Many of the programs we've been using, like `echo` and `ls`, are executable files that live in the `/bin` directory along with many others, including `bash`, our shell program.

```
oneils@atmosphere ~$ cd /bin
oneils@atmosphere /bin$ ls -l
total 7384
-rwxr-xr-x 1 root root 959120 Mar 28  2013 bash
-rwxr-xr-x 1 root root  31112 Dec 14  2011 bunzip2
-rwxr-xr-x 1 root root  31112 Dec 14  2011 bzcat
...
```

If we like, we can even attempt to take a look at one of these with `less`. For example, we can try to examine the contents of the `bash` program with `less /bin/bash`; even though `less` reports a warning that the file is binary encoded, it won't hurt to try.

```
^?   ^B^A^A^@^@^@^@^@^@^@^@^B^@ ^@^A^@^@^@ <D4> ^@^@^@^@ ^@^@^@^@^@^@<90>
<9B>^N^@^@^@^@^@^@^@^@ ^@ ^@   ^@ ^@^\^@ESC^@^F^@^@^@^E^@^@^@ ^@^@^@^@^@^@
^@ ^@^@^@^@^@ ^@ ^@^@^@^@^@<F8>^A^@^@^@^@^@<F8>^A^@^@^@^@^@^H^@^@^@^@^@^@
^C^@^@^@^D^@^@^@ ^B^@^@^@^@^@ ^B ^@^@^@^@ ^B ^@^@^@^@^@^\^@^@^@^@^@^@^\^@
^@^@^@^@^@^A^@^@^@^@^@^@^A^@^@^@^E^@^@^@^@^@^@^@^@^@^@^@ ^@^@^@^@^@^@ ^
<FE>^M^@^@^@^@^@^@^@^@^@^@^@^A^@^@^@^F^@^@
^@^H^N^N^@^@^@^@^@^H^N ^@^@^@^@^@^H^N ^@^@^@^@^@<88><8C>^@^@^@^@^@ <E8>^@^@^@
^@^@^@^@^@ ^@^@^@^@^@^B^@^@^@^F^@^@^@ ^N^N^@^@^@^@^@ ^N ^@^@^@^@^@ ^N ^@^@^@^@^@
bash
```

A binary-encoded file doesn't look like much when we attempt to convert it to text and view it with `less`. In any case, here's an "execution rule" we're going to break almost immediately: *to get the shell to run a program (executable file), we specify the absolute or relative path to it.*

In this case, our execution rule would indicate that to run the `echo` program, we would specify the absolute path `/bin/echo hello`, as the `echo` program lives in `/bin`, or `../../../../bin/echo hello` for the relative path (because `/bin` is four folders above our present working directory).

```
oneils@atmosphere ~/apcb/intro$ /bin/echo hello
hello
oneils@atmosphere ~/apcb/intro$ ../../../../bin/echo hello
hello
```

Now for the rule-breaking part: we already know that we can just run `echo` *without* specifying a path to the program. This means that when we attempt to run a program, the shell must be able to find the executable file somehow. How is this done?

The answer, as with so many questions involving the shell, is an environment variable called `$PATH`. Let's check the contents of this variable:[13]

```
oneils@atmosphere ~/apcb/intro$ echo $PATH
/usr/local/sbin:/usr/local/bin:/usr/sbin:/usr/bin:/sbin:/bin:/usr/games
```

The `$PATH` environment variable contains a simple string, describing a list of absolute paths separated by `:` characters. When we specify what looks to the shell like the name of a program, it searches this list of paths, in order, for an executable file of that name. When we type `echo`, it tries `/usr/local/sbin/echo`, then `/usr/local/bin/echo`, and so on, until it finds it in `/bin/echo`.

The first matching executable file the shell finds in the directories listed in `$PATH` is the one that is executed. This could lead to some mischief: if a coworker with a penchant for practical jokes could modify your `$PATH` variable, he could add his own home directory as the first entry. From there, he could create an executable file called, say, `ls` that did whatever he wanted, and you would unknowingly be running that! It is possible for anyone with access to your account to modify your `$PATH`, so it's a good idea not to leave your terminal window open around anyone with a dastardly sense of humor.

13. The `tcsh` and `csh` shells do not use the `$PATH` environment variable. Instead, they look in a shell variable called `$path`.

If there are multiple executable files with the same name in this list of paths, can we discover which one the shell will execute? Yes: in `bash` we can make this determination using the `which` command.[14]

```
oneils@atmosphere ~$ which echo
/bin/echo
```

What about a command like `cd`? We can try to use `which` to locate a program called `cd`, but we'll find that nothing is reported.

```
oneils@atmosphere ~/apcb/intro$ which cd
oneils@atmosphere ~/apcb/intro$
```

This is because `cd` is not a program (executable file), but rather a "command," meaning the shell notices that it's a special keyword it should handle, rather than searching for an executable file of that name. Said another way, `bash` is performing the action, rather than calling an external executable program. Knowing about the difference between commands handled by the shell and programs that are executable files is a minor point, but one that could be confusing in cases like this.

Making Files Executable

Let's do something decidedly weird, and then come back and explain it. First, we'll use `nano` to create a new file called `myprog.sh`, using the `-w` flag for `nano` to ensure that long lines are not automatically wrapped (`nano -w myprog.sh`). In this file, we'll make the first two characters `#!`, followed immediately by the absolute path to the `bash` executable file. On later lines, we'll put some commands that we might run in `bash`, like two `echo` calls.

14. In `tcsh` and `csh`, the closest approximation to `which` is `where`, though `which` may also work.

```
 GNU nano 2.2.6              File: myprog.sh                              Modified

#!/bin/bash

echo "Hello!"
echo "This is a bit weird..."

^G Get Help  ^O WriteOut  ^R Read File ^Y Prev Page ^K Cut Text  ^C Cur Pos
^X Exit      ^J Justify   ^W Where Is  ^V Next Page ^U UnCut Text^T To Spell
```

Although it looks like the `#!` (pronounced "shebang," rhyming with "the bang") line starts on the second line, it is actually the first line in the file. This is important. Notice that `nano` has realized we are writing a file that is a bit odd, and has turned on some coloring. Your `nano` may not be configured for this syntax highlighting. If not, don't worry: we are creating a simple text file.

After we save the file (`Control-o`, then Enter confirm the file name to write) and exit `nano` (`Control-x`), we can add execute permissions to the file (for everyone, perhaps) with `chmod +x myprog.sh`.

```
oneils@atmosphere ~/apcb/intro$ ls
myprog.sh  p450s.fasta  temp
oneils@atmosphere ~/apcb/intro$ chmod +x myprog.sh
oneils@atmosphere ~/apcb/intro$ ls -l
total 24
-rwxr-xr-x 1 oneils iplant-everyone    57 Oct 20 20:49 myprog.sh
-rw-rw-rw- 1 oneils community       15891 Oct 20 17:42 p450s.fasta
drwxr-xr-x 2 oneils iplant-everyone  4096 Oct 20 17:40 temp
```

It would appear that we might have created a program—we do have an executable file, and you might have guessed that the special syntax we've used makes the file executable in a meaningful way. Let's try it out: according to our execution rule, we can specify the absolute path to it to run it.

```
oneils@atmosphere ~/apcb/intro$ pwd
/home/oneils/apcb/intro
oneils@atmosphere ~/apcb/intro$ /home/oneils/apcb/intro/myprog.sh
Hello!
This is a bit weird...
```

It ran! What we've created is known as a script to be run by an interpreter; in this case, the interpreter is `bash`. A *script* is a text file with execute permissions set, containing commands that may be run by an interpreter, usually specified through the absolute path at the top of the script with a `#!` line. An *interpreter* is a program that can execute commands, sometimes specified in a script file.

What is happening here is that the shell has noticed that the user is attempting to run an executable file, and passes the execution off to the operating system. The operating system, in turn, notices the first two bytes of the file (the `#!` characters), and rather than having the CPU run the file as binary machine code, executes the program specified on the `#!` line, passing to that program the contents of the file as "code" to be run by that program. Because in this case the interpreting program is `bash`, we can specify any commands that we can send to our shell, `bash`. Later, we'll see that we can create scripts that use much more sophisticated interpreters, like `python`, to run more sophisticated code.

According to our execution rule, we can also run our program by specifying a relative path to it, like `./myprog.sh` (which specifies to run the `myprog.sh` file found in the present working directory).

```
oneils@atmosphere ~/apcb/intro$ ./myprog.sh
Hello!
This is a bit weird...
```

This is the most common way to run files and programs that exist in the present working directory.

If we change to another present working directory, like our home directory, then in order to run the program according to the execution rule, we have to again specify either the absolute or relative path.

```
oneils@atmosphere ~/apcb/intro$ cd $HOME
oneils@atmosphere ~$ /home/oneils/apcb/intro/myprog.sh
Hello!
This is a bit weird...
oneils@atmosphere ~$ apcb/intro/myprog.sh
Hello!
This is a bit weird...
```

This process is tedious; we'd *like* to be able to specify the name of the program, but because the location of our program isn't specified in a directory listed in `$PATH`, we'll get an error.

```
oneils@atmosphere ~$ myprog.sh
-bash: myprog.sh: command not found
```

Installing a Program

To add our own programs to the system so that we can run them at will from any location, we need to:

1. Obtain or write an executable program or script.
2. Place it in a directory.
3. Ensure the absolute path to that directory can be found in `$PATH`.

Traditionally, the location to store one's own personal executables is in one's home directory, inside a directory called `local`, inside a directory called `bin`. Let's create these directories (creating them was also part of a previous exercise, so you may not need to), and move our `myprog.sh` file there.

```
oneils@atmosphere ~$ cd $HOME
oneils@atmosphere ~$ mkdir local
oneils@atmosphere ~$ mkdir local/bin
oneils@atmosphere ~$ ls
apcb     Documents  local  Pictures  Templates      Videos
Desktop  Downloads  Music  Public    todo_list.txt
oneils@atmosphere ~$ mv apcb/intro/myprog.sh local/bin
```

This accomplishes steps 1 and 2. For step 3, we need to make sure that our `local/bin` directory can be found in `$PATH`. Because `$PATH` is an environment variable, we can set it with `export`, making use of the fact that environment variables inside of double quotes (but not single quotes) are expanded to their contents.

```
oneils@atmosphere ~$ export PATH="$HOME/local/bin:$PATH"
oneils@atmosphere ~$ echo $PATH
/home/oneils/local/bin:/usr/local/sbin:/usr/local/bin:/usr/sbin:/usr/bin:/sbin:/b
:/usr/games
```

Because the right-hand side of the = is evaluated before the assignment happens, the `$PATH` variable now contains the full path to the `local/bin` directory, followed by the previous contents of `$PATH`.[15] If we type a program name without specifying a path to it, the shell will search our own install location first!

15. The corresponding command to set the `tcsh` or `csh` `$path` variable is: `set path = ("$HOME/local/bin" $path)`.

```
oneils@atmosphere ~$ myprog.sh
Hello!
This is a bit weird...
```

There's only one problem: every time we log out and log back in, modifications of environment variables that we've made are forgotten. Fortunately, `bash` looks for two important files when it starts:[16] (1) commands in the `.bash_login` file (in your home directory) are executed whenever `bash` starts as a consequence of a login session (e.g., when entering a password causes `bash` to start), and (2) commands in the `.bashrc` file (in your home directory) are executed whenever `bash` starts (e.g., on login *and* when `bash` is executed via a `#!` script).

If you want to see a friendly greeting every time you log in, for example, you might add the line `echo "Hello $USER, nice to see you again!"` to your `.bash_login` file. Because we want our `$PATH` to be modified even if `bash` somehow starts without our logging in, we'll add the `export` command to the `.bashrc` file.

The `.bashrc` file may have information in it already, representing a default shell configuration placed there when the administrator created the account. While we can add our own commands to this file, we should do so at the end, and we should be careful to not disturb the other configuration commands that are likely there for a good reason. Also, the commands in this file should be free of errors and typos—some errors are bad enough to prevent you from logging in! Using the `-w` when editing the file with `nano` will help ensure that the editor does not attempt to autowrap long commands that shouldn't be broken over multiple lines.

```
oneils@atmosphere ~$ ls -a
.               .dbus         .gnome2            Public          .vnc
..              Desktop       .gstreamer-0.10    .pulse          .Xauthority
apcb            Documents     .gvfs              .pulse-cookie   .Xdefaults
.bash_history   Downloads     .ICEauthority      .ssh            .xscreensaver
.bash_logout    .gconf        local              Templates       .xsession-errors
.bashrc         .git          Music              todo_list.txt
.cache          .gitconfig    Pictures           Videos
.config         .gitignore    .profile           .vim
oneils@atmosphere ~$ nano -w .bashrc
```

At the bottom of this file, we'll add the `export line`:

16. The corresponding files for `tcsh` and `csh` shells are `.login` and `.cshrc`, respectively.

```
## ...
export IDS_HOME="/irods/data.iplantc.org/iplant/home/oneils"
alias ids_home="cd $IDS_HOME"

## End Atmosphere System

## Added by Shawn O'Neil, for my local/bin
export PATH="$HOME/local/bin:$PATH"
```

Because lines starting with `#` are "comments" (unexecuted, aside from the `#!` line, of course), we can use this feature to remind our future selves how that line got in the file. Because commands in these files are only executed when the shell starts, in order to activate the changes, it suffices to log out and log back in.

Exercises

1. Suppose a file has the following permissions listed by `ls -l`: `-rwxrw-r--`. What does this permission string indicate the about the file?
2. What is the difference between `export PATH="$HOME/local/bin:$PATH"` and `export PATH="$PATH:$HOME/local/bin"`? In what situations might the former be preferred over the latter?
3. Carefully add the line `export PATH="$HOME/local/bin:$PATH"` to your `.bashrc` (assuming you have a `local/bin` directory in your home directory, and your default shell is `bash`). Be sure not to alter any lines already present, or to create any typos that might prevent you from logging in.
4. Create an executable script in your `local/bin` directory called `myinfo.sh`, which runs `echo` on the `$HOME`, `$PWD`, and `$USER` environment variables, and also runs the `date` utility. Try running it by just running `myinfo.sh` from your home directory (you may need to log out and back in first, to get the shell to recognize the change in the contents of paths listed in `$PATH` if you modified your `.bashrc`).
5. Executable `bash` scripts that start with `#!/bin/bash` will work fine, provided that the `bash` program lives in the `/bin` directory. On any system where this isn't the case (probably a rare occurrence), it won't.

Try creating a `bash` script where the first line is `#!/usr/bin/env bash`. The `env` program uses the `$PATH` variable to locate the `bash` executable, and passes off interpretation of the script to the located `bash` interpreter. This poses the same problem: what if `env` is not located in `/usr/bin`? Fortunately, this has been an agreed-upon location for the `env` program for decades, so scripts written in this manner are portable across more machines.

Chapter 6

Installing (Bioinformatics) Software

Ideally, the computational infrastructure to which you have access already includes a host of specialized software packages needed for your work, and the software installations are kept up to date as developers make improvements. If this isn't the case, you might consider bribing your local system administrator with sweets and caffeine. Failing that, you're likely to have to install the software you need yourself.

Installing more sophisticated software than the simple scripts described in chapter 5, "Permissions and Executables," will follow the same basic pattern: (1) obtain executable files, (2) get them into `$HOME/local/bin`, and (3) ensure that `$HOME/local/bin` is present in the `$PATH` environment variable. Chapter 5 covered step 3, which needs to be done only once for our account. Steps 2 and 3, however, are often quite different depending on how the software is distributed.

In this chapter, we're going to run through an example of installing and running a bioinformatics suite known as HMMER. This software searches for protein sequence matches (from a set of sequences) based on a probabilistic hidden Markov model (HMM) of a set of similar protein sequences, as in orthologous proteins from different species. The motivation for choosing this example is not so we can learn about HMM modeling or this software suite specifically, but rather that it is a representative task requiring users to download files, install software in different ways, and obtain data from public repositories.

The first step to installing HMMER is to find it online. A simple web search takes us to the homepage:

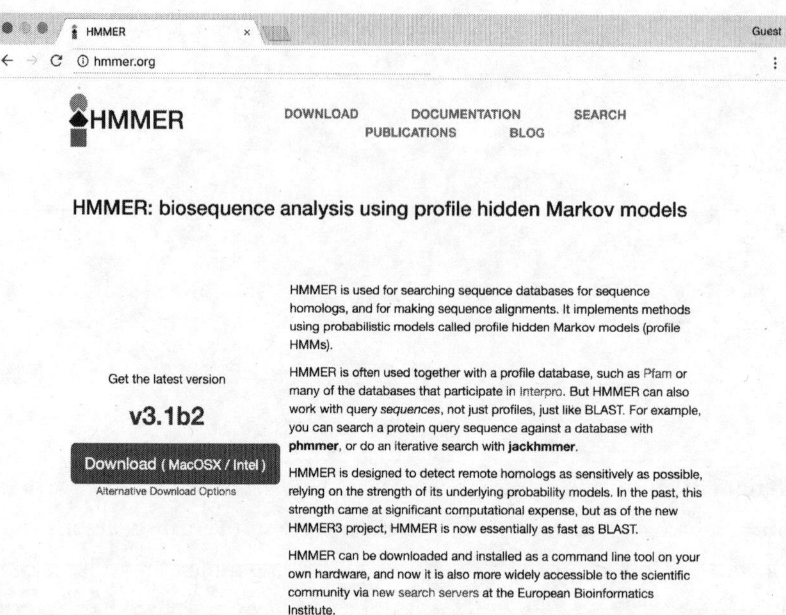

Conveniently, we see a nice large “Download” button, but the button indicates that the download is made for MacOS X/Intel, the operating system running on my personal laptop. Because we are remotely logged in to a Linux computer, this download won’t work for us. Clicking the “Alternative Download Options” link reveals options that might work for the Linux system we’re using.

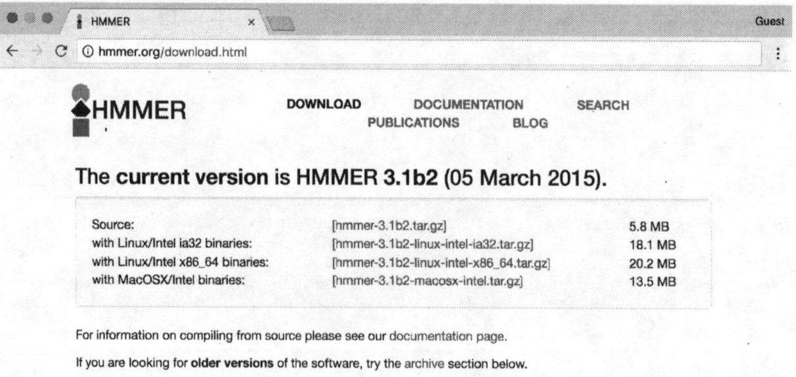

In this screenshot, we see a number of interesting download options, including one for "Source," two for "Linux binaries," and below a suggestion of some documentation, to which we'll return later.

Source or Binary?

Some bioinformatics software is created as a simple script of the kind discussed in chapter 5: a text file with a `#!` line referencing an interpreting program (that is hopefully installed on the system) and made executable with `chmod`.

But it turns out that such interpreted programs are slow owing to the extra layer of execution, and for some applications, the convenience and relative ease aren't worth the loss in speed. In these cases, software may be written in a *compiled* language, meaning that the program code starts as human-readable "source code" but is then processed into machine-readable binary code. The trick is that the process of compilation needs to be independently performed for each type of CPU. Although there are fewer CPU types in common use than in days past, both 32- and 64-bit x86 CPU architectures are still common, and software compiled for one won't work on the other. If the developer has made available compiled binaries compatible with our system, then so much the better: we can download them, ensure they are executable, and place them in `$HOME/local/bin`. Alternatively, we may need to download the source code files and perform the compilation ourselves. In some cases, developers distribute binaries, but certain features of the program can be customized in the compilation process.

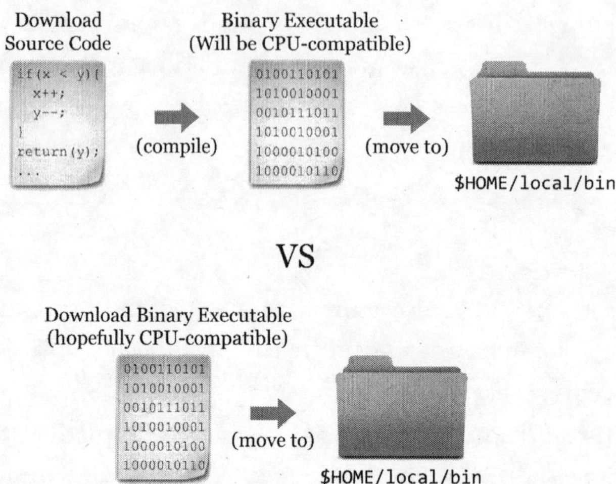

For the sake of completeness, we'll do a source install of HMMER; later, we'll get some other software as binaries.[17]

Downloading and Unpacking

We're going to download the source files for HMMER; first, we are going to create a new directory to store downloads, called `downloads`, in our home directory (you may already have such a directory).

```
oneils@atmosphere ~$ cd $HOME
oneils@atmosphere ~$ mkdir downloads
oneils@atmosphere ~$ cd downloads/
oneils@atmosphere ~/downloads$
```

If we were to click on the link in the HMMER download page, the web browser would attempt to download the file located at the corresponding URL (http://eddylab.org/software/hmmer3/3.1b2/hmmer-3.1b2.tar.gz) to the local desktop. Because we want the file downloaded to the remote system, clicking on the download button won't work. What we need is a tool called `wget`, which

17. If you have administrator privileges on the machine, software repositories curated with many packages are also available. Depending on the system, if you log in as `root`, installing HMMER may be as simple as running `apt-get install hmmer` or `yum install hmmer`.

can download files from the Internet on the command line.[18] The `wget` utility takes at least one important parameter, the URL, to download. It's usually a good idea to put URLs in quotes, because they often have characters that confuse the shell and would need to be escaped or quoted. Additionally, we can specify `-O <filename>`, where `<filename>` is the name to use when saving the file. Although not required in this instance, it can be useful for URLs whose ending file names aren't reasonable (like `index.php?query=fasta&search=drosophila`).

```
oneils@atmosphere ~/downloads$ wget 'http://eddylab.org/software/hmmer3/3.1b1/h
mmer-3.1b1.tar.gz' -O hmmer-3.1b1.tar.gz
```

At this point, we have a file ending in `.tar.gz`, known as a "gzipped tarball," representing a collection of files that have first been combined into a single file (a tarball), and then compressed (with the `gzip` utility).

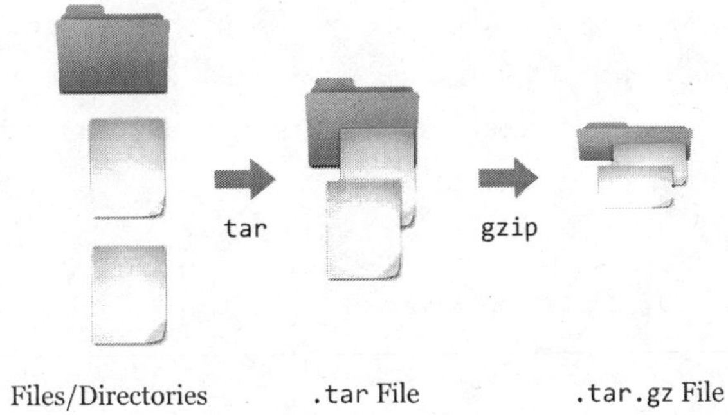

To get the contents out, we have to reverse this process. First, we'll un-gzip the file with `gzip -d hmmer-3.1b1.tar.gz`, which will replace the file with the un-gzipped `hmmer-3.1b1.tar`.[19] From there, we can un-tar the tarball with `tar -xf hmmer-3.1b1.tar` (the `-x` indicates extract, and the `f` indicates that the data will be extracted from the specified file name).

18. A similar tool called `curl` can be used for the same purpose. The feature sets are slightly different, so in some cases `curl` is preferred over `wget` and vice versa. For the simple downloading tasks in this book, either will suffice.

19. The `gzip` utility is one of the few programs that care about file extensions. While most programs will work with a file of any extension, `gzip` requires a file that ends in `.gz`. If you are unsure of a file's type, the `file` utility can help; for example, `file hmmer-3.1b1.tar.gz` reports that the file is `gzip`-compressed data, and would do so even if the file did not end in `.gz`.

```
oneils@atmosphere ~/downloads$ ls
hmmer-3.1b1.tar.gz
oneils@atmosphere ~/downloads$ gzip -d hmmer-3.1b1.tar.gz
oneils@atmosphere ~/downloads$ ls
hmmer-3.1b1.tar
oneils@atmosphere ~/downloads$ tar -xf hmmer-3.1b1.tar
oneils@atmosphere ~/downloads$ ls
hmmer-3.1b1  hmmer-3.1b1.tar
```

It looks like the gzipped tarball contained a directory, called `hmmer-3.1b1`.

Other Download and Compression Methods

Before continuing to work with the downloaded source code, there are a couple of things to note regarding compressed files and downloading. First, although gzipped tarballs are the most commonly used compression format for Unix-like systems, other compression types may also be found. They can usually be identified by the file extension. Different tools are available for each type, though there is also a generic `uncompress` utility that can handle most common types.

Extension	Decompress Command
`file.bz2`	`bunzip2 file.bz2`
`file.zip`	`unzip file.zip`
`file.tgz`	Same as for `.tar.gz`

The most common syntax for creating a gzipped tarball uses the `tar` utility, which can do both jobs of tarring and gzipping the inputs. As an example, the command `tar -cvzf hmmer_compress_copy.tar.gz hmmer-3.1b1` would create (`c`), with verbose output (`v`), a gzipped (`z`) tarball in a file (`f`) called `hmmer_compress_copy.tar.gz` from the input directory `hmmer-3.1b1`.

Traditionally, zipped files of source code were the most common way to distribute software. More recently, version control systems (used by developers to track changes to their software over time) have become web-enabled as a way to distribute software to end-users. One such system is `git`, which allows users to download entire directories of files using a "clone URL" over the web. GitHub is a similarly popular page for hosting these projects. Here's a screenshot of the GitHub page for Sickle, a fast-quality trimmer for high-throughput sequence data.

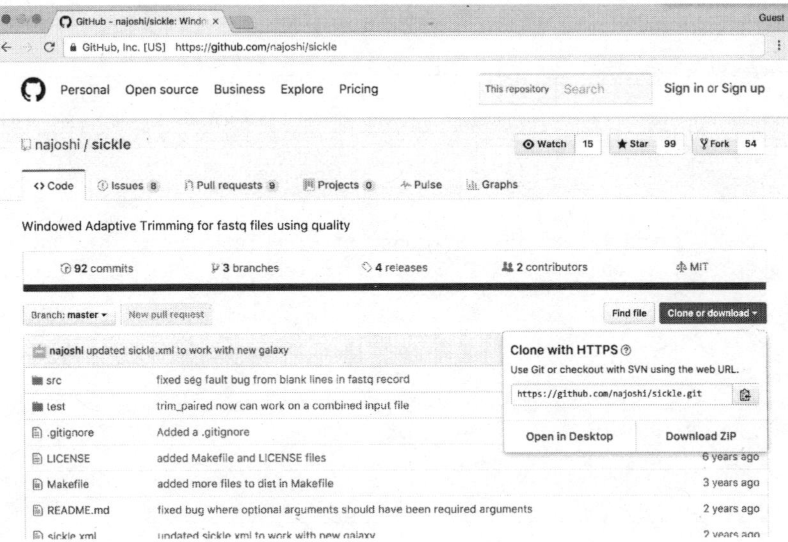

The GitHub "HTTPS clone URL" is shown after clicking on the green "Clone or download" link; to download the Sickle source on the command line, one need only run a command like `git clone https://github.com/najoshi/sickle.git`, provided the `git` program is installed. (The `git` program can also be used on the command line for tracking "snapshots" of files in a directory over time, providing a log of the work done. Such logged directories can then be synced to GitHub, making it relatively easy to share projects with others.)

Compiling the Source

Having downloaded and unpacked the HMMER source code, the first step is to check the contents of the directory and look for any `README` or `INSTALL` files. Such files are often included and contain important information from the software developer.

```
oneils@atmosphere ~/downloads$ cd hmmer-3.1b1/
oneils@atmosphere ~/downloads/hmmer-3.1b1$ ls
aclocal.m4     configure.ac   INSTALL        Makefile.in    src
config.guess   COPYRIGHT      install-sh     profmark       testsuite
config.sub     documentation  libdivsufsort  README         tutorial
configure      easel          LICENSE        RELEASE-NOTES  Userguide.pdf
```

Taking a look at the contents of the `hmmer-3.1b1` directory, there is an `INSTALL` file, which we should read with `less`. Here's the top part of the file:

```
Brief installation instructions
HMMER 3.1b1; May 2013
-------------------------------------------------------------

These are quick installation instructions. For complete documentation,
including customization and troubleshooting, please see the
Installation chapter in the HMMER User's Guide (Userguide.pdf).

Starting from a source distribution, hmmer-3.1b1.tar.gz:
 uncompress:                  uncompress hmmer-3.1b1.tar.gz
 unpack:                      tar xf hmmer-3.1b1.tar
 move into new directory:     cd hmmer-3.1b1
 configure:                   ./configure
 build:                       make
 automated tests:             make check
 automated install:           make install
```

The installation documentation describes a number of commands, including many we've already run (for extracting the data from the gzipped tarball). There are also four more commands listed: `./configure`, `make`, `make check`, and `make install`. Three of these comprise the "canonical install process"—`make check` is an optional step to check the success of the process midway through. The three important steps are: (1) `./configure`, (2) `make`, and (3) `make install`.

1. The contents of the directory (above) include `configure` as an executable script, and the command `./configure` executes the script from the present working directory. This script usually verifies that all of the prerequisite libraries and programs are installed on the system. More importantly, this step may set some environment variables or create a file called `Makefile`, within which will be instructions detailing how the compilation and installation process should proceed, customized for the system.
2. Actually, `make` is an interpreting program much like `bash` (`which make` is likely to return `/usr/bin/make`—it's a binary program). When running `make`, its default behavior is to look for a file called `Makefile` in the current directory, and run a default set of commands specified in the `Makefile` in the order specified. In this case, these default commands run the compilation programs that turn the source code into executable binaries.
3. The `make install` command again executes `make`, which looks for the `Makefile`, but this time we are specifying that the "install" set of commands in the `Makefile` should run. This step

copies the binary executable files (and other supporting files, if necessary) to the install location.

This final step, `make install`, may lead us to ask: what is the install location? By default, it will be something like `/usr/bin`—a system-wide location writable to by only the administrator. So, unless we are logged in as `root` (the administrator), the final step in the process will fail. We must specify the install location, and although the install itself happens in the third step, the entire process is configured in the first step. There may be many options that we can specify in the `./configure` step, though the install location (known as the `PREFIX`) is by far the most commonly used. Running `./configure --help` prints a lot of information; here's the relevant section:

```
Installation directories:
  --prefix=PREFIX         install architecture-independent files in PREFIX
                          [/usr/local]
  --exec-prefix=EPREFIX   install architecture-dependent files in EPREFIX
                          [PREFIX]
```

The `--prefix` option is the one we'll use to determine where the binaries should be located. Although our executable binaries will eventually go in `$HOME/local/bin`, for this option we're going to specify `$HOME/local`, because the `bin` portion of the path is implied (and other directories like `lib` and `share` might also be created alongside the `bin` directory). Finally, our modified canonical install process will consist of three steps: `./configure --prefix=$HOME/local`, `make`, and `make install`.

```
oneils@atmosphere ~/downloads/hmmer-3.1b1$ ./configure --prefix=$HOME/local
configure: Configuring HMMER for your system.
checking build system type... x86_64-unknown-linux-gnu
checking host system type... x86_64-unknown-linux-gnu
...
```

```
oneils@atmosphere ~/downloads/hmmer-3.1b1$ make
     SUBDIR easel
make[1]: Entering directory `/home/oneils/downloads/hmmer-3.1b1/easel'
     CC easel.o
     CC esl_alphabet.o
...
```

```
oneils@atmosphere ~/downloads/hmmer-3.1b1$ make install
/usr/bin/install -c -d /home/oneils/local/bin
/usr/bin/install -c -d /home/oneils/local/lib
/usr/bin/install -c -d /home/oneils/local/include
...
```

At this point, if we navigate to our `$HOME/local` directory, we will see the added directories and binary files.

```
oneils@atmosphere ~/downloads/hmmer-3.1b1$ cd $HOME/local
oneils@atmosphere ~/local$ ls
bin  include  lib  share
oneils@atmosphere ~/local$ cd bin
oneils@atmosphere ~/local/bin$ ls
alimask   hmmc2       hmmfetch  hmmpress   hmmsim     myprog.sh  phmmer
hmmalign  hmmconvert  hmmlogo   hmmscan    hmmstat    nhmmer
hmmbuild  hmmemit     hmmpgmd   hmmsearch  jackhmmer  nhmmscan
```

Because these executable files exist in a directory listed in the `$PATH` variable, we can, as always, type their names on the command prompt when working in any directory to run them. (Though, again, we may need to log out and back in to get the shell to see these new programs.[20])

Installation from Binaries

Our objective is to run HMMER to search for a sequence-set profile in a larger database of sequences. For details, the HMMER documentation (available on the website) is highly recommended, particularly the "Tutorial" section, which describes turning a multiple alignment of sequences into a profile (with `hmmbuild`) and searching that profile against the larger set (with `hmmsearch`). It is also useful to read the peer-reviewed publication that describes the algorithms implemented by HMMER or any other bioinformatics software. Even if the material is outside your area of expertise, it will reveal the strengths and weaknesses of software.

20. It's not strictly necessary to log back out and back in; when working in bash, running `hash -r` will cause the shell to update its list of software found in `$PATH`.

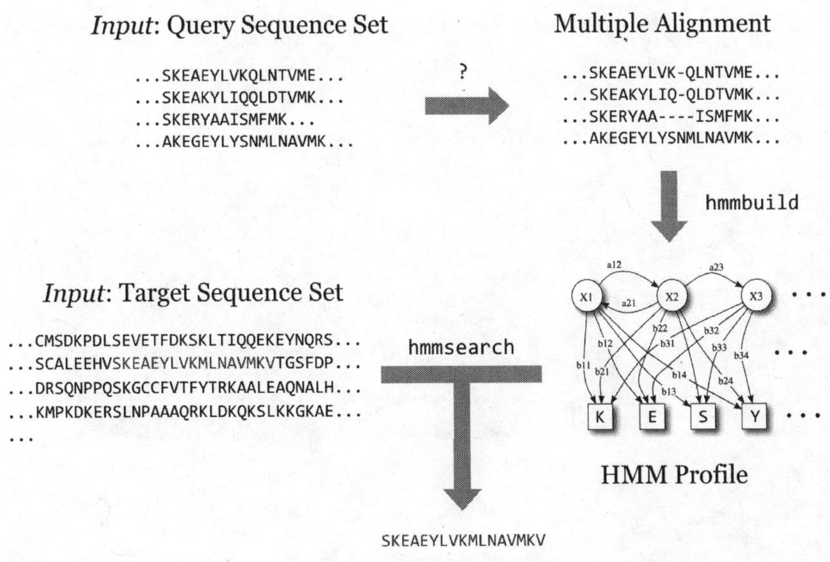

We'll soon get to downloading query and target sequence sets, but we'll quickly come to realize that although the programs in the HMMER suite can produce the profile and search it against the target set, they cannot produce a multiple alignment from a set of sequences that are similar but not all the same length. Although there are many multiple-alignment tools with different features, we'll download the relatively popular `muscle`. This time, we'll install it from binaries.

It's worth discussing how one goes about discovering these sequences of steps, and which tools to use. The following strategies generally work well, though creativity is almost always rewarded.

1. Read the methods sections of papers with similar goals.
2. Ask your colleagues.
3. Search the Internet.
4. Read the documentation and published papers for tools you are already familiar with, as well as those publications that cite them.
5. Don't let the apparent complexity of an analysis prevent you from taking the first steps. Most types of analyses employ a number of steps and many tools, and you may not have a clear picture of what the final procedure will be. Experiment with alternative tools, and look for

help when you get stuck. Be sure to document your work, as you will inevitably want to retrace your steps.

If we visit the `muscle` homepage, we'll see a variety of download options, including binaries for our system, Linux.

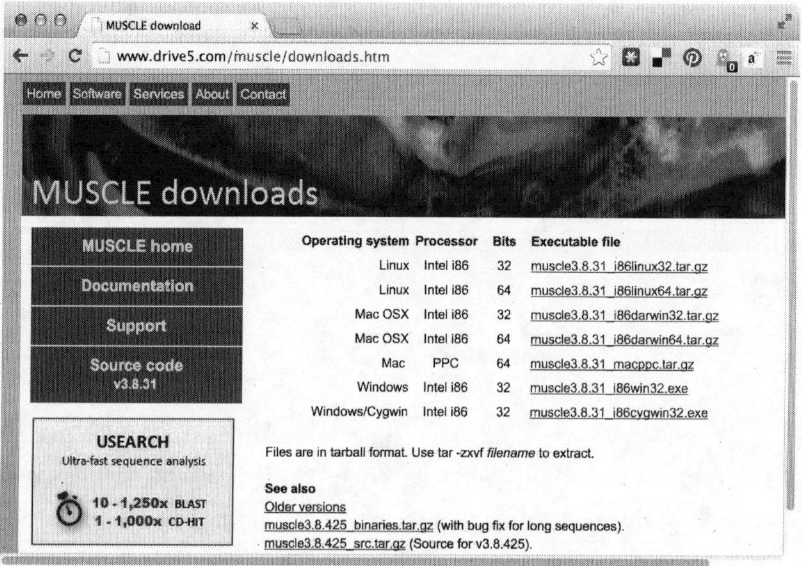

Unfortunately, there appear to be two options for Linux binaries: 32-bit and 64-bit. How do we know which of these we want? We can get a hint by running the `uname` program, along with the `-a` parameter to give as much information as possible.

```
oneils@atmosphere ~$ uname -a
Linux 172.31.88.3 3.2.0-37-virtual #58-Ubuntu SMP Thu Jan 24 15:48:03 UTC 2013
x86_64 x86_64 x86_64 GNU/Linux
```

The `uname` program gives information about the operating system, which in this case appears to be GNU/Linux for a 64-bit, x86 CPU. If any of the binaries are likely to work, it will be the "i86linux64" set. We'll `wget` that gzipped tarball in the `downloads` directory.

```
oneils@atmosphere ~$ cd downloads/
oneils@atmosphere ~/downloads$ wget 'http://www.drive5.com/muscle/downloads3.8.31
8.31_i86linux64.tar.gz'
```

Note that in this case we haven't used the -O option for wget, because the file name described by the URL (muscle3.8.31_i86linux64.tar.gz) is what we would like to call the file when it is downloaded anyway. Continuing on to unpack it, we find it contains only an executable that we can attempt to run.

```
oneils@atmosphere ~/downloads$ ls
hmmer-3.1b1  hmmer-3.1b1.tar  muscle3.8.31_i86linux64.tar.gz
oneils@atmosphere ~/downloads$ gzip -d muscle3.8.31_i86linux64.tar.gz
oneils@atmosphere ~/downloads$ tar -xf muscle3.8.31_i86linux64.tar
oneils@atmosphere ~/downloads$ ls
hmmer-3.1b1      muscle3.8.31_i86linux64
hmmer-3.1b1.tar  muscle3.8.31_i86linux64.tar
oneils@atmosphere ~/downloads$ ./muscle3.8.31_i86linux64

MUSCLE v3.8.31 by Robert C. Edgar

http://www.drive5.com/muscle
This software is donated to the public domain.
...
```

Because it didn't report an execution error, we can install it by copying it to our $HOME/local/bin directory. While doing so, we'll give it a simpler name, muscle.

```
oneils@atmosphere ~/downloads$ cp muscle3.8.31_i86linux64 $HOME/local/bin/muscle
oneils@atmosphere ~/downloads$
```

Now our multiple aligner, muscle, is installed!

Exercises

1. Follow the steps above to install the HMMER suite (from source) as well as muscle (from binaries) in your $HOME/local/bin directory. Ensure that you can run them from anywhere (including from your home directory) by running muscle --help and hmmsearch --help. Both commands should display help text instead of an error. Further, check that the versions being found by the shell are from your home directory by running which hmmsearch and which muscle.

2. Determine whether you have the "NCBI Blast+" tools installed by searching for the blastn program. If they are installed, where are they located? If they are not installed, find them and install them from binaries.

3. Install `sickle` from the `git` repo at https://github.com/najoshi/sickle. To install it, you will need to follow the custom instructions inside of the `README.md` file. If you don't have the `git` program, it is available for binary and source install at http://git-scm.com.

Getting Data

Now that we've got the software installed for our example analysis, we'll need to get some data. Supposing we don't have any novel data to work with, we'll ask the following question: can we identify, using HMMER and `muscle`, homologues of P450-1A1 genes in the *Drosophila melanogaster* protein data set? (This is a useless example, as the *D. melanogaster* genome is already well annotated.)

The first step will be to download the *D. melanogaster* data set, which we can find on http://flybase.org, the genome repository for *Drosophila* genomes. Generally, genome repositories like FlyBase provide full data sets for download, but they can be difficult to find. To start, we'll navigate to "Files," then "Releases (FTP)."

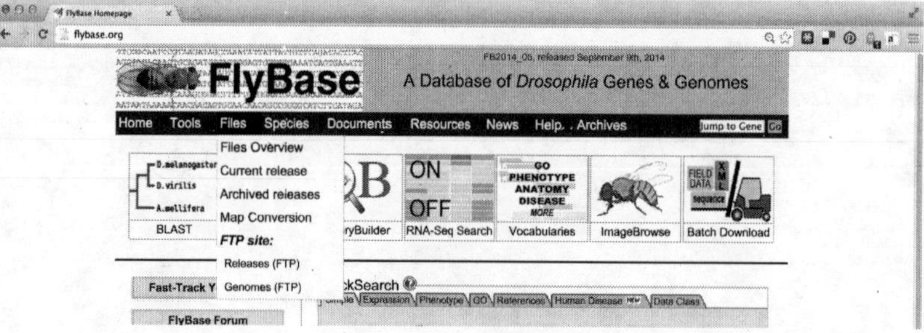

From there, we'll navigate to a recent release, like FB2014_05. Because genomic information is often being updated as better information becomes available, newer versions are indicated by newer releases.

FB2014_02/		3/18/14, 12:00:00 AM
FB2014_03/		5/9/14, 3:23:00 PM
FB2014_04/		7/21/14, 6:51:00 PM
FB2014_05/		9/9/14, 8:26:00 PM
current	0 B	9/9/14, 8:28:00 PM

Next we'll see links for specific species; we'll work with `dmel_r6.02`. It is often a good idea to note specific release numbers or release dates for data sets you download, for eventual description in the methods sections of any papers you write that are based on those data.

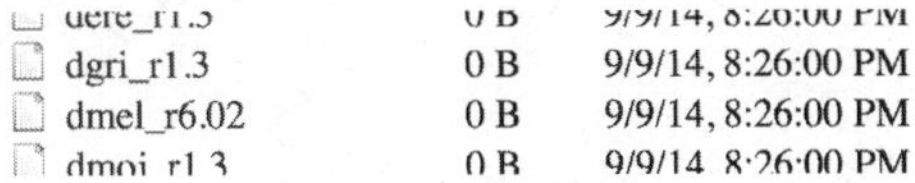

dgri_r1.3	0 B	9/9/14, 8:26:00 PM
dmel_r6.02	0 B	9/9/14, 8:26:00 PM

The most common format for sequence information is known as FASTA and is the format we want, so the next step is to navigate to the `fasta` directory. Other potentially interesting options include the `gff` and `gtf` directories, which hold text-based annotation files describing the location and function of genes in the genome. (These formats are often useful for RNA-seq analysis.)

dna/	9/9/14, 5:24:00 PM
fasta/	9/9/14, 5:59:00 PM
gff/	9/9/14, 6:00:00 PM
gtf/	9/9/14, 5:36:00 PM

Finally, we see a variety of gzipped files we can `wget` on the command line. Because we are interested in the full protein data set for this species, we will use the URL for `dmel-all-translation-r6.02.fasta.gz`.

dmel-all-translation-r6.02.fasta.gz	7.6 MB	9/9/14, 1:23:00 PM

Before running `wget` on the file, we'll create a `projects` directory in our home directory, and a `p450s` directory inside there to work in.

```
oneils@atmosphere ~/downloads$ cd $HOME
oneils@atmosphere ~$ mkdir projects
oneils@atmosphere ~$ cd projects/
oneils@atmosphere ~/projects$ mkdir p450s
oneils@atmosphere ~/projects$ cd p450s/
oneils@atmosphere ~/projects/p450s$ wget 'ftp://ftp.flybase.net/releases/FB2014_
05/dmel_r6.02/fasta/dmel-all-translation-r6.02.fasta.gz'
```

Because the file is gzipped, we can use `gzip -d` to decompress it, and then use `less -S` to view the results without the long lines wrapped in the terminal window.

```
oneils@atmosphere ~/projects/p450s$ gzip -d dmel-all-translation-r6.02.fasta.gz
oneils@atmosphere ~/projects/p450s$ less -S dmel-all-translation-r6.02.fasta
```

The result illustrates the standard format for a FASTA file. Each sequence record begins with line starting with a > character, and the first non-whitespace-containing word following that is considered the sequence ID.

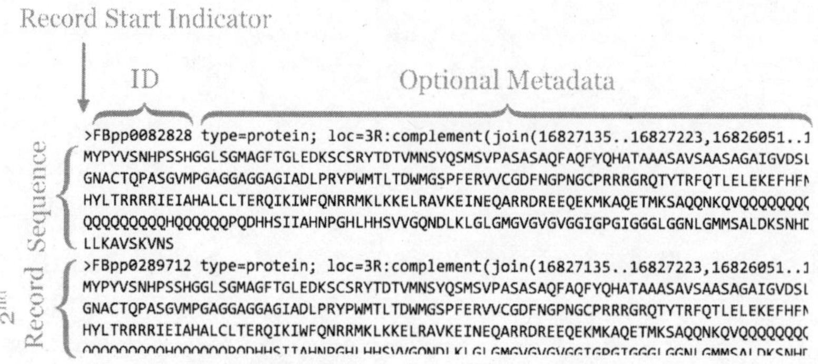

This line might then contain *whitespace* characters and metadata. Whitespace comprises a sequence of one or more spaces, tabs (represented in Unix/Linux as a special character sometimes written as `\t`), or newlines (represented in Unix/Linux as a special character sometimes written as `\n`) in a row.

Lines following the header line contain the sequence information, and there is no specific format for the number of lines over which the sequence may be broken, or how long those lines should be. After the last sequence line for a record, a new sequence record may start.

Depending on the source of the FASTA file, the IDs or metadata may represent multiple pieces of data; in this example, the metadata are separated by spaces and have a `<label>=<value>;` format that is specific to protein sequences from FlyBase.

For our next trick, we'll download some P450-1A1 protein sequences from Uniprot.org. Uniprot.org is a well-known protein database, and it is composed of the "TrEMBL" database and the subset of TrEMBL, known as "Swiss-Prot." While the former contains many sequences with annotations, many of those annotations have been assigned by automated homology searches and have not been reviewed. The latter, Swiss-Prot, contains only sequences whose annotations have been manually reviewed.

For the download, we'll enter "p450 1A1" into the search field, and we'll filter the results to only those in Swiss-Prot by clicking on the "Reviewed" link, resulting in 28 matches. Next, we can click on the "Download" button to download a "FASTA (canonical)" (rather than with all isoforms included) file.

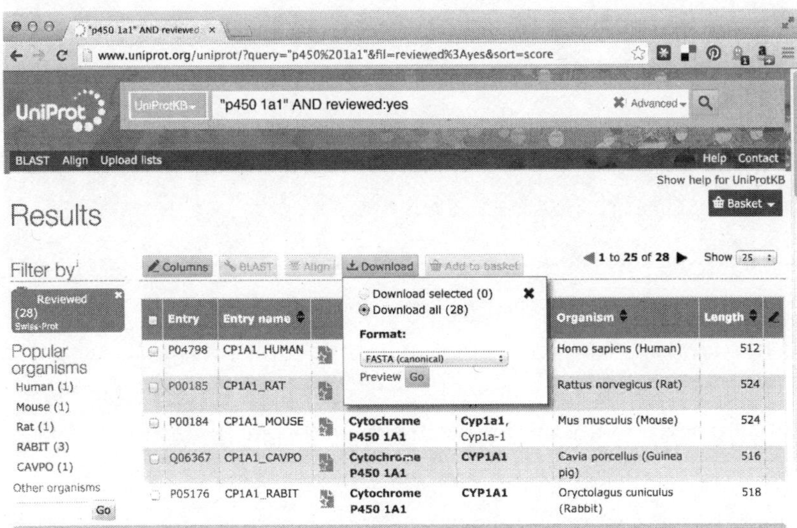

The Uniprot website recently underwent a redesign, such that the downloaded file is transferred directly to the web browser, rather than presented as a URL that could be accessed with `wget`. This isn't a problem, as it gives us a chance to discuss how to transfer files between the remote system and local desktop via SFTP.

Like SSH, SFTP is a common client/server protocol. Provided the server is running on the remote computer (in fact, it uses the same port as SSH, port 22, because SSH provides the secure connection), we just need to install and run an SFTP client on our desktop. There are many SFTP clients available for Microsoft Windows (e.g., Core-FTP), OS X (e.g., Cyberduck), and Linux systems (e.g., `sftp` on the command line or the graphical FileZilla). The client discussed here is called FireFTP, and it is available as an extension for the Mozilla Firefox web browser (itself available for Windows, OS X, and Linux). To get it requires installing and running Firefox, navigating to Tools → Addons, and searching for "FireFTP."

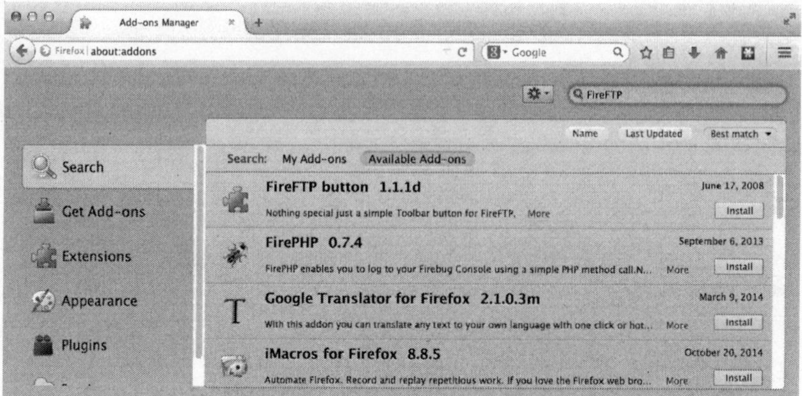

Once the plugin is installed (which requires a restart of Firefox), we can access it from the Tools → Developer submenu. Connecting the client to a remote computer requires that we first configure the connection by selecting "Create an account." The basic required information includes an account name, the host to connect to (e.g., an IP address like `128.196.64.120` or a host name like `files.institution.edu`), as well as our login name and password.

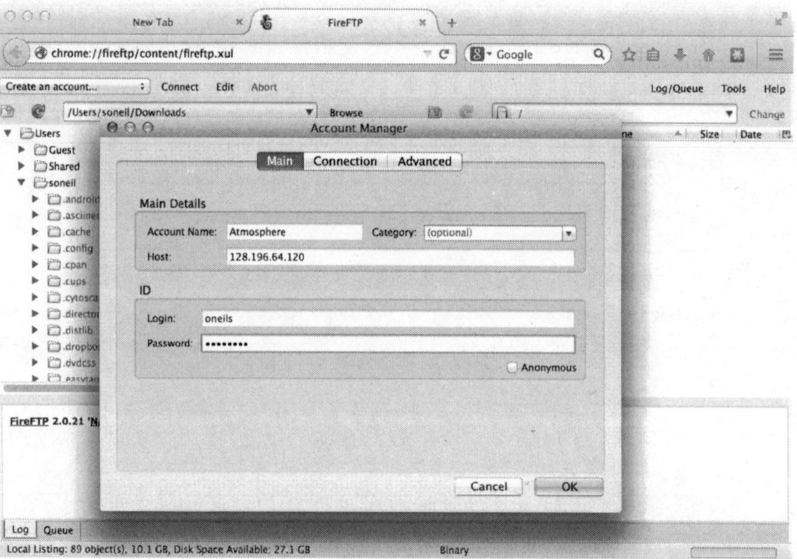

We also need to tell the client which protocol to connect with, which is done on the "Connection" tab; we want SFTP on port 22.

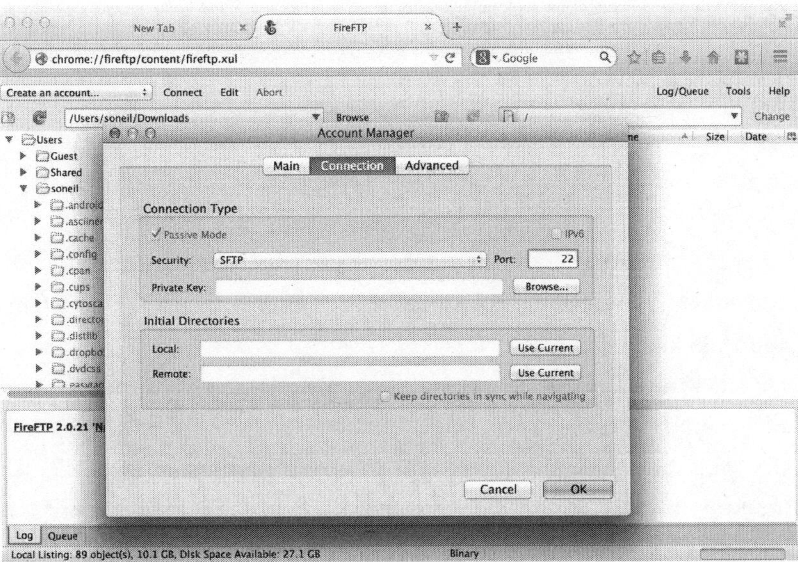

With that accomplished, we can transfer any file back and forth using the green arrows in the interface, where the remote filesystem is shown on the right and the local filesystem is shown on the left. Here's the result after transferring our `p450s.fasta` file.

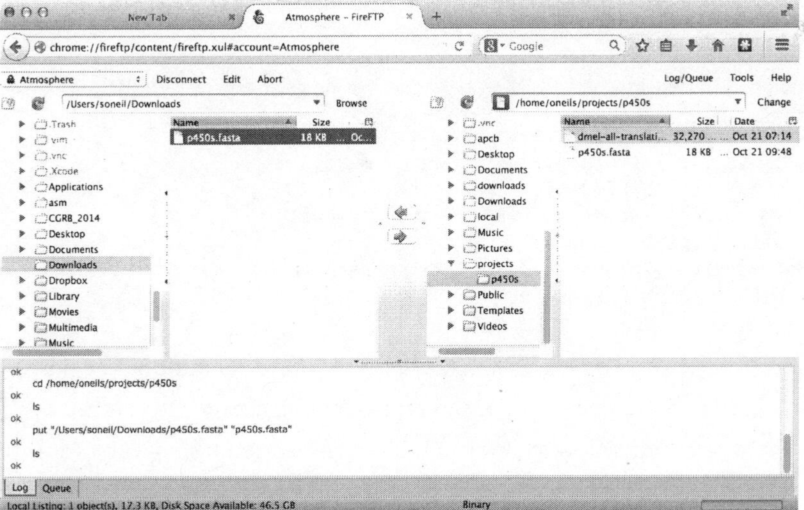

DOS/Windows and Unix/Linux Newlines

For the most part, the way text is encoded on Microsoft operating systems (like DOS and Windows) and on Unix-like systems (like Linux and OS X) is similar. But there is one difference: how the ends of lines, or "newline characters" are represented. In Unix-like systems, a newline is represented by a single 8-bit byte (the "Line Feed" (NF) character): `00001010`. On Microsoft systems, they are represented by a pair of 8-bit bytes ("Carriage Return" (CR) followed by NF): `0000110100001010`. This means that text files created on Microsoft operating files may not be readable on Unix-like systems, and vice versa.

Fortunately, there are utilities available for converting between these formats. On the command line, the utilities `dos2unix` and `unix2dos` convert to and from Unix-like format, respectively. This isn't often an issue, as most file transfer programs (FireFTP included) automatically perform the appropriate conversion. The command-line utility `file` can also be used to determine the type of a file, including its newline type.

Putting It All Together

At this point, we've obtained the data we wish to process, and we've successfully installed the software we intend to run on that data.

```
oneils@atmosphere ~/projects/p450s$ ls
dmel-all-translation-r6.02.fasta  p450s.fasta
oneils@atmosphere ~/projects/p450s$ which muscle
/home/oneils/local/bin/muscle
oneils@atmosphere ~/projects/p450s$ which hmmbuild
/home/oneils/local/bin/hmmbuild
oneils@atmosphere ~/projects/p450s$ which hmmsearch
/home/oneils/local/bin/hmmsearch
```

The first step is to run `muscle` on the `p450s.fasta` file to produce a multiple alignment. One quick way to identify how a program like `muscle` should be run (what parameters it takes) is to run it without any parameters. Alternatively, we could try the most common options for getting help: `muscle -h`, `muscle --help`, `muscle --h`, or `muscle -help`.

```
oneils@atmosphere ~/projects/p450s$ muscle

MUSCLE v3.8.31 by Robert C. Edgar

http://www.drive5.com/muscle
This software is donated to the public domain.
Please cite: Edgar, R.C. Nucleic Acids Res 32(5), 1792-97.


Basic usage

    muscle -in <inputfile> -out <outputfile>

Common options (for a complete list please see the User Guide):

    -in <inputfile>    Input file in FASTA format (default stdin)
    -out <outputfile>  Output alignment in FASTA format (default stdout)
    -diags             Find diagonals (faster for similar sequences)
    -maxiters <n>      Maximum number of iterations (integer, default 16)
    -maxhours <h>      Maximum time to iterate in hours (default no limit)
    -html              Write output in HTML format (default FASTA)
...
```

The most important line of this help text is the usage information: `muscle -in <inputfile> -out <outputfile>`; parameters in angle brackets indicate that the parameter is required (nonrequired parameters often appear in straight brackets). Further help text indicates other parameters that we could opt to add. Presumably, they could be placed before or after the input or output specifiers. So, we'll run `muscle` on our `p450s.fasta` file, and produce a file whose file name indicates its pedigree in some way:

```
oneils@atmosphere ~/projects/p450s$ muscle -in p450s.fasta -out p450s.fasta.aln
```

Once the command has finished executing, we can view the alignment file with `less -S p450s.fasta.aln`.

```
>sp|P05176|CP1A1_RABIT Cytochrome P450 1A1 OS=Oryctolagus cuniculus GN=CYP1A1 PE
------------------------------------------------------------
---------------------------------------------------MVSDFGLPT
FISATELLLASAVFCLVFWVAGASKPRVPKGLKRLPGPWGWPLLGHVLTLGK---NPHV-
---ALARLSRRYGDVFQIRLGSTPVVVLSGLDTIKQALVRQGDDFKGRPDLYSFSFVTK-
-----------------------------GQSMIFGSDSGPV---------WAARRRLAQ
NALNSFSVAS--------------------------------------------------
----------------DPASSSSCYLEEHV---SQEAENLISK-FQELMAAVGH------
```

With further inspection, we'd see that the sequences have been made the same length by the insertion of gap characters. The next step is to run `hmmbuild` to produce the HMM profile. Again, we'll run `hmmbuild` without any options to get information on what parameters it needs.

```
oneils@atmosphere ~/projects/p450s$ hmmbuild
Incorrect number of command line arguments.
Usage: hmmbuild [-options] <hmmfile_out> <msafile>

where basic options are:
  -h     : show brief help on version and usage
  -n <s> : name the HMM <s>
  -o <f> : direct summary output to file <f>, not stdout
  -O <f> : resave annotated, possibly modified MSA to file <f>

To see more help on other available options, do:
  hmmbuild -h
```

The help output for `hmmbuild` is shorter, though the command also notes that we could run `hmmbuild -h` for more detailed information. The usage line, `hmmbuild [-options] <hmmfile_out> <msafile>`, indicates that the last two parameters are required, and are the name of the output file (for the profile HMM) and the multiple sequence alignment input file. The brackets indicate that, before these last two parameters, a number of optional parameters may be given, described later in the help output. In this case, `<hmmfile_out>` and `<msafile>` are *positional*: the second-to-last argument must specify the output, and the last must specify the input.

```
oneils@atmosphere ~/projects/p450s$ hmmbuild p450s.fasta.aln.hmm p450s.fasta.aln
```

After this operation finishes, it may be interesting to take a look at the resulting HMM file with `less -S p450s.fasta.aln.hmm`. Here's a snippet:

```
CKSUM 797170701
STATS LOCAL MSV       -12.2089  0.69650
STATS LOCAL VITERBI   -13.2160  0.69650
STATS LOCAL FORWARD    -6.4971  0.69650
HMM          A        C        D        E        F        G        H        I
            m->m     m->i     m->d     i->m     i->i     d->m     d->d
  COMPO   2.56763  4.18442  2.94826  2.71355  3.22576  2.92226  3.70840  2.94423
          2.68504  4.42336  2.77569  2.73176  3.46465  2.40496  3.72606  3.29362
          0.66606  1.72211  1.17901  3.05405  0.04832  0.00000        *
      1   3.15916  4.51751  4.70912  4.16707  3.23103  4.39232  4.76778  1.44083
```

With some documentation reading, we may even be able to decode how the probabilistic profile is represented in this matrix of letters and numbers. As a reminder, our project directory now contains the original sequence file, a multiple-alignment file, and the HMM profile file, as well as the *D. melanogaster* protein file in which we wish to search for the profile.

```
oneils@atmosphere ~/projects/p450s$ ls
dmel-all-translation-r6.02.fasta  p450s.fasta.aln
p450s.fasta                       p450s.fasta.aln.hmm
```

At this point, we are ready to search for the profile in the *D. melanogaster* protein set with `hmmsearch`. As usual, we'll first inspect the usage for `hmmsearch`.

```
oneils@atmosphere ~/projects/p450s$ hmmsearch
Incorrect number of command line arguments.
Usage: hmmsearch [options] <hmmfile> <seqdb>

where most common options are:
  -h : show brief help on version and usage

To see more help on available options, do hmmsearch -h
```

This brief help text indicates that `hmmsearch` may take a number of optional parameters (and we'd have to run `hmmsearch -h` to see them), and the last two parameters are required. These last two parameters constitute the HMM profile file we are searching for, as well as the `<seqdb>` in which to search. It doesn't say what format `<seqdb>` should be, so we'll try it on our *D. melanogaster* FASTA file and hope for the best (if it fails, we'll have to read more help text or documentation).

```
oneils@atmosphere ~/projects/p450s$ hmmsearch p450s.fasta.aln.hmm dmel-all-trans
lation-r6.02.fasta
```

Note that there was no required option for an output file. Running this command causes quite a lot of information to be printed to the terminal, including lines like:

```
...
  Alignments for each domain:
  == domain 1  score: 2.4 bits;  conditional E-value: 0.19
  p450s.fasta 433 sdekivpivndlfganfdtisvalswslpylvaspeigkklkke 476
                  + +++       f a+f+ is  l++   +l  +p ++ +l++e
  FBpp0086933 305 TADDLLAQCLLFFFAGFEIISSSLCFLTHELCLNPTVQDRLYEE 348
                  555566666667899***************************99 PP
...
```

And, when we run `ls`, we find that no output file has been created.

```
oneils@atmosphere ~/projects/p450s$ ls
dmel-all-translation-r6.02.fasta  p450s.fasta.aln
p450s.fasta                       p450s.fasta.aln.hmm
```

It seems that `hmmsearch`, by default, prints all of its meaningful output to the terminal. Actually, `hmmsearch` is printing its output to the *standard output* stream. Standard output is the primary output mechanism for command-line programs (other than writing files directly). By default, standard output, also known as "standard out" or "stdout," is printed to the terminal.

Fortunately, it is possible to *redirect* the standard output stream into a file by indicating this to our shell with a > redirect operator, and specifying a file name or path. In this case, we'll redirect the output of standard out to a file called `p450s_hmmsearch_dmel.txt`.

```
oneils@atmosphere ~/projects/p450s$ hmmsearch p450s.fasta.aln.hmm dmel-all-trans
lation-r6.02.fasta > p450s_hmmsearch_dmel.txt
```

When this command executes, nothing is printed, and instead our file is created. When using the > redirect, the file will be *overwritten* if it already exists. If, instead, we wished to append to an existing file (or create a new file if there is no file to append to), we could have used the >> redirect.

Here are the contents of our final analysis, a simple text file with quite a bit of information, including some nicely formatted row and column data, as displayed by `less -S p450s_hmmsearch_dmel.txt`.

```
Query:       p450s.fasta  [M=730]
Scores for complete sequences (score includes all domains):
   --- full sequence ---    --- best 1 domain ---    -#dom-
    E-value  score  bias     E-value  score  bias     exp  N  Sequence     Descript
    ------- ------ -----     ------- ------ -----    ---- --  --------     --------
    1.2e-98  332.1   0.1       2e-98  331.3   0.1     1.2  1  FBpp0081483  type=pr
    1.2e-98  332.1   0.1       2e-98  331.3   0.1     1.2  1  FBpp0307731  type=pr
      7e-67  227.0   0.0       7e-67  227.0   0.0     1.5  1  FBpp0082768  type=pr
    9.5e-67  226.6   0.0     9.5e-67  226.6   0.0     2.1  1  FBpp0297168  type=pr
```

Reproducibility with Scripts

It is highly unlikely that an analysis of this type is performed only once. More often than not, we'll wish to adjust or replace the input files, compare to different protein sets, or adjust the parameters for the programs ran. Thus it make sense to capture the analysis we just performed as an executable script, perhaps called `runhmmer.sh`.

```
  GNU nano 2.2.6              File: runhmmer.sh

#!/bin/bash

muscle -in p450s.fasta -out p450s.fasta.aln
hmmbuild p450s.fasta.aln.hmm p450s.fasta.aln
hmmsearch p450s.fasta.aln.hmm dmel-all-translation-r6.02.fasta \
   > p450s_hmmsearch_dmel.txt

^G Get Help  ^O WriteOut  ^R Read File ^Y Prev Page ^K Cut Text  ^C Cur Pos
^X Exit      ^J Justify   ^W Where Is  ^V Next Page ^U UnCut Text^T To Spell
```

Note in the above that we've broken the long `hmmsearch` line into two by ending it midway with a backslash and continuing it on the next line. The backslash lets `bash` know that more of the command is to be specified on later lines. (The backslash should be the last character on the line, with no spaces or tabs following.) After making this script executable with `chmod`, we could then rerun the analysis by navigating to this directory and running `./runhmmer.sh`.

What if we wanted to change the input file, say, to `argonase-1s.fasta` instead of `p450s.fasta`? We could create a new project directory to work in, copy this script there, and then change all instances of `p450s.fasta` in the script to `argonase-1s.fasta`.

Alternatively, we could use the power of environment variables to architect our script in such a way that this process is easier.

```
GNU nano 2.2.6                File: runhmmer.sh

#!/bin/bash

export query=p450s.fasta
export db=dmel-all-translation-r6.02.fasta
export output=p450s_hmmsearch_dmel.txt

muscle -in $query -out $query.aln
hmmbuild $query.aln.hmm $query.aln
hmmsearch $query.aln.hmm $db \
   > $output


^G Get Help  ^O WriteOut  ^R Read File ^Y Prev Page ^K Cut Text  ^C Cur Pos
^X Exit      ^J Justify   ^W Where Is  ^V Next Page ^U UnCut Text^T To Spell
```

Now the file names of interest are specified only once, near the top of the script, and from then on the script uses its own identifiers (as environment variables) to refer to them. Reusing this script would be as simple as changing the file names specified in three lines.

We can go a step further. It turns out that shell scripts can take parameters from the command line. The first parameter given to a script on the command line will be automatically stored in a variable accessible to the script called `$1`, the second parameter will be stored in `$2`, and so on. We can thus further generalize our script:

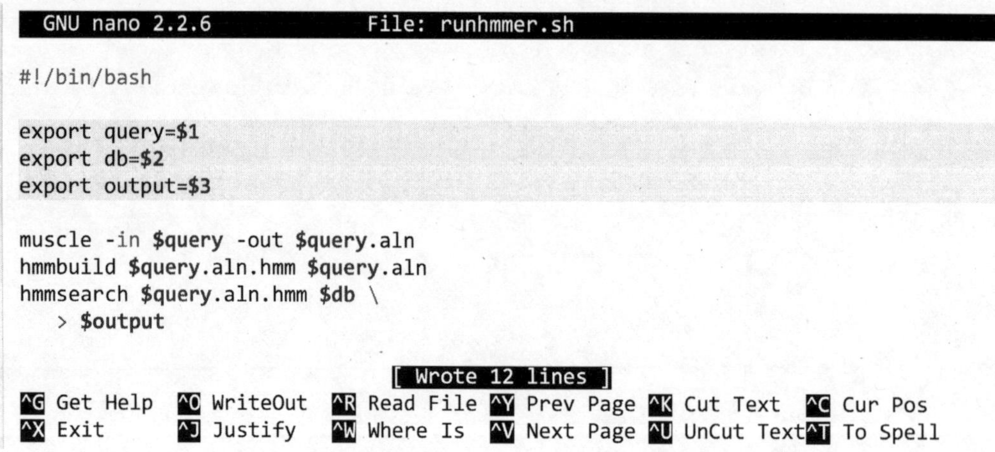

```
GNU nano 2.2.6                File: runhmmer.sh

#!/bin/bash

export query=$1
export db=$2
export output=$3

muscle -in $query -out $query.aln
hmmbuild $query.aln.hmm $query.aln
hmmsearch $query.aln.hmm $db \
   > $output

                              [ Wrote 12 lines ]
^G Get Help  ^O WriteOut  ^R Read File ^Y Prev Page ^K Cut Text  ^C Cur Pos
^X Exit      ^J Justify   ^W Where Is  ^V Next Page ^U UnCut Text^T To Spell
```

We could have replaced all instances of `$query` with `$1`, but this organization makes our script easier to read in the future, an important consideration when programming. Now we can run a full analysis by specifying the three relevant file names on the command line, as in: `./runhmmer.sh p450s.fasta dmel-all-translation-r6.02.fasta p450s_hmmsearch_dmel.txt`.

This `runhmmer.sh` is a good candidate for inclusion in our `$HOME/local/bin` so that we can run it from anywhere, though we may want to add lines immediately following the `#!` line, to provide some help text for anyone who attempts to run the script without providing the correct inputs:

```
 GNU nano 2.2.6                 File: runhmmer.sh

#!/bin/bash

# Check number of input parameters:
if [ $# -ne 3 ]; then
    echo "Wrong number of parameters."
    echo "Usage: runhmmer.sh <query_fasta> <db_fasta> <output_name>"
    exit
fi

export query=$1
export db=$2
export output=$3

muscle -in $query -out $query.aln
hmmbuild $query.aln.hmm $query.aln
hmmsearch $query.aln.hmm $db \
   > $output


^G Get Help  ^O WriteOut  ^R Read File ^Y Prev Page ^K Cut Text  ^C Cur Pos
^X Exit      ^J Justify   ^W Where Is  ^V Next Page ^U UnCut Text^T To Spell
```

The "if block" above will only execute if the number of parameters given (`$#`) is not equal to 3. Although languages like Python provide much nicer facilities for this sort of logic-based execution, the ability to conditionally provide usage information for scripts is important. As usual for `bash`, the interpreter ignores lines that start with `#`.

Exercises

1. Create a new folder in your `projects` folder called `c_elegans`. Locate the FASTA file for the

reference genome of Caenorhabditis elegans from http://wormbase.org, and download it to this folder using `wget`. The file you are searching for will be named something like `c_elegans.PRJNA13758.WS244.genomic.fa.gz`. After it is downloaded, decompress it and view it with `less -S`.

2. Install an SFTP client on your desktop, like FireFTP or CyberDuck, and attempt to connect to the same machine you log in to via SFTP. Download a FASTA file of some potentially homologous sequences from Uniprot to your local desktop, and transfer it to your remote `c_elegans` directory.
3. Try running `muscle` and HMMER on the sequences you downloaded from uniprot.org against the C. elegans genome.
4. If you have access to more than one Unix-based machine (such as an OS X desktop and a remote Linux computer, or two remote Linux computers), read the man page for `scp` with `man scp`, and also read about it online. Try to transfer a file and a directory between machines using `scp` on the command line.
5. Write an executable `bash` script that automates a process of some kind, and install it in your `$HOME/local/bin`. Test it after logging out and back in.

Chapter 7

Command Line BLAST

While the previous chapters covered installing and using a few bioinformatics tools as examples of the process, there is one nearly ubiquitous tool: BLAST, or Basic Local Alignment Search Tool.[21]

Given one or more *query* sequences (usually in FASTA format), BLAST looks for matching sequence regions between them and a *subject* set.

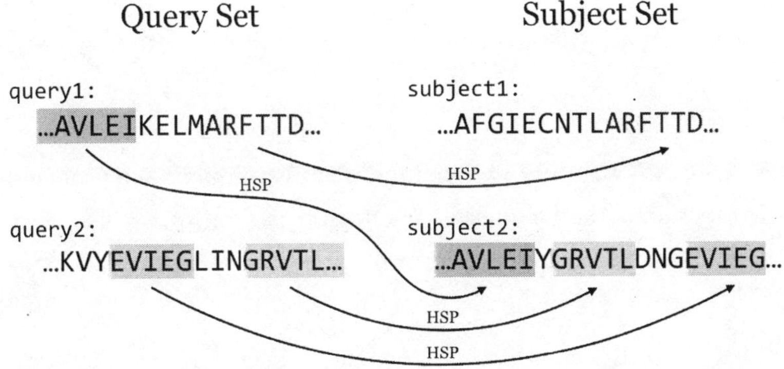

A sufficiently close match between subsequences (denoted by arrows in the figure above, though matches are usually longer than illustrated here) is called a high-scoring pair (HSP), while a query sequence is said to *hit* a target sequence if they share one or more HSPs. Sometimes, however, the term "hit" is used loosely, without differentiating between the two. Each HSP is associated with a

21. The original BLAST paper is by Stephen F. Altschul, Warren Gish, Webb Miller, Eugene W. Myers, and David J. Lipman, "Basic Local Alignment Search Tool," *Journal of Molecular Biology* 215 (1990): 403–404. I'm convinced that naming bioinformatics tools exciting verbs like BLAST and HMMER results in greater usage.

"bitscore" that is based on the similarity of the subsequences as determined by a particular set of rules. Because in larger subject sets some good matches are likely to be found by chance, each HSP is also associated with an "*E* value," representing the expected number of matches one might find by chance in a subject set of that size with that score or better. For example, an *E* value of 0.05 means that we can expect a match by chance in 1 in 20 similar searches, whereas an *E* value of 2.0 means we can expect 2 matches by chance for each similar search.

BLAST is not a single tool, but rather a suite of tools (and the suite grows over the years as more features and related tools are added). The most modern version of the software, called BLAST+, is maintained by the National Center for Biotechnology Information (NCBI) and may be downloaded in binary and source forms at ftp://ftp.ncbi.nlm.nih.gov/blast/executables/blast+/LATEST/.

This chapter only briefly covers running BLAST on the command line in simple ways. Reading the help information (e.g., with `blastn --help`) and the NCBI BLAST Command Line Applications User Manual at http://www.ncbi.nlm.nih.gov/books/NBK1763/ is highly recommended. The NCBI manual covers quite a few powerful and handy features of BLAST on the command line that this book does not.

BLAST Types

The programs in the BLAST+ suite can search for and against sequences in protein format (as we did for the HMMER example) and in nucleotide format (A's, C's, T's, and G's). Depending on what type the query and subject sets are, different BLAST programs are used.

Program	Query Type	Subject Type	Computation
`blastn`	N	N	~ 1X
`blastp`	P	P	~ 1X
`blastx`	N	P	~ 6X
`tblastn`	P	N	~ 6X
`tblastx`	N	N	~36X

(other BLAST types not listed: `psiblast, deltablast, rpsblast`)

While two nucleotide sequences (N comparisons in the figure above) may be compared directly (as may two protein sequences, represented by P), when we wish to compare a nucleotide sequence to a protein sequence, we need to consider which reading frame of the nucleotide sequence corresponds to a protein. The `blastx` and `tblastn` programs do this by converting nucleotide sequences into protein sequences in all six reading frames (three on the forward DNA strand and three on the reverse) and comparing against all of them. Generally such programs result in six times as much work to be done. The `tblastx` program compares nucleotide queries against nucleotide subjects, but it does so in protein space with all six conversions compared to all six on both sides.

Other more exotic BLAST tools include `psiblast`, which produces an initial search and tweaks the scoring rules on the basis of the results; these tweaked scoring rules are used in a second search, generally finding even more matches. This process is repeated as many times as the user wishes, with more dissimilar matches being revealed in later iterations. The `deltablast` program considers a precomputed database of scoring rules for different types of commonly found (conserved) sequences. Finally, `rpsblast` searches for sequence matches against sets of profiles, each representing a collection of sequences (as in HMMER, though not based on hidden Markov models).

All this talk of scoring rules indicates that the specific scoring rules are important, especially when comparing two protein sequences. When comparing protein sequences from two similar species, for example, we might wish to give a poor score to the relatively unlikely match of a nonpolar valine (V) to a polar tyrosine (Y). But for dissimilar species separated by vast evolutionary time, such a mismatch might not be as bad relative to other possibilities. Scoring matrices representing these rule sets with names like BLOSUM and PAM have been developed using a variety of methods to capture these considerations. A discussion of these details can be found in other publications.

Each of the various programs in the BLAST suite accepts a large number of options; try running `blastn -help` to see them for the `blastn` program. Here is a summary of a few parameters that are most commonly used for `blastn` et al.:

`-query <fasta file>`

The name (or path) of the FASTA-formatted file to search for as query sequences.

`-subject <fasta file>`

The name (or path) of the FASTA-formatted file to search in as subject sequences.

`-evalue <real number>`

Only HSPs with *E* values smaller than this should be reported. For example: `-evalue 0.001` or `-evalue 1e-6`.

`-outfmt <integer>`

How to format the output. The default, `0`, provides a human-readable (but not programmatically parseable) text file. The values `6` and `7` produce tab-separated rows and columns in a text file, with `7` providing explanatory comment lines. Similarly, a value of `10` produces comma-separated output; `11` produces a format that can later be quickly turned into any other with another program called `blast_formatter`. Options `6`, `7`, and `10` can be highly configured in terms of what columns are shown.

`-max_target_seqs <integer>`

When the output format is `6`, `7`, or `10` for each query sequence, only report HSPs for the best `<integer>` different subject sequences.

`-max_hsps <integer>`

For each query/target pair, only report the best `<integer>` HSPs.

`-out <output file>`

Write the output to `<output file>` as opposed to the default of standard output.

BLAST Databases

No doubt readers familiar with BLAST have been curious: aren't there *databases* of some kind involved in BLAST searches? Not necessarily. As we've seen, simple FASTA files will suffice for both the query and subject set. It turns out, however, that from a computational perspective, simple FASTA files are not easily searched. Thus BLAST+ provides a tool called `makeblastdb` that converts a subject FASTA file into an indexed and quickly searchable (but not human-readable) version of the same information, stored in a set of similarly named files (often at least three ending in `.pin`, `.psq`, and `.phr` for protein sequences, and `.nin`, `.nsq`, and `.nhr` for nucleotide sequences). This set of files represents the "database," and the database name is the shared file name prefix of these files.

Running `makeblastdb` on a FASTA file is fairly simple: `makeblastdb -in <fasta file> -out <database name> -dbtype <type> -title <title> -parse_seqids`, where `<type>` is one of `prot` or `nucl`, and `<title>` is a human-readable title (enclosed in quotes if necessary). The `-parse_seqids` flag indicates that the sequence IDs from the FASTA file should be included in the database so that they can be used in outputs as well as by other tools like `blastdbcmd` (discussed below).

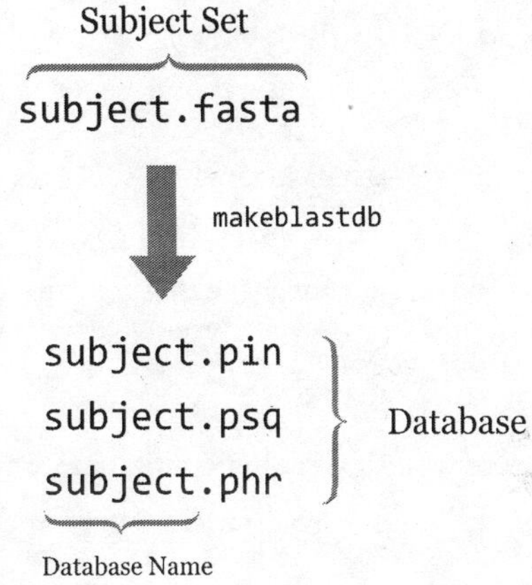

Once a BLAST database has been created, other options can be used with `blastn` et al.:

`-db <database name>`
The name of the database to search against (as opposed to using `-subject`).

`-num_threads <integer>`
Use `<integer>` CPU cores on a multicore system, if they are available.

When using the `-db` option, the BLAST tools will search for the database files in three locations: (1) the present working directory, (2) your home directory, and (3) the paths specified in the `$BLASTDB` environment variable.

The tool blastdbcmd can be used to get information about BLAST databases—for example, with blastdbcmd -db <database name> -info—and can show the databases in a given path with blastdbcmd -list <path> (so, blastdbcmd -list $BLASTDB will show the databases found in the default search paths). This tool can also be used to extract sequences or information about them from databases based on information like the IDs reported in output files. As always, reading the help and documentation for software like BLAST is highly recommended.

Running a Self-BLAST

To put these various tools and options to use, let's consider using blastp to look for proteins that are similar in sequence to other proteins in the yeast exome. First, we'll need to use wget to download the protein data set (after locating it at http://yeastgenome.org), and then gzip -d to decompress it, calling it **orf_trans.fasta**.

```
oneils@atmosphere ~/apcb/intro/blast$ wget 'http://downloads.yeastgenome.org/seq
uence/S288C_reference/orf_protein/orf_trans.fasta.gz' -O orf_trans.fasta.gz
...
oneils@atmosphere ~/apcb/intro/blast$ gzip -d orf_trans.fasta.gz
oneils@atmosphere ~/apcb/intro/blast$ ls
orf_trans.fasta
```

In order to find sequences that are similar to others, we're going to want to blastp this file against *itself.* So, we'll start by creating a database of these sequences.

```
oneils@atmosphere ~/apcb/intro/blast$ makeblastdb -in orf_trans.fasta -out orf_t
rans -dbtype prot -title "Yeast Open Reading Frames" -parse_seqids
...
oneils@atmosphere ~/apcb/intro/blast$ ls
orf_trans.fasta  orf_trans.pin  orf_trans.psd  orf_trans.psq
orf_trans.phr    orf_trans.pog  orf_trans.psi
```

Now we need to determine what options we will use for the blastp. In particular, do we want to limit the number of HSPs and target sequences reported for each query? Because we're mostly interested in determining which proteins match others, we probably only need to keep one hit. But each protein's best hit will likely be to itself! So we'd better keep the top two with `-max_target_seqs 2` and only the best HSP per hit with `-max_hsps 1`. We'll also use an `-evalue 1e-6`, a commonly used cutoff.[22]

For the output, we'll create a tab-separated output with comment lines (`-outfmt 7`), creating columns for the query sequence ID, subject sequence ID, HSP alignment length, percentage identity of the alignment, subject sequence length, query sequence length, start and end positions in the query and subject, and the *E* value. (The coded names—`qseqid`, `sseqid`, `length`, etc.—can be found by running `blastp -help`.) Finally, we'll call the output file **`yeast_blastp_yeast_top2.txt`** and use four processors to speed the computation (which will only really help if the machine we are logged in to has at least that many).

```
oneils@atmosphere ~/apcb/intro/blast$ blastp -query orf_trans.fasta -db orf_tran
s -max_target_seqs 2 -max_hsps 1 -evalue 1e-6 -outfmt '7 qseqid sseqid length ql
en slen qstart qend sstart send evalue' -out yeast_blastp_yeast_top2.txt -num_th
reads 4
```

It's a long command, to be sure! This operation takes several minutes to finish, even with `-num_threads 4` specified. When it does finish, we can see with `less` that the output file contains the columns we specified interspersed with the comment lines provided by `-outfmt 7`.

```
...
# BLASTP 2.2.30+
# Query: YAL003W EFB1 SGDID:S000000003, Chr I from 142174-142253,142620-143160,
# Database: orf_trans
# Fields: query id, subject id, alignment length, query length, subject length,
# 1 hits found
YAL003W YAL003W 207     207     207     1       207     1       207     2e-148
# BLASTP 2.2.30+
# Query: YAL005C SSA1 SGDID:S000000004, Chr I from 141431-139503, Genome Release
# Database: orf_trans
# Fields: query id, subject id, alignment length, query length, subject length,
# 2 hits found
YAL005C YAL005C 643     643     643     1       643     1       643     0.0
YAL005C YLL024C 643     643     640     1       643     1       640     0.0
...
```

In the output snippet above, YAL0005C has an HSP with itself (naturally), but also one with YLL024C. We'll consider basic analyses containing this sort of data—rows and columns stored in text files, interspersed with extraneous lines—in later chapters.

22. Using an *E*-value cutoff of `1e-6` may not be the best choice; as with all analyses, some evaluations or literature guidance may be necessary, depending on the task at hand.

Exercises

1. If you don't already have the NCBI Blast+ tools installed, install them. Once you do, check the contents of the `$BLASTDB` environment variable. If it's not empty, use `blastdbcmd` to determine whether you have the "nr" database available, and any information you can determine about it (when it was downloaded, how many sequences it has, etc.)

2. Create a new folder in your `projects` folder called `blast`. In this directory, download the **`p450s.fasta`** file and the yeast exome **`orf_trans.fasta`** from the book website. Create a database called `orf_trans` using `makeblastdb`, and use `blastp` to search the `p450s.fasta` file against it. When doing the search, use an *E*-value cutoff of `1e-6`, keep the top one target sequences, and produce an output file called `p450s_blastp_yeast_top1.blast` in output format `11`.

3. Use the `blast_formatter` tool to convert the output format `11` file above into an output format `6` called `p450s_blastp_yeast_top1.txt`, with columns for: (1) Query Seq-id, (2) Subject Seq-id, (3) Subject Sequence Length, (4) Percentage of Identical Matches, (5) *E* Value, (6) Query Coverage per Subject, and (7) Subject title. (You may find browsing the NCBI BLAST+ manual and the output of `blast_formatter -help` to be informative.) The output, when viewed with `less -S`, should look something like this:

```
sp|Q3LFU0|CP1A1_BALAC   YHR007C 531     23.83   2e-12   49      ERG11 SGDID:S000
sp|P56590|CP1A1_CANFA   YHR007C 531     24.10   1e-14   48      ERG11 SGDID:S000
sp|Q06367|CP1A1_CAVPO   YDR402C 490     28.03   7e-13   49      DIT2 SGDID:S0000
sp|Q92039|CP1A1_CHACA   YHR007C 531     23.81   8e-09   40      ERG11 SGDID:S000
...
```

   What do these various output columns represent?

4. The file **`yeast_selected_ids.txt`** contains a column of 25 IDs identified as interesting in some way. Use `blastdbcmd` to extract just those sequence records from the `orf_trans` database as a FASTA file named `yeast_selected_ids.fasta`. (Again, browsing the BLAST+ manual and the output of `blastdbcmd -help`will be useful.)

Chapter 8

The Standard Streams

In previous chapters, we learned that programs may produce output not only by writing to files, but also by printing them on the standard output stream. To further illustrate this feature, we've created a simple program called **`fasta_stats`** that, given a FASTA file name as its first parameter, produces statistics on each sequence. We'll also look at the file **`pz_cDNAs.fasta`**, which contains a set of 471 de novo assembled transcript sequences from *Papilio zelicaon*, and **`pz_cDNAs_sample.fasta`**, which contains only the first two.

```
oneils@atmosphere ~/apcb/intro$ mkdir fasta_stats
oneils@atmosphere ~/apcb/intro$ cd fasta_stats/
oneils@atmosphere ~/apcb/intro/fasta_stats$ wget http://library.open.oregonstate
.edu/doc/computationalbiology/fasta_stats
Resolving library.open.oregonstate.edu (library.open.oregonstate.edu)...
...
oneils@atmosphere ~/apcb/intro/fasta_stats$ chmod +x fasta_stats
oneils@atmosphere ~/apcb/intro/fasta_stats$ wget http://library.open.oregonstate
.edu/doc/computationalbiology/pz_cDNAs.fasta
Resolving library.open.oregonstate.edu (library.open.oregonstate.edu)...
...
oneils@atmosphere ~/apcb/intro/fasta_stats$ http://library.open.oregonstate
.edu/doc/computationalbiology/pz_cDNAs_sample.fasta
Resolving library.open.oregonstate.edu (library.open.oregonstate.edu)...
...
oneils@atmosphere ~/apcb/intro/fasta_stats$ ls
fasta_stats  pz_cDNAs.fasta  pz_cDNAs_sample.fasta
```

We can run the `fasta_stats` program (after making it executable) with `./fasta_stats pz_cDNAs_sample.fasta`.

```
oneils@atmosphere ~/apcb/intro/fasta_stats$ ./fasta_stats pz_cDNAs_sample.fasta
# Column 1: Sequence ID
# Column 2: GC content
# Column 3: Length
# Column 4: Most common 5mer
# Column 5: Count of most common 5mer
# Column 6: Repeat unit of longest simple perfect repeat (2 to 10 chars)
# Column 7: Length of repeat (in characters)
# Column 8: Repeat type (dinucleotide, trinucleotide, etc.)
Processing sequence ID PZ7180000031590
PZ7180000031590 0.378 486 ACAAA 5 unit:ATTTA 10 pentanucleotide
Processing sequence ID PZ7180000000004_TX
PZ7180000000004_TX 0.279 1000 AAATA 12 unit:TAA 12 trinucleotide
```

Based on the information printed, it appears that sequence `PZ7180000031590` has a GC content (percentage of the sequence composed of G or C characters) of 37.8%, is 486 base pairs long, the most common five-base-pair sequence is `ACAAA` (occurring 5 times), and the longest perfect repeat is 10 base pairs long, caused by the pentanucleotide `ATTTA`, occurring twice.

Much like `hmmsearch`, this program writes its output to standard output. If we would like to save the results, we know that we can redirect the output of standard out with the > redirect.

```
oneils@atmosphere ~/apcb/intro/fasta_stats$ ./fasta_stats pz_cDNAs_sample.fasta
> pz_sample_stats.txt
Processing sequence ID PZ7180000031590
Processing sequence ID PZ7180000000004_TX
oneils@atmosphere ~/apcb/intro/fasta_stats$ ls
fasta_stats  pz_cDNAs.fasta  pz_cDNAs_sample.fasta  pz_sample_stats.txt
```

When we run this command, however, we see that even though the output file has been created, text is still printed to the terminal! If we use `less -S` to view the `pz_sample_stats.txt` file, we see that *some* of the output has gone to the file.

```
# Column 1: Sequence ID
# Column 2: GC content
# Column 3: Length
# Column 4: Most common 5mer
# Column 5: Count of most common 5mer
# Column 6: Repeat unit of longest simple perfect repeat (2 to 10 chars)
# Column 7: Length of repeat (in characters)
# Column 8: Repeat type (dinucleotide, trinucleotide, etc.)
PZ7180000031590 0.378    486     ACAAA    5       unit:ATTTA      10      pentanuc
PZ7180000000004_TX       0.279   1000     AAATA   12      unit:TAA        12
pz_sample_stats.txt (END)
```

So what is going on? It turns out that programs can produce output (other than writing to files) on *two* streams. We are already familiar with the first, standard output, which is by default printed to the terminal but can be redirected to a file with `>`. The second, called *standard error*, is also by default printed to the terminal but is not redirected with `>`.

By default, like standard output, standard error (also known as "standard err" or "stderr") is printed to the terminal.

Because standard error usually contains diagnostic information, we may not be interested in capturing it into its own file. Still, if we wish, `bash` can redirect the standard error to a file by using the `2>` redirect.[23]

```
oneils@atmosphere ~/apcb/intro/fasta_stats$ ./fasta_stats pz_cDNAs_sample.fasta
> pz_sample_stats.txt 2> pz_sample_stats.err.txt
oneils@atmosphere ~/apcb/intro/fasta_stats$ ls
fasta_stats     pz_cDNAs_sample.fasta    pz_sample_stats.txt
pz_cDNAs.fasta  pz_sample_stats.err.txt
```

We might pictorially represent programs and their output as alternative information flows:

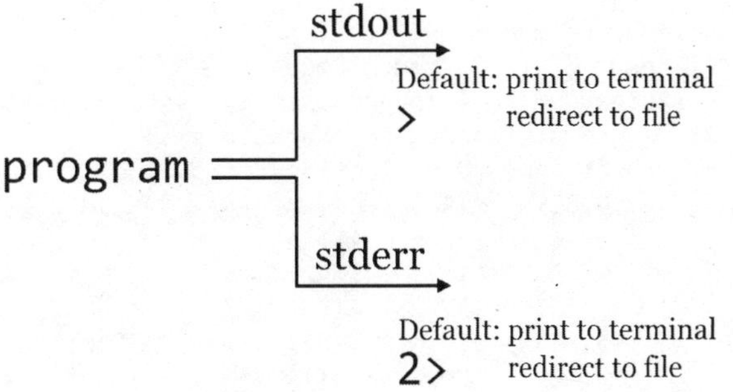

23. The `tcsh` and `csh` shells unfortunately cannot natively separately redirect stdout and stderr to files. A potential workaround looks like: `( ./fasta_stats pz_cDNAs_sample.fasta > pz_sample_stats.txt ) > & pz_sample_stats.err.txt`. This command runs two independent redirects; using parentheses causes the redirect of stdout to happen first, then the further redirect of stderr can occur next. How `bash`-compatible shells handle standard output and standard error is one of the primary reasons they are preferred over the older `csh`-compatible shells.

Filtering Lines, Standard Input

It can often be useful to extract lines from a file based on a pattern. For example, the `pz_sample_stats.txt` file contains information on what each column describes, as well as the data itself. If we want to extract all the lines that match a particular pattern, say, `unit:`, we can use the tool `grep` (for Global search for Regular Expression and Print), which prints to standard output lines that match a given pattern (or don't match a given pattern, if using the `-v` flag): `grep '<pattern>' <file>`. To illustrate, we'll first run `fasta_stats` on the full input file, redirecting the standard output to a file called `pz_stats.txt`.

```
oneils@atmosphere ~/apcb/intro/fasta_stats$ ./fasta_stats pz_cDNAs.fasta > pz_st
ats.txt
```

Looking at the file with `less -S pz_stats.txt`, we can see that informational lines as well as data-containing lines are stored in the file:

```
# Column 1: Sequence ID
# Column 2: GC content
# Column 3: Length
# Column 4: Most common 5mer
# Column 5: Count of most common 5mer
# Column 6: Repeat unit of longest simple perfect repeat (2 to 10 chars)
# Column 7: Length of repeat (in characters)
# Column 8: Repeat type (dinucleotide, trinucleotide, etc.)
PZ832049        0.321   218     CTTAA   4       unit:CGT        6       trinucle
PZ21878_A       0.162   172     ATTAA   8       unit:ATT        6       trinucle
PZ439397        0.153   111     TTAAT   5       unit:GAAAT      10      pentanuc
PZ16108_A       0.157   191     ATTAA   7       unit:ATT        6       trinucle
PZ21537_A       0.158   82      TTATT   3       unit:ATT        6       trinucle
PZ535325        0.108   120     AATTA   6       unit:TA 6       dinucleotide
...
```

To get rid of the informational lines, we can use `grep` to extract the other lines by searching for some pattern they have in common; in this case, the pattern `unit:` will do. Because `grep` prints its results to standard output, we will need to redirect the modified output to a file called, perhaps, `pz_stats.table` to indicate its tabular nature.

```
oneils@atmosphere ~/apcb/intro/fasta_stats$ grep 'unit:' pz_stats.txt > pz_stats
.table
```

This time, `less -S pz_stats.table` reveals only the desired lines.

```
PZ832049        0.321   218     CTTAA   4       unit:CGT        6       trinucle
PZ21878_A       0.162   172     ATTAA   8       unit:ATT        6       trinucle
PZ439397        0.153   111     TTAAT   5       unit:GAAAT      10      pentanuc
PZ16108_A       0.157   191     ATTAA   7       unit:ATT        6       trinucle
PZ21537_A       0.158   82      TTATT   3       unit:ATT        6       trinucle
PZ535325        0.108   120     AATTA   6       unit:TA 6       dinucleotide
...
```

Rather than viewing the file with `less`, we can also count the number of lines present in the file with the `wc` tool, which counts the number of lines, words, and characters of input: `wc <file>`.

Working with the cleaned data table reveals that our program produced 21,131 characters broken into 3,297 words among 471 lines of data output.

```
oneils@atmosphere ~/apcb/intro/fasta_stats$ wc pz_stats.table
  471  3297 21131 pz_stats.table
```

This sort of command-line-driven analysis can be quite powerful, particularly because many of these programs—like `less`, `grep`, and `wc`—can both print their results on standard output and read input from *standard input* rather than from a file. Standard input is the secondary input mechanism for command-line programs (other than reading from files directly). By default, standard input, or "stdin," is unused.

How can we get input to a program on its standard input? It turns out that the easiest way to do so is to *redirect* the standard output of another program to it using the `|`, also known as the "pipe," redirect (found above the Enter key on most keyboards). In this case, the data come in "on the left":

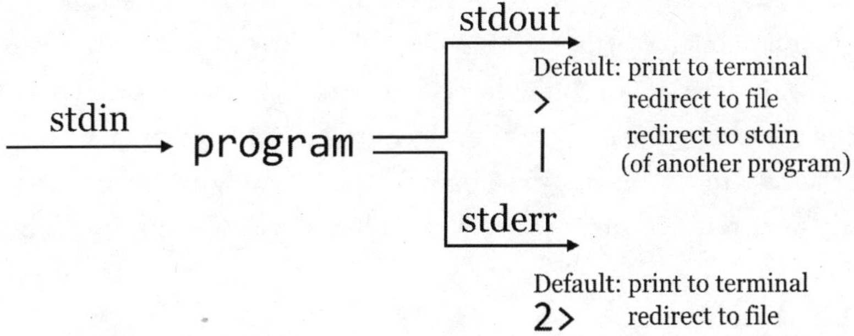

To drive this home, we'll first remove our `pz_stats.table` file, and then rerun our `grep` for `unit:` on the `pz_stats.txt` file, but rather than send the result of `grep` to a file with the `>` redirect, we'll direct it straight to the standard input of `wc` with a `|` redirect.

```
oneils@atmosphere ~/apcb/intro/fasta_stats$ grep 'unit:' pz_stats.txt | wc
    471    3297   21131
```

In this example, we've neither created a new file nor specified a file for `wc` to read from; the data are stored in a temporary buffer that is handled automatically by the shell and operating system. The `less` program can also read from standard input, so if we wanted to see the contents of the `grep` without creating a new file, we could run `grep 'unit:' pz_stats.txt | less -S`.

Recall that the `fasta_stats` program wrote its output to standard out, and because `grep` can read from standard input as well, we can process the entire FASTA file without needing to create any new files by using multiple such buffers:

```
oneils@atmosphere ~/apcb/intro/fasta_stats$ ./fasta_stats pz_cDNAs.fasta | gre
p 'unit:' | wc
```

When this command runs, the results printed by `fasta_stats` on standard error will still be printed to the terminal (as that is the default and we didn't redirect standard error), but the standard output results will be filtered through `grep` and then filtered through `wc`, producing the eventual output of 471 lines.

At this point, the longish nature of the commands and the fact that our terminal window is only so wide are making it difficult to read the commands we are producing. So, we'll start breaking the commands over multiple lines by ending partial commands with backslashes. Just as in the shell scripts we wrote at the end of chapter 6, "Installing (Bioinformatics) Software," using backslashes will let the shell know that we aren't finished entering the command. However, the `bash` shell indicates that a command spans multiple lines by showing us a `>`, which shouldn't be confused with the redirect character that we might type ourselves. The following example shows the exact same command in a more readable form broken over multiple lines, but the highlighted characters have *not* been typed.

```
oneils@atmosphere ~/apcb/intro/fasta_stats$ ./fasta_stats pz_cDNAs.fasta | \
> grep 'unit:' | \
> wc
```

A chain of commands like the above, separated by pipe characters, is often called a "pipeline." More generally, though, a pipeline can describe any series of steps from input data to output data (as in the Muscle/HMMER series covered in chapter 6).

Counting Simple AT Repeats

Let's expand on the small pipeline above to inspect just the "simple" AT repeats, that is, those that are the string "AT" repeated one or more times. We can start with what we have, but rather than just searching for `unit:`, we'll modify the pattern to find `unit:AT`, and see what we get:

```
oneils@atmostphere ~/apcb/intro/fasta_stats$ ./fasta_stats pz_cDNAs.fasta | \
> grep 'unit:AT' | \
> less -S
```

The resulting output is close to what we hoped for, but not quite complete, as this pattern also matches things like `unit:ATT` and `unit:ATG`.

```
PZ21878_A         0.162   172     ATTAA   8       unit:ATT        6       trinucle
PZ16108_A         0.157   191     ATTAA   7       unit:ATT        6       trinucle
PZ21537_A         0.158   82      TTATT   3       unit:ATT        6       trinucle
PZ7180000031590 0.378   486     ACAAA   5       unit:ATTTA      10      pentanuc
PZ7180000031597 0.287   403     ATTAT   6       unit:ATTTTG     12      hexanucl
PZ7180000025478 0.516   829     TGATG   18      unit:ATG        18      trinucle
...
```

We probably want to further filter the output, but based on what pattern? In this case, those lines that match not only `unit:AT`, but also the term `dinucleotide`. Rather than attempt to produce a single complicated pattern that does this job in a single `grep`, we can add another `grep` call into the pipeline.

```
oneils@atmosphere ~/apcb/intro/fasta_stats$ ./fasta_stats pz_cDNAs.fasta | \
> grep 'unit:AT' | \
> grep 'dinucleotide' | \
> less -S
```

This command results in the output we want:

```
PZ7180000031598 0.209    81       AATAT    5        unit:AT 6        dinucleotide
PZ463243        0.226    97       TTGTA    3        unit:AT 4        dinucleotide
PZ7180000000106_T        0.246    1044     AAAAA    22       unit:AT 10       dinucleo
PZ17593_A       0.157    76       ATTAA    5        unit:AT 4        dinucleotide
PZ492422        0.144    90       ATTAA    5        unit:AT 4        dinucleotide
PZ22453_A       0.267    269      ATTAA    8        unit:AT 4        dinucleotide
...
```

Rather than run the results through `less -S`, we could instead use `wc` to count the simple (dinucleotide) AT repeats. There is an important concept at play here, that of *iterative development*, the idea that as we get closer to a solution, we inspect the results and repeat as necessary. Iterative development is a good strategy for many areas in life, but it is essential and pervasive in computing.

Once we've decided that we like the small computational process we have created, we might decide to encapsulate it and make it repeatable as a shell script, perhaps called **`count_ATs.sh`**.

```
  GNU nano 2.2.6               File: count_ATs.sh

#!/bin/bash

if [ $# -ne 1 ]; then
    echo "Wrong number of parameters."
    echo "Usage: count_ATs.sh <fasta_file>"
    exit
fi

export file=$1

fasta_stats $file | \
    grep 'unit:AT' | \
    grep 'dinucleotide' | \
    wc

^G Get Help  ^O WriteOut  ^R Read File ^Y Prev Page ^K Cut Text  ^C Cur Pos
^X Exit      ^J Justify   ^W Where Is  ^V Next Page ^U UnCut Text^T To Spell
```

The above script will need to be made executable and placed in a location referenced by `$PATH`, as will the `fasta_stats` program.

Exercises

1. Use `grep` and `wc` to determine how many sequences are in the file **`orf_trans.fasta`** without

creating any temporary files.

2. How many sequence headers in the file **`orf_trans.fasta`** have the term "polymerase"?
3. Some of the sequence headers in **`orf_trans.fasta`** have the phrase "Verified ORF" to indicate that the open reading frame has been verified experimentally. Some also have the term "reverse complement" to indicate that the ORF (open reading frame) is present on the reverse complement sequence of the canonical genome description. How many sequences are verified ORFs *and* are not on the reverse complement?
4. The sequence headers in **`orf_trans.fasta`** have information on the chromosome from which they originate, such as `Chr I` or `Chr II`. How many sequences are present on chromosome I?

Chapter 9

Sorting, First and Last Lines

Continuing with the `fasta_stats` examples from chapter 8, "The Standard Streams," the seventh column of the output contains the length of the longest perfect repeat, in characters.

```
oneils@atmosphere ~/apcb/intro/fasta_stats$ ./fasta_stats pz_cDNAs_sample.fasta
# Column 1: Sequence ID
# Column 2: GC content
# Column 3: Length
# Column 4: Most common 5mer
# Column 5: Count of most common 5mer
# Column 6: Repeat unit of longest simple perfect repeat (2 to 10 chars)
# Column 7: Length of repeat (in characters)
# Column 8: Repeat type (dinucleotide, trinucleotide, etc.)
Processing sequence ID PZ7180000031590
PZ7180000031590 0.378 486 ACAAA 5 unit:ATTTA 10 pentanucleotide
Processing sequence ID PZ7180000000004_TX
PZ7180000000004_TX 0.279 1000 AAATA 12 unit:TAA 12 trinucleotide
```

Which sequence contains the longest perfect repeat? To answer this question, we could consider sorting the lines according to this seventh column. (First, we'll have to remove the header lines themselves, which we can accomplish by filtering out lines matching # using the `-v` flag of grep, or by grepping for `unit:`, as in chapter 8.) Enter `sort`, which sorts lines from a text file (or from standard input) by specified columns: `sort <file>` or `... | sort`.

By default, `sort` sorts by all columns, by comparing the entire lines in "dictionary," or lexicographic, order. To sort by specific columns, we need to use a rather sophisticated syntax. We'll illustrate with a figure.

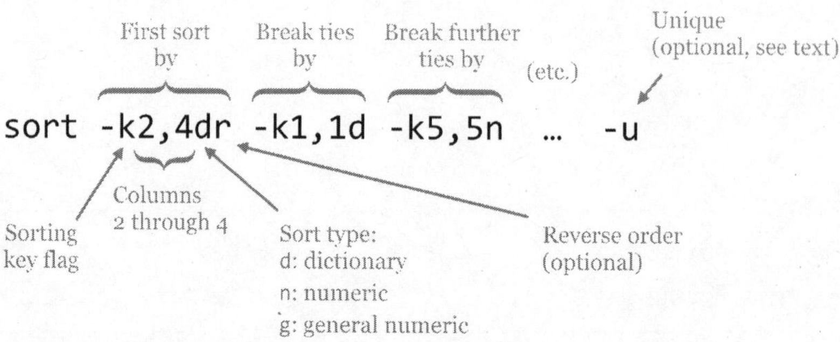

The `sort` utility takes many potential parameters, though the most important are the `-k` parameters that specify the columns by which to sort and how that sorting should be done, and occasionally the `-u` flag. The `-k` (key) parameters are considered in order; the above specifies that the sorting should be done on columns 2 through 4 (conglomerated into a single "column"), considering them in dictionary order, and sorting them in reverse. In the case of ties, only the first column is considered in normal dictionary order, and in the case of further ties, the fifth column is considered in numeric order.[24] (The difference between `n` and `g` ordering is that `g` can handle entries in scientific notation like `1e-6`, but generally `n` is preferred because it is faster and not subject to small rounding errors.)

The optional `-u` flag (which may be specified before or after the keys, or even mixed in) specifies that after all the keys are considered, if there are still any ties between rows, then only the first row should be output. It outputs only "unique" lines according to the overall sorting order.

By default, `sort` uses whitespace as the column separator, though it can be changed (run `man sort` for more information). To view information about the longest perfect repeat, we will use `sort -k7,7nr`, indicating that we wish sort on the seventh column only, in reverse numeric order.

```
oneils@atmosphere ~/apcb/intro/fasta_stats$ ./fasta_stats pz_cDNAs.fasta | \
> grep -v '#' | \
> sort -k7,7nr | \
> less -S
```

The first few lines of output indicate that the longest perfect repeat is 94 bases long and occurs in sequence `PZ805359` (this sequence's GC content is `0`, because it's composed entirely of a long `AT` repeat).

24. A word of caution: if a column contains an entry that cannot be interpreted as an integer or general number, it will be treated as `0` in the sorting order.

```
PZ805359         0.0     101     ATATA   47      unit:AT 94      dinucleotide
PZ796092         0.365   361     TACGT   9       unit:GTACGT     48      hexanucl
PZ7180000019700 0.375   564     GAGTG   12      unit:GAGTG      30      pentanuc
PZ7180000028921 0.31    561     TGTAA   8       unit:CTGTG      30      pentanuc
PZ851952         0.399   338     TATAT   12      unit:AT 24      dinucleotide
PZ7180000000664_B        0.3     652     TTTTT   18      unit:TAAAATTAT  18
PZ7180000023622 0.31    687     TTAAT   9       unit:TGA        18      trinucle
PZ7180000023665_ATQ      0.401   508     ACTGA   5       unit:TGACACTGA  18
PZ7180000025478 0.516   829     TGATG   18      unit:ATG        18      trinucle
PZ7180000030412 0.461   258     TGATG   8       unit:ATG        18      trinucle
PZ7180000036892 0.268   548     AATAA   16      unit:TAA        18      trinucle
PZ801814         0.262   255     TTACA   5       unit:TATTTACAT  18      enneanuc
...
```

The results also indicate that there are a number of ties; several sequences contain perfect repeats of length 18 base pairs. If we only wanted one sequence reported per different repeat length, we could try `sort -k7,7nr -u`. We could alternatively modify our sort to secondarily sort by GC content (second column), `sort -k7,7nr -k2,2g`.

A useful trick is to perform two sorts: one that initially sorts the data on whatever criteria are wanted, and a second that gets only the first line of a group based on secondary criteria. We may wish report only the highest GC content sequence per different repeat length, for example. We can initially use a `sort -k7,7nr -k2,2gr` to sort by repeat length and break ties by GC content, leaving the highest-GC-content sequences on top. From there, we can use a `sort -k7,7nr -u` to *re-sort* the data (even though they are already sorted!) by the seventh column, keeping only the top line per repeat length.

```
oneils@atmosphere ~/apcb/intro/fasta_stats$ ./fasta_stats pz_cDNAs.fasta | \
> grep -v '#' | \
> sort -k7,7nr -k2,2gr | \
> sort -k7,7nr -u | \
> less -S
```

Output:

```
PZ805359         0.0     101     ATATA   47      unit:AT 94      dinucleotide
PZ796092         0.365   361     TACGT   9       unit:GTACGT     48      hexanucl
PZ7180000019700 0.375   564     GAGTG   12      unit:GAGTG      30      pentanuc
PZ851952         0.399   338     TATAT   12      unit:AT 24      dinucleotide
PZ7180000025478 0.516   829     TGATG   18      unit:ATG        18      trinucle
PZ7180000000447_B        0.484   578     ATCCA   7       unit:TCCA       16
...
```

There is one small concern, however: how can we be sure that our careful ordering by GC content wasn't undone in the second sort? After all, the second sort would technically be free to reorder ties according to the seventh column, resulting in incorrect output. There is an additional flag for `sort`, the `-s` flag, indicating that *stable* sorting should be used. Stable sorting means that, in the case of ties, elements are left in their original order. So, to be safe, we could use a secondary sort of `sort -k7,7nr -u -s`, though a careful reading of the documentation for `sort` indicates that on most systems the `-u` flag implies the `-s` flag.

First and Last Lines

Often we wish to extract from a file (or from standard input) the first or last few lines. The tools `head` and `tail` are designed to do exactly this, and in combination with other tools are surprisingly handy. The `head` tool extracts the first lines of a file or standard input: `head -n <number> <file>` or `... | head -n <number>`. The `tail` tool extracts the last lines of a file or standard input: `tail -n <number> <file>` or `... | tail -n <number>`.

The `head` and `tail` utilities, like the others covered previously, write their output to standard output, and so they can be used within pipelines. They are often employed to inspect the beginning or end of a file (to check results or formatting). They also commonly extract test data sets. For example, `head -n 40000 input.fastq > test.fastq` would extract the first 10,000 sequence records from `input.fastq` and produce `test.fastq` (because every four lines of a FASTQ sequence file represents information for a single sequence).

```
@DB775P1:229:C1JDAACXX:3:1101:17936:7565 1:N:0:
CTATTACTGCCTGCAACAACATAAGGATACCATAAATTGTAATTCTTAACAAAGCAGAGATCGGAAGAGCGGTTCAGCAG
+
CCCFFFFFHHHHGJIJJJJJGGJJJJJJJIJIJJJJGIGIIIJJJJJJJJGIJJIIIJJIHIIIJJIJJIHHEDFCEEEE
@DB775P1:229:C1JDAACXX:3:1101:17888:7593 1:N:0:
TACATCAGCAAACCACCAGTGGTACGCAAATACATGGTGCTTCTTTAATCTTCAGTCCTCGGCTTTCCTCATATATCACT
+
CCCFFFFFHHHHHJJJJJHIJHHIJJJJJJJJJJJJFHIJJJJJJJJJJJJJJJIJJJJIIJJJHHHHHHFFFFFFFEE
@DB775P1:229:C1JDAACXX:3:1101:17820:7597 1:N:0:
GTAACTGCACCATATGCTGGATTCTGGACAATGAATATCGAAGTAAATAAGAAAACCTCATACCTCATCTTTCTCAATGG
+
BCCDFFEFHHHHHJJJJIJIJJIJJJJJIIGIIIDFIIJIJIFHIJGGIGIJJJIJHJIGHGDIJCEEGHIGIGCACEF
```

The above shows the first 12 lines of a FASTQ file generated on an Illumina HiSeq 2000. The first line in each set of four represents an identifier for the sequence read, the second line contains the sequence itself, the third line is often unused (containing only a `+`, though it may be followed by the identifier and other optional data), and the fourth line contains the "quality" of each base in the sequence encoded as a character. (The encoding has varied in the past, but in recent years, sequencing companies have standardized the encoding used by Sanger sequencing machines.)

With a bit of modified syntax, `tail` can also be used to extract all lines of a file starting at a given line. As an example, `tail -n +5 input.fastq > test.fastq` would result in `test.fastq` having all but the first sequence record of `input.fastq` (i.e., starting at the fifth line). This feature is especially useful for stripping off header lines of output or files before further processing, as in `./fasta_stats pz_cDNAs.fasta | tail -n +9`, rather than using `grep -v '#'` above.

Exercises

1. Running **`fasta_stats`** on **`pz_cDNAs.fasta`**, the seventh column represents the length of the longest perfect repeat found in each sequence. Use only `grep`, `sort`, `wc`, `head`, and `tail` to determine the median value in this column. (You may need to run multiple commands or pipelines.)

2. Running **`fasta_stats`** on **`pz_cDNAs.fasta`**, how many different perfect repeat units (column six) are found?

3. The file **`pz_blastx_yeast_top10.txt`** is the result of running `blastx -query ../fasta_stats/pz_cDNAs.fasta -db orf_trans -evalue 1e-6 -max_target_seqs 10 -max_hsps 1 -outfmt 7 -out pz_blastx_yeast_top1.txt`. Aside from the "comment" lines that start with `#`, the first column is the query ID, the second the target (yeast) ID, the third the percentage identity of the HSP, the eleventh the *E* value, and the twelfth the "bitscore." Which query ID had the largest bitscore? How many different query sequences (entries in the first column) had one or more HSPs against the database?

4. Extract from **`pz_blastx_yeast_top10.txt`** a file called `pz_blastx_yeast_top1.txt` containing only the smallest *E*-valued HSP line per query ID. (You may remove comment lines starting with `#` altogether.)

Chapter 10

Rows and Columns

Let's return to the output of the yeast proteins versus yeast proteins self-BLAST we performed previously, from the file **`yeast_blastp_yeast_top2.txt`**.

```
...
# BLASTP 2.2.30+
# Query: YAL003W EFB1 SGDID:S000000003, Chr I from 142174-142253,142620-143160,
# Database: orf_trans
# Fields: query id, subject id, alignment length, query length, subject length,
# 1 hits found
YAL003W YAL003W 207     207     207     1       207     1       207     2e-148
# BLASTP 2.2.30+
# Query: YAL005C SSA1 SGDID:S000000004, Chr I from 141431-139503, Genome Release
# Database: orf_trans
# Fields: query id, subject id, alignment length, query length, subject length,
# 2 hits found
YAL005C YAL005C 643     643     643     1       643     1       643     0.0
YAL005C YLL024C 643     643     640     1       643     1       640     0.0
...
```

Let's ask the following: how many sequences in this data set had a match to some other sequence? To start with, we would probably use a `grep -v '#'` to remove all of the comment lines, but then what? We could try using `wc` to count the lines, but only after also removing the self-hits, where the ID in the first column is equal to the ID in the second column. None of the utilities we've seen so far—`grep`, `sort`, `head`, or `tail`—can perform this task. We need a new tool, `awk`, which is a line-by-line and column-by-column processing tool for text files: `awk '<program>' <file>` or `... | awk '<program>'`.

Written in the late 1970s and named after its authors (Alfred Aho, Peter Weinberger, and Brian Kernigan), awk provides a sophisticated programming language that makes it easy to parse tabular data like the BLAST results above. The syntax for awk can be fairly complex, but much of the complexity can be ignored in regular use.

First, let's answer our specific question, of how many sequences had matches to other sequences, and then we'll look at some awk syntax more generally. The awk command that we want, printing only those lines where the first two columns are not equal, is `awk '{if($1 != $2) {print $0}}'`.

```
oneils@atmosphere ~/apcb/intro/blast$ cat yeast_blastp_yeast_top2.txt | \
> grep -v '#' | \
> awk '{if($1 != $2) {print $0}}' | \
> less -S
```

Breaking down the awk command, the "program" that is executed is delimited by the single quotes (which collate their contents into a single command line parameter that is sent to the awk program). The code inside the outer pair of curly brackets is executed for each line. For each line, if the contents of the first column (represented by `$1`) are not equal to (`!=`) the second column (`$2`), then the code in the following pair of curly brackets is executed, and the line (`$0`) is printed (to standard output).

```
YDR534C YOR383C 134     529     205     1       129     1       134     7e-33
YDR536W YEL069C 525     570     565     21      532     50      553     6e-59
YDR541C YOL151W 342     345     343     4       345     2       343     6e-150
YDR542W YGL261C 121     121     121     1       121     1       121     3e-83
YDR545W YOR396W 1797    1797    1797    1       1797    1       1797    0.0
YDR545W YLR467W 1797    1797    1797    1       1797    1       1797    0.0
YEL006W YIL006W 330     336     374     15      330     41      365     4e-130
...
```

In theory, at this point we should be able to replace the `less -S` with a `wc` to count the lines, and thus counting the number of sequences that had matches to other sequences. Unfortunately, in this case, theory doesn't align with reality: inspecting the output above reveals that ID `YDR545W` is still represented by two lines, so this sequence would be counted twice.

Why? In the BLAST command, we requested the top two HSPs per query with `-max_target_seqs 2` and `-max_hsps 1`, so we expected that the best HSP would be to the sequence itself, with the second best (if it existed) to be to a non-self-hit. But in this case, `blastx` decided to report two non-self-hits. In fact, if we were to inspect `YDR545W`, `YOR396W`, and `YLR467W`, we'd find that their sequences are identical, and so BLAST needed to pick two HSPs out of the three-way tie.

In order to get the correct number of sequences that had matches to others, we need to remove any duplicates that might still be found in the first column. We can do this by adding a `sort -k1,1d -u`, for a final answer of 2,884.

```
oneils@atmosphere ~/apcb/intro/blast$ cat yeast_blastp_yeast_top2.txt | \
> grep -v '#' | \
> awk '{if($1 != $2) {print $0}}' | \
> sort -k1,1d -u | \
> wc
   2884   28840  138048
```

For any sufficiently complex data set, it is a good idea to check as many assumptions about it as possible when performing an analysis. In the above, using `wc` on the lines to count the number of sequences that had hits to others implied that in the first column no ID was listed twice. In this case, comparing the counts with and without the `sort -k1,1d -u` would serve to verify or reject this assumption. In later chapters, we'll learn more techniques for this kind of "sanity checking."

Basic Syntax for awk

Although `awk` can be used as a full-featured programming language (complete with loops, arrays, and so on), for sophisticated programming needs, other languages like Python and R are usually better suited. Let's take a look at a practical subset of its syntax.

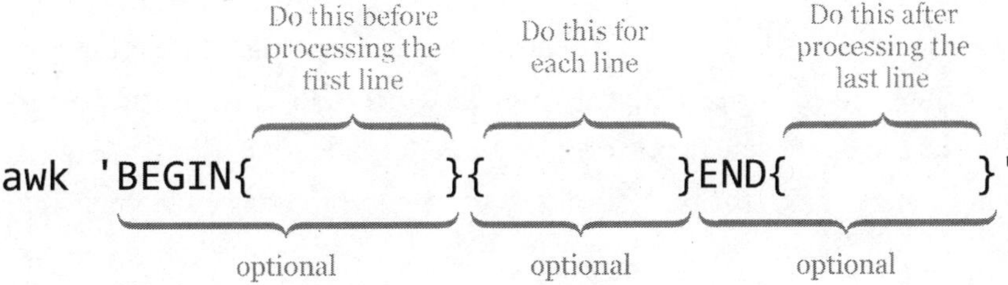

Statements in the `BEGIN` block are executed before any line of the input is processed, statements in the unadorned middle block are executed for every line, and statements in the `END` block are executed after the last line is processed. Each of these blocks is optional, though the middle "for each line" block is rarely omitted in practice.

When processing a line, a number of variables are available for use. Although many of them start with a `$`, they are not environment variables (because we've placed the "program" inside single quotes, they are sent to `awk` as unaltered strings, `$` signs intact).

`$0`

Holds the contents of the entire line that is being processed.

`$1`, `$2`, etc.

`$1` holds the contents of the first column of the current line, `$2` holds the contents of the second column of the current line, and so on. Like `sort`, `awk` by default assumes that columns are separated by whitespace.

`NF`

The special variable `NF` holds the number of columns (also known as fields) in the current line (`awk` does not require that all lines contain the same number of columns).

`NR`

`NR` holds the number of lines that have been processed so far, including the current line. Thus in the `BEGIN` line, `NR` holds `0`; during the processing of the first line, `NR` holds 1; and in the `END` block, `NR` holds the total number of lines in the input.

Note that both the `NF` and `NR` lack a `$` prefix. The text placed in the three blocks can consist of a wide variety of things, including conditionals, print statements, logic and mathematical manipulations, and so on. Perhaps a collection of examples will help illustrate the utility of `awk`. All of these examples are based on the BLAST output above, after filtering out comment lines with `grep -v '#'`.

This command prints only the first two columns of the table, separated by a space (the default when a comma is included in a `print` statement):

```
oneils@atmosphere ~/apcb/intro/blast$ cat yeast_blastp_yeast_top2.txt | \
> grep -v '#' | \
> awk '{print $1,$2}'
YAL001C YAL001C
YAL002W YAL002W
YAL003W YAL003W
...
```

Instead of separating the two output columns by a space, we can instead separate them by a string like :::, producing only a single conglomerated column of output.

```
oneils@atmosphere ~/apcb/intro/blast$ cat yeast_blastp_yeast_top2.txt | \
> grep -v '#' | \
> awk '{print $1":::"$2}'
YAL001C:::YAL001C
YAL002W:::YAL002W
YAL003W:::YAL003W
...
```

If we'd like to add a new first column that simply contains the line number, we can use the `NR` variable in conjunction with the `$0` variable:

```
oneils@atmosphere ~/apcb/intro/blast$ cat yeast_blastp_yeast_top2.txt | \
> grep -v '#' | \
> awk '{print NR,$0}'
1 YAL001C       YAL001C 1161    1161    1161    1       1161    1       1161
2 YAL002W       YAL002W 1275    1275    1275    1       1275    1       1275
3 YAL003W       YAL003W 207     207     207     1       207     1       207
...
```

If-statements allow `awk` to execute other statements conditionally; the syntax is `if( <logical expression> ) { <statements to execute> }`. Additionally, if a column contains numeric values, `awk` can work with them as such, and `awk` even understands scientific notation. Here's an example where only lines with HSP *E* values (the tenth column in our example) of less than `1e-10` are printed.

```
oneils@atmosphere ~/apcb/intro/blast$ cat yeast_blastp_yeast_top2.txt | \
> grep -v '#' | \
> awk '{if($10 < 1e-10) {print $0}}'
YAL001C YAL001C 1161    1161    1161    1       1161    1       1161    0.0
YAL002W YAL002W 1275    1275    1275    1       1275    1       1275    0.0
YAL003W YAL003W 207     207     207     1       207     1       207     2e-148
...
```

Notice that the organization of the curly brackets produces a nested block structure; although for this simple case the inside set of brackets could be omitted, it's usually best practice to include them, as they illustrate exactly which statement is controlled by the preceding `if`.[25]

If-statements can control multiple conditions, and sometimes it helps to break `awk` programs over multiple lines to help with their readability, especially when they are included in executable scripts. This sophisticated statement adds a new first column that categorizes each HSP as either "great," "good," or "ok," depending on the *E* value, printing only the two IDs and the *E* value (columns 1, 2, and 10):

```
oneils@atmosphere ~/apcb/intro/blast$ cat yeast_blastp_yeast_top2.txt | \
> grep -v '#' | \
> awk '{ \
> if($10 < 1e-30) {print "great",$1,$2,$10} \
> else if($10 < 1e-20) {print "good",$1,$2,$10} \
> else {print "ok",$1,$2,$10} \
> }'
...
great YAL018C YAL018C 0.0
good YAL018C YOL048C 4e-25
great YAL019W YAL019W 0.0
great YAL019W YOR290C 1e-84
great YAL020C YAL020C 0.0
great YAL021C YAL021C 0.0
good YAL021C YOL042W 6e-27
great YAL022C YAL022C 0.0
...
```

It is easy enough to determine whether a particular column is equal to a given string, for example, to pull out all lines where the first column is `YAL054C`:

```
oneils@atmosphere ~/apcb/intro/blast$ cat yeast_blastp_yeast_top2.txt | \
> grep -v '#' | \
> awk '{if($1 == "YAL054C") {print $0}}'
YAL054C YAL054C 714     714     714     1       714     1       714     0.0
YAL054C YLR153C 645     714     684     73      712     37      674     0.0
```

25. This nested construct, a controlled block inside of another block that is executed for each element of a set (in this case, for each line), is one of our first examples of programming! One hint for reducing confusion when producing such structures is to fill in their structure from the outside in, adding pairs of symbols and then "backing up" into them as needed. In this example, we might have started with `awk ''`, and then added the curly brackets to produce `awk '{}'`, next `awk '{if() {}}'`, and finally filled in the logic with `awk '{if($10 < 1e-10) {print $0}}'`.

Mathematical computations are a nice feature of awk. For example, columns 4 and 5 contain the total length of the query sequence and subject sequence, respectively, so we might wish to print the ratio of these two as an additional column at the end.

```
oneils@atmosphere ~/apcb/intro/blast$ cat yeast_blastp_yeast_top2.txt | \
> grep -v '#' | \
> awk '{print $1,$2,$4/$5}'
...
YAL017W YAL017W 1
YAL017W YOL045W 1.2314
YAL018C YAL018C 1
YAL018C YOL048C 0.950437
YAL019W YAL019W 1
YAL019W YOR290C 0.664319
YAL020C YAL020C 1
...
```

We could then pipe the result to a `sort -k3,3g | tail -n 5` to see the five HSPs with the largest ratios. Beware, however, that when performing mathematical operations or comparisons with columns, any contents that can't be parsed as a number (`1.5` can be, as can `2` and `4e-4`, but not `i5` or `NA`) may be truncated (e.g., `10x1` is treated as just `10`) or treated as `0`. Using `sort` on columns with `-g` can reveal such potential problems, as the same underlying method is used for parsing.

There are a variety of mathematical functions built into awk. Here's a sample:

`log()`

Returns the natural logarithm of its argument, as in `print $10 * log($3 * $4)` for printing the log of the multiplication of the third and fourth columns times the tenth column.[26]

`length()`

The `length()` function returns the number of characters in its argument, as in `length($1)` for the character length of the first column, and `length($0)` for the character length of the whole line (spaces and tab characters included).

26. If you are concerned about where spaces are allowed in awk statements, try not to be: for the most part, they are allowed anywhere, and you may feel free to use them to enhance the readability of your statements. They are not allowed in keywords and variables: `i f($ 1 > $2) {print N R}` would be an invalid expression because `i f`, `$ 1`, and `N R` have erroneous spaces.

`**`

This operator returns the left-hand side raised to the power of the right-hand side, as in `$1**2` for the square of the first column.

`%`

The modulus operator, returning the remainder after dividing the left-hand side by the right-hand side. For example, `NR%4` will be 1 on the first line, 2 on the second, 3 on the third, 0 on the fourth, 1 on the fifth, and so on.

`exp()`

This function returns its argument raised to the natural power *e*. For example, `log(exp($1))` returns the first column's value.

`int()`

Returns the integer portion of its argument. For example, `int(6.8)` returns `6`, and `int(-3.6)` returns `-3`.

`rand()`

When given no arguments, returns a random number between 0 (inclusive) and 1 (exclusive). By default, every time `awk` is run, the series of random numbers produced by multiple calls to `rand()` is the same. To get a "random" random series, run `srand()` (which "seeds" the random number generator) in the `BEGIN` block, as in `BEGIN{srand()}{print rand(),$0}`.

Logical expressions may be combined with Boolean operators, including `&&` for "and" and `||` for "or" (which produces true if either or both sides are true), and grouping can be accomplished with parentheses. For instance, we might wish to print only those lines where the first column is not equal to the second, and either the tenth column is less than `1e-30` or the second column is `YAL044C`.

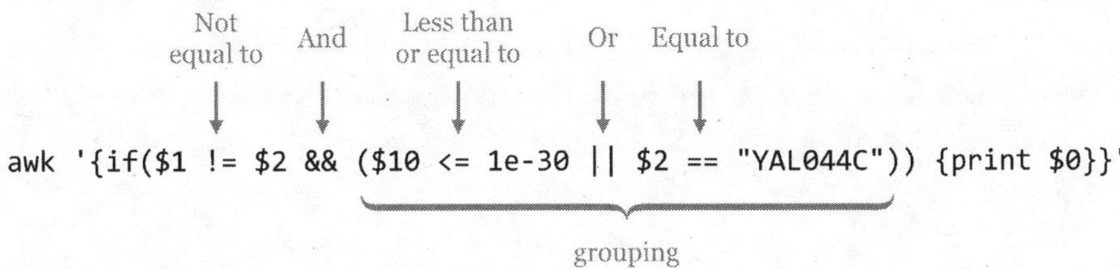

Thus far, we haven't made much use of the `BEGIN` or `END` blocks, which are especially handy when we define and update our own variables. We can accomplish this task with an `=` assignment (not to be confused with the `==` comparison). This command prints the average *E* values in our example BLAST result file.

```
oneils@atmosphere ~/apcb/intro/blast$ cat yeast_blastp_yeast_top2.txt | \
> grep -v '#' | \
> awk 'BEGIN{sumeval = 0} {sumeval = sumeval + $10} END{print sumeval/NR}'
3.00206e-09
```

This command works because the right-hand side of an assignment to a variable with `=` is evaluated before the assignment happens. Thus, in the `BEGIN` block, the `sumeval` variable is initialized to `0`, then for each line the value of `sumeval` is added to the contents of the tenth column (the *E* value of that line), and the result is stored in `sumeval`. Finally, in the `END` block, `sumeval` contains the total sum of *E* values, and we can divide this result by the number of lines processed, `NR`.

We can execute multiple statements within a single block if we separate them with semicolons. In the above example, the average *E* value computed includes self-hits. We can filter them out with an if-statement before modifying `sumeval`, but then we won't want to divide the result by `NR`, because that will include the self-hit counts as well. To solve this problem, we'll need to keep *two* variables.

```
oneils@atmosphere ~/apcb/intro/blast$ cat yeast_blastp_yeast_top2.txt | \
> grep -v '#' | \
> awk 'BEGIN{sumeval = 0; count = 0} \
> {if($1 != $2){sumeval = sumeval + $10; count = count + 1}} \
> END{print sumeval/count}'
9.06228e-09
```

As before, some IDs are still present more than one time in the first column with this solution, so it may make better sense to first filter the desired lines by using the `awk` and `sort -k1,1d -u` solution from above, and then use another `awk` for the average computation.

Exercises

1. In the file **`pz_blastx_yeast_top10.txt`**, how many HSPs (lines) have an *E* value that is less than `1e-30` *or* have an identity value of greater than 50%? Use `awk`, `wc`, and `grep` if needed to compute the answer.
2. The file **`contig_stats.txt`** describes statistics for contigs from a de novo genome assembly (a contig is an assembled "piece" of the genome). The fourth column describes the GC content of the various contigs. Other analyses indicate that the majority of correctly assembled genes have an average coverage (second column) of between 80.0 and 150.0. Use `awk` to determine the average GC content for contigs with coverage between 80.0 and 150.0. Then use another invocation of `awk` to determine the average GC content for all other contigs. (Do not count the header line in the file in your computations.)
3. The file **`PZ.annot.txt`** is the result of a gene ontology (GO) analysis for the full set of assembled *Papilio zelicaon* cDNA sequences. Owing to the incomplete nature of the annotation process, not all sequences were assigned GO terms. How many different sequence IDs are represented in this file?
4. Some versions of the `sort` program can sort lines in "random" order by using the `-R` flag. Rather than using this flag, however, use `grep`, `awk` (with the `rand()` feature) `sort` (without the `-R` flag) and `head` to select five random IDs from **`pz_cDNAs.fasta`**. An example output might look like:

```
>PZ725649
>PZ442821
>PZ805359
>PZ760258
>PZ7180000019707
>PZ725649
```

The same command should produce a different list of five IDs each time it is run.

Chapter 11

Patterns (Regular Expressions)

In previous chapters, we used a simple **`fasta_stats`** program to perform basic analyses on a FASTA file called **`pz_cDNAs.fasta`**, mostly as an excuse to learn about the standard streams and tools like `grep` and `sort`. It turns out that the information in the `pz_cDNAs.fasta` file provides us with many potential questions to ponder.

The sequences in this file are actually a subset of putative transcripts, produced from a de novo transcriptome assembly for the butterfly *Papilio zelicaon*. Each sequence header line encodes a variety of information: the second and third columns of the header lines reveal the number of reads contributing to each assembled sequence and the average coverage of the sequence (defined as the total number of bases contributed by reads, divided by the assembled sequence length). Even the sequence IDs encode some information: they all start with a pseudorandom identifier, but some have a suffix like `_TY`.

```
...
>PZ456916 nReads=1 cov=1
AAACTGTCTCTAATTAATTTATAAAATTTAATTTTTTAGTAAAAAAGCTAAAATAATTTTAAAAGACGAG
AAGACCCTATAGAGTTTTATAATTTATTTAATTATTATTAATATATAAATTTTAAAATTAAAATTAGGTA
AATTATTTTGTTGGGGTGACAG
>PZ7180000037718 nReads=9 cov=6.26448
ACTTTTTTTTTAATTTATTTAATTATATTAACTAATAAATCCGTTGTAATTGTGAGTTTATATGCAATTT
CGAAGTAGAACCGTTTCACTGGAAAGCGTTGTGTTGTCAGTTCGGTCGCTCTTTCGTATTTTTAAATATA
AGTAGGCTTATAAATTGAAGCGTTTTGCTTCTTGACAATTTATCTTACTGCATATGTGATAAGTATCAGA
ATTGCCCGCAGTATTCCCGAAGCGAGCGACCGAAGCCGGTCAATGTGAAAAACGAAAAACATTTTTTTTA
TATAAGCAACAAAAAAAACCTCTTTATACGTTTAACTTAGATAGTTATTATTAATTTTAGCTTTTAATAG
GTGTTTCGATGATTTTCACGAATTTTTTTTTGTTTTCTCGCATTTAGCTAGCGATTGCAAGAGTCGCAGT
GTACATAATATAATAGTTAGACATGATATGGACAATACCTAACAAGTGAAAAGAAAAAAAATATCATTTT
ATTGAACAAATACATTCAGTGGAATTAT
>PZ7180000000004_TY nReads=86 cov=36.4238
CCAAAGTCAACCATGGCGGCCGGGGTACTTTATACTTATCCGACAAACTTCCGTGCTTATAAGGCGTTGA
TCGCCGCACAGTACTCCGGGGTCGATCTTAAAGTTGCAACGGGTTTCGTATTTGGCGAGACAAATAAATC
TGAAGAGTTCCTGAAAAAATTCCCTGCGGGCAAAGTACCGGCTTATGAAAGTGCTGATGGAAAAGTGGTG
...
```

Groups of sequences that share the same _ suffix were previously identified as having shared matches using a self-BLAST. Sequence IDs without such a suffix had no matches. We might ask: how many sequences are in such a group? This could be easily answered by first using `grep` to extract lines that match `>` (the header lines), and then using another `grep` to extract those with the pattern _ (those in a group) before sending the result to `wc`.

A more complex question would be to ask how many different groups are represented in the file. If the group information was stored in a separate column (say, the second column), this question could be answered with the same process as above, followed by a `sort -k2,2d -u` to remove duplicate group identifiers. But how can we coerce the group information into its own column? We could do this by *substituting* instances of _ with spaces. The `sed` (Stream EDitor) tool can help us. Here's the overall pipeline we'll use:

```
oneils@atmosphere ~/apcb/intro/fasta_stats$ cat pz_cDNAs.fasta | \
> grep '>' | \
> grep '_' | \
> sed -r 's/_/ /' | \
> sort -k2,2d -u | \
> less -S
```

And here is some of the output, where only sequences in groups are represented and each group is represented only once (replacing the `less -S` with `wc` would thus count the number of groups):

```
...
>PZ7180000000067 AF nReads=16 cov=12.0608
>PZ7180000028269 AFW nReads=3 cov=2.97992
>PZ7180000036480 AIJ nReads=11 cov=4.61416
>PZ783221 AOC nReads=1 cov=1
...
```

The `sed` tool is a sophisticated program for modifying input (either from a file or standard input) and printing the results to standard output: `sed '<program>' <file>` or `... | sed '<program>'`.

Like `awk`, `sed` hails from the 1970s and provides a huge variety of powerful features and syntax, only a tiny fraction of which we'll cover here. In particular, we'll focus on the `s`, or substitution, operation.

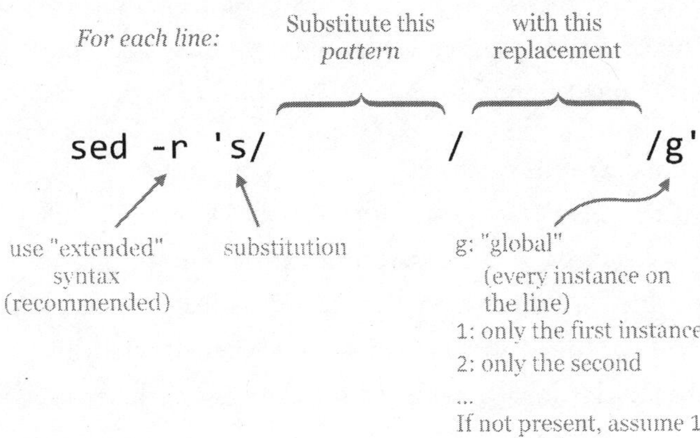

The `-r` option that we've used lets `sed` know that we want our pattern to be specified by "POSIX extended regular expression" syntax.[27] The general pattern of the program for substitution is `s/<pattern>/<replacement>/g`, where the `g` specifies that, for each line, each instance of the pattern should be replaced. We can alternatively use `1` in this spot to indicate that only the first instance should be replaced, `2` to indicate only the second, and so on. Often, `s/<pattern>/<replacement>/` is used, as it has the same meaning as `s/<pattern>/<replacement>/1`.[28]

27. POSIX, short for Portable Operating System Interface, defines a base set of standards for programs and operating systems so that different Unix-like operating systems may interoperate.

28. We should also note that the `/` delimiters are the most commonly used, but most characters can be used instead; for example, `s|<pattern>|<replacement>|g` may make it easier to replace `/` characters.

Regular Expressions

The true power of sed comes not from its ability to replace text, but from its utility in replacing text based on "patterns" or, more formally, *regular expressions.* A regular expression is a syntax for describing pattern matching in strings. Regular expressions are described by the individual characters that make up the pattern to search for, and "meta-operators" that modify parts of the pattern for flexibility. In [ch]at, for example, the brackets function as a meta-operator meaning "one of these characters," and this pattern matches both cat and hat, but not chat. Regular expressions are often built by chaining smaller expressions, as in [ch]at on the [mh]at, matching cat on the hat, cat on the mat, hat on the hat, and hat on the mat.

In the example above, the entire pattern was specified by _, which is not a meta-operator of any kind, and so each instance of _ was replaced by the replacement (a space character). The meta-operators that are supported by regular expressions are many and varied, but here's a basic list along with some biologically inspired examples:

non-meta-operator characters or strings

Most characters that don't operate in a meta-fashion are simply matched. For example, _ matches _, A matches A, and ATG matches a start codon. (In fact, ATG is three individual patterns specified in a row.) When in doubt, it is usually safe to escape a character (by prefixing it with a backslash) to ensure it is interpreted literally. For example, \[_\] matches the literal string [_], rather than making use of the brackets as meta-operators.

.

A period matches any single character. For example, CC. matches any P codon (CCA, CCT, CCG, CCC), but also strings like CCX and CC%.

[<charset>]

Matches any single character specified in <charset>. For example, TA[CT] matches a Y codon (TAC or TAT).

[^<charset>]

Placing a ^ as the first character inside charset brackets negates the meaning, such that any single character *not* named in the brackets is matched. TA[^CT] matches TAT, TAG, TA%, and so on, but not TAC or TAT.

^ (outside of [])

Placing a ^ outside of charset brackets matches the start of the input string or line. Using `sed -r 's/^ATG/XXX/g'`, for example, replaces all instances of start codons with `XXX`, but only if they exist at the start of the line.

$

Similar to ^, but $ matches the end of the string or line. So, `sed -r 's/ATG$/XXX/g'` replaces all start codons that exist at the end of their respective lines.

So far our patterns aren't really all that flexible, because most of the pieces covered to this point match a single character. The next five meta-operators resolve that limitation.

`{x,y}`

Modifies the preceding pattern so that it matches if it occurs between `x` and `y` times in a row, inclusive. For example, `[GC]{4,8}` matches any string of C's and/or G's that is four to eight characters long (shooting for eight characters, if possible). So, `sed -r 's/[GC]{4,8}/_X_/g'` would result in the following substitutions:

`ATCCGTCT` to `ATCCGTCT` (no replacement)
`ATCCGCGGCTC` to `AT_X_TC`
`ATCGCGCGGCCCGTTCGGGCCT` to `AT_X_CCGTT_X_T`

Using `{0,1}` has the effect of making what it follows optional in the pattern, and `{x,}` has the effect of allowing the pattern to match `x` or more times with no upper limit.

*

An asterisk modifies the preceding pattern so that it matches if it occurs zero or more times; thus it is equivalent to `{0,}`.

The usage of * deserves a detailed example. Consider the pattern `ATG[ATGC]*TGA`, where `ATG` is the pattern for a start codon, `[ATGC]*` indicates zero or more DNA bases in a row, and `TGA` is one of the canonical stop codons. This pattern matches `ATGTACCTTGA`, and also matches `ATGTGA` (where the middle part has been matched zero times).

+

The most prominent repetition modifier, a plus sign modifies the preceding pattern so that it is matched one or more times; it is equivalent to `{1,}`. In contrast with the example above, `ATG[ATGC]+TGA` matches `ATGTACCTTGA` and `ATGCTGA`, but not `ATGTGA`.

`(<pattern>)`

Parentheses may be used to group an expression or series of expressions into a single unit so that they may be operated on together. Because `AT` is the pattern `A` followed by `T`, for example, `AT+` matches `AT`, `ATT`, `ATTT`, and so on. If we wanted to instead match `AT` repeats, we might wish to specify a pattern like `(AT)+`, which matches `AT`, `ATAT`, `ATATAT`, and so on. Parentheses also "save" the string that was matched within them for later use. This is known as back-referencing, discussed below.

`<pattern x>|<pattern y>`

Match either the pattern `<pattern x>` or the pattern `<pattern y>`. Multiple such patterns or operations can be chained; for example, `TAA|TAG|TGA` matches any one of the three canonical stop codons. This example is a bit ambiguous, though: does this pattern read "TA (A or T) A (G or T) GA," or "TAA or TAG or TGA"? To make it concrete, we'd probably want to specify it as `((TAA)|(TAG)|(TGA))`.

Using these pieces, we can put together a regular expression that serves as a simple (and not actually useful in practice) open reading frame finder. For prokaryotic sequences (where introns are not a consideration), we'll define these as a start codon `ATG`, followed by one or more codons, followed by one of the three canonical stop codons `TAA`, `TAG`, or `TGA`. The pattern for the start is `ATG`, and we've seen how we can encode a stop above, with `((TAA)|(TAG)|(TGA))`. How about "one or more codons?" Well, "one or more" is embodied in the `+` operator, and a codon is any three A's, T's, C's, or G's. So, "one or more codons" is encoded as `([ACTG]{3,3})+`. Thus the regular expression for our simple open reading frame finder is:

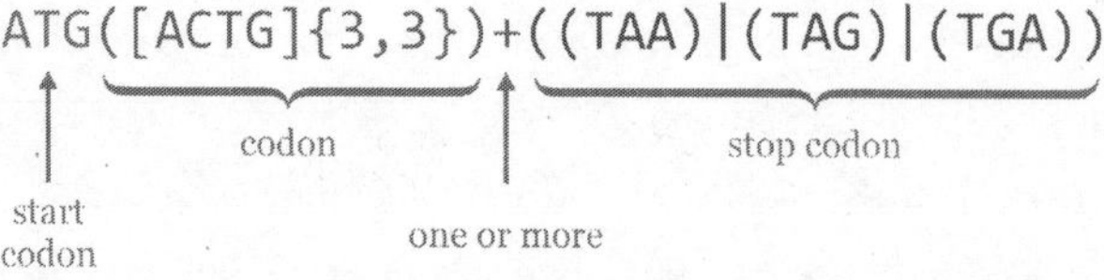

In reality, regular expressions are not often used in searching for coding regions (though they are sometimes used for identifying smaller motifs). Part of the reason is that regular expressions are, by default, *greedy*: they match the first occurring pattern they can, and they seek to match as much of the string as they can. (The cellular machinery that processes open reading frames is not greedy in this way.) Consider the following sequence, which has three open reading frames according to our simple definition and regular expression above.

```
Input:   TATGCATGTTTAGTAGCTTTTAG
ORF 1:    ATGCATGTTTAGTAG
          start        stop
ORF 2:    ATGCATGTTTAG
          start     stop
ORF 3:        ATGTTTAGTAGCTTTTAG
              start          stop
```

Notice that the string `TAG` is both a type of codon in general (`[ACTG]{3,3}`) and a stop, so technically both of the first two options are valid according to the regular expression. By the rules of greediness, the first will be matched, which we can verify with a simple `echo` and `sed`.

```
oneils@atmosphere ~/apcb/intro/fasta_stats$ echo "TATGCATGTTTAGTAGCTTTTAG" | \
> sed -r 's/ATG([ACTG]{3,3})+((TAA)|(TAG)|(TGA))/_ORF_/g'
T_ORF_CTTTTAG
```

The regular expression syntax used by `sed` is similar to the syntax used in languages such as Perl, Python, and R. In fact, all of the examples we've seen so far would work the same in those languages (though they are applied by their own specific functions rather than a call to `sed`). One helpful feature provided by more modern regular expression engines like these is that operators like `*` and `+` can be made nongreedy (though I prefer the clearer term "reluctant") by following them with a question mark. In Python, the regular expression `ATG([ACTG]{3,3})+?((TAA)|(TAG)|(TGA))` would match the second option. (When not following a `*`, or `+`, it makes the previous optional; thus `TG(T)?CC` is equivalent to `TG(T){0,1}CC`.) More sophisticated features allow the user to access all the matches of a pattern, even if they overlap, so that the most satisfying one can be pulled out by some secondary criteria. Unfortunately, `sed` does not support nongreedy matching and several other advanced regular expression features.

Character Classes and Regular Expressions in Other Tools

We often wish to use charset brackets to match any one of a "class" of characters; for example, `[0123456789]` matches any single digit. Most regular expression syntaxes (including that used by `sed`) allow a shorthand version `[0-9]` (if we wanted to match only a 0, 9, or -, we could use `[09-]`). Similarly, `[a-z]` matches any single lowercase letter, and `[A-Z]` any uppercase letter. These can even be combined: `[A-Za-z0-9]` matches any digit or letter. In the POSIX extended syntax used by `sed`, `0-9` can also be specified as `[:digit:]`. Notice the lack of brackets in the former—to actually match any single digit, the regular expression is `[[:digit:]]` (which, yes, is annoying). To match any nondigit, we can negate the bracketed set as `[^[:digit:]]`.

These POSIX character classes are especially useful when we want to match character types that are difficult to type or enumerate. In particular, `[[:space:]]` matches one of any whitespace character (spaces, tabs, newlines), and `[[:punct:]]` matches any "punctuation" character, of which there are quite a few. The `[[:space:]]` character class is particularly helpful when you are reformatting data stored in rows and columns but are not sure whether the column separators are spaces, tabs, or some combination.

In many regular expression syntaxes (including those used by Perl, Python, R, and some versions of `sed`), even shorter shortcuts for character classes are available. In these, `\d` is equivalent to `[[:digit:]]`, `\D` is equivalent to `[^[:digit:]]`, `\s` for `[[:space:]]`, `\S` for `[^[:space:]]`, among others.

As it turns out, regular expressions can be utilized by `grep` as well as `awk`. When using `grep`, we can specify that the pattern should be treated as an extended regular expression by adding the flag `-E` (as opposed to the `-r` used for `sed`.) Thus `grep -E '[[:digit:]]+'` extracts lines that contain an integer.

In `awk`, we can use the `~` comparator instead of the `==` comparator in an if-statement, as in `awk '{if($1 ~ /PZ718[[:digit:]]+/) {print $3}}'`, which prints the third column of each line where the first column matches the pattern `PZ718[[:digit:]]+`.

Back-Referencing

According to the definition above for the header lines in the `pz_cDNAs.fasta` file, the IDs should be characterizable as a pseudorandom identifier followed by, optionally, an underscore and a set of capital letters specifying the group. Using `grep '>'` to extract just the header lines, we can inspect this visually:

```
...
>PZ7180000037718 nReads=9 cov=6.26448
>PZ7180000000004_TY nReads=86 cov=36.4238
>PZ7180000000067_AF nReads=16 cov=12.0608
>PZ7180000031591 nReads=4 cov=3.26022
>PZ7180000024036 nReads=14 cov=5.86079
>PZ15501_A nReads=1 cov=1
...
```

If we send these results through `wc`, we see that this file contains 471 header lines. How can we verify that each of them follows this pattern? By using a regular expression with `grep` for the pattern, and comparing the count to 471. Because IDs must begin immediately after the > symbol in a FASTA file, that will be part of our pattern. For the pseudorandom section, which may or may not start with `PZ` but should at least *not* include an underscore or a space, we can use the pattern `[^_[:space:]]+` to specify one or more nonunderscore, nonwhitespace characters. For the optional group identifier, the pattern would be `(_[A-Z]+){0,1}` (because `{0,1}` makes the preceding optional). Putting these together with `grep -E` and counting the matches should produce 471.

```
oneils@atmosphere ~/apcb/intro/fasta_stats$ cat pz_cDNAs.fasta | \
> grep -E '>[^_[:space:]]+(_[A-Z]+){0,1}' | \
> wc
    471    1413   15279
```

All of the headers matched the pattern we expected. What if they hadn't? We could inspect which ones didn't by using a `grep -v -E` to print the lines that didn't match the pattern.

Now, hypothetically, suppose a (stubborn, and more senior) colleague has identified a list of important gene IDs, and has sent them along in a simple text file.

```
id      annotation      expression      sample
LQ00X000020     Hypothetical protein    5.024142433     C6_control_A1
LQ00X000020     Hypothetical protein    4.646697026     C6_control_A3
LQ00X000020     Hypothetical protein    4.986591902     C6_control_B1
LQ00X000020     Hypothetical protein    5.164291761     C6_control_B2
LQ00X000020     Hypothetical protein    5.348809843     C6_control_B3
LQ00X000020     Hypothetical protein    5.519392911     C6_control_C1
LQ00X000020     Hypothetical protein    5.305463561     C6_control_C2
LQ00X000020     Hypothetical protein    5.535279863     C6_control_C3
LQ00X000020     Hypothetical protein    4.831259167     C6_chemical_A1
LQ00X000020     Hypothetical protein    4.784184435     C6_chemical_A3
LQ00X000020     Hypothetical protein    5.577740866     C6_chemical_B1
```

Unfortunately, it looks like our colleague has decided to use a slightly altered naming scheme, appending `-gene` to the end of each pseudorandom identifier, before the _, if it is present. In order to continue with peaceful collaborations, it might behoove us to modify *our* sequence file so that it corresponds to this scheme. We can do this with `sed`, but it will be a challenge, primarily because we want to perform an insertion, rather than a substitution. Actually, we'll be performing a substitution, but we'll be substituting matches with contents from themselves!

Regarding back-references, in a regular expression, matches or submatches that are grouped and enclosed in parentheses have their matching strings saved in variables `\1`, `\2`, and so on. The contents of the first pair of parentheses is saved in `\1`, the second in `\2` (some experimentation may be needed to identify where nested parentheses matches are saved). The entire expression match is saved in `\0`.

To complete the example, we'll modify the pattern used in the `grep` to capture both relevant parts of the pattern, replacing them with `\1-gene\2`.

```
sed -r 's/(>[^_[:space:]]+)(_[A-Z]+){0,1}/\1-gene\2/'
```

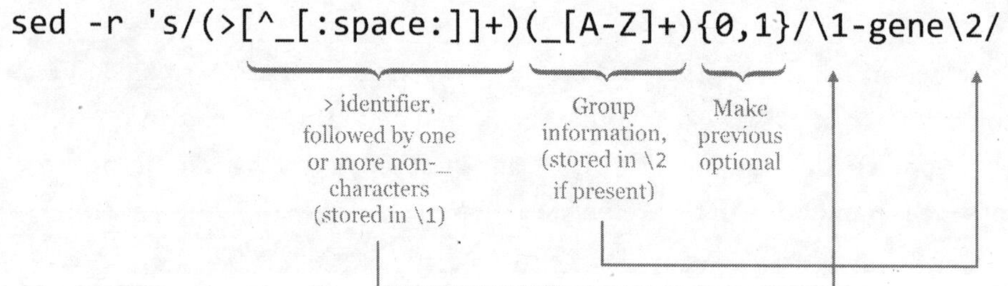

The contents of `pz_cDNAs.fasta`, after running it through the `sed` above, are as follows:

```
...
>PZ456916-gene nReads=1 cov=1
AAACTGTCTCTAATTAATTTATAAAATTTAATTTTTTAGTAAAAAAGCTAAAATAATTTTAAAAGACGAG
AAGACCCTATAGAGTTTTATAATTTATTTAATTATTATTAATATATAAATTTTAAAATTAAAATTAGGTA
AATTATTTTGTTGGGGTGACAG
>PZ7180000037718-gene nReads=9 cov=6.26448
ACTTTTTTTTTAATTTATTTAATTATATTAACTAATAAATCCGTTGTAATTGTGAGTTTATATGCAATTT
CGAAGTAGAACCGTTTCACTGGAAAGCGTTGTGTTGTCAGTTCGGTCGCTCTTTCGTATTTTTAAATATA
AGTAGGCTTATAAATTGAAGCGTTTTGCTTCTTGACAATTTATCTTACTGCATATGTGATAAGTATCAGA
ATTGCCCGCAGTATTCCCGAAGCGAGCGACCGAAGCCGGTCAATGTGAAAAACGAAAAACATTTTTTTTA
TATAAGCAACAAAAAAAACCTCTTTATACGTTTAACTTAGATAGTTATTATTAATTTTAGCTTTTAATAG
GTGTTTCGATGATTTTCACGAATTTTTTTTTGTTTTCTCGCATTTAGCTAGCGATTGCAAGAGTCGCAGT
GTACATAATATAATAGTTAGACATGATATGGACAATACCTAACAAGTGAAAAGAAAAAAAATATCATTTT
ATTGAACAAATACATTCAGTGGAATTAT
>PZ7180000000004-gene_TY nReads=86 cov=36.4238
CCAAAGTCAACCATGGCGGCCGGGGTACTTTATACTTATCCGACAAACTTCCGTGCTTATAAGGCGTTGA
TCGCCGCACAGTACTCCGGGGTCGATCTTAAAGTTGCAACGGGTTTCGTATTTGGCGAGACAAATAAATC
TGAAGAGTTCCTGAAAAAATTCCCTGCGGGCAAAGTACCGGCTTATGAAAGTGCTGATGGAAAAGTGGTG
...
```

Back-references may be used within the pattern itself as well. For example, a `sed -r 's/([A-Za-z]+) \1/\1/g'` would replace "doubled" words `([A-Za-z]+) \1` with a single copy of the word `\1`, as in `I like sed very very much`, resulting in `I like sed very much`. But beware if you are thinking of using substitution of this kind as a grammar checker, because this syntax does not search across line boundaries (although more complex `sed` programs can). This example would not modify the following pair of lines (where the word `the` appears twice):

> The quick sed regex modifies the
> the lazy awk output.

A few final notes about `sed` and regular expressions in general will help conclude this chapter.

1. Regular expressions, while powerful, lend themselves to mistakes. Work incrementally, and regularly check the results.
2. It is often easier to use multiple invocations of regular expressions (e.g., with multiple `sed` commands) rather than attempt to create a single complex expression.
3. Use regular expressions where appropriate, but know that they are not always appropriate. Many problems that might seem a natural fit for regular expressions are also naturally fit by other strategies, which should be taken into consideration given the complexity that regular

expressions can add to a given command or piece of code.

> Some people, when confronted with a problem, think, "I know, I'll use regular expressions." Now they have two problems.
>
> ~Jamie Zawinski

Exercises

1. In the de novo assembly statistics file **`contig_stats.txt`**, the contig IDs are named as `NODE_1`, `NODE_2`, and so on. We'd prefer them to be named `contig1`, `contig2`, and the like. Produce a `contig_stats_renamed.txt` with these changes performed.
2. How many sequences in the file **`pz_cDNAs.fasta`** are composed of only one read? You will likely need to use both `awk` and `sed` here, and be sure to carefully check the results of your pipeline with `less`.
3. A particularly obnoxious colleague insists that in the file **`pz_cDNAs.fasta`**, sequences that are not part of any group (i.e., those that have no `_` suffix) should have the suffix `_nogroup`. Appease this colleague by producing a file to this specification called `pz_cDNAs_fixed.fasta`.
4. The headers lines in the yeast protein set **`orf_trans.fasta`** look like so when viewed with `less -S` after grepping for `>`:

   ```
   >YAL001C TFC3 SGDID:S000000001, Chr I from 151006-147594,151166-151097, Genome R
   >YAL002W VPS8 SGDID:S000000002, Chr I from 143707-147531, Genome Release 64-2-1,
   >YAL003W EFB1 SGDID:S000000003, Chr I from 142174-142253,142620-143160, Genome R
   ...
   ```

   Notably, header lines contain information on the locations of individual exons; sequence YAL001C has two exons on chromosome I from locations 151006 to 147594 and 151166 to 151097 (the numbers go from large to small because the locations are specified on the forward strand, but the gene is on the reverse complement strand). By contrast, sequence YAL002W has only one exon on the forward strand.

   How many sequences are composed of only a single exon? Next, produce a list of sequence IDs in a file called `multi_exon_ids.txt` containing all those sequence IDs with more than one exon as a single column.

5. As a continuation of question 4, which sequence has the most exons? Which single-exon sequence is the longest, in terms of the distance from the first position to the last position noted in the exon list?

Chapter 12

Miscellanea

Tools like `sort`, `head` and `tail`, `grep`, `awk`, and `sed` represent a powerful toolbox, particularly when used with the standard input and standard output streams. There are many other useful command line utilities, and we'll cover a few more, but we needn't spend quite as much time with them as we have for `awk` and `sed`.

Manipulating Line Breaks

All of the features of the tools we've covered so far assume the line as the basic unit of processing; `awk` processes columns within each line, `sed` matches and replaces patterns in each line (but not easily across lines), and so on. Unfortunately, sometimes the way data break over lines isn't convenient for these tools. The tool `tr` translates sets of characters in its input to another set of characters as specified: `... | tr '<set1>' '<set2>'`[29]

As a brief example, `tr 'TA' 'AT' pz_cDNAs.fasta` would translate all `T` characters into `A` characters, and vice versa (this goes for every `T` and `A` in the file, including those in header lines, so this tool wouldn't be too useful for manipulating FASTA files). In a way, `tr` is like a simple `sed`. The major benefit is that, unlike `sed`, `tr` does not break its input into a sequence of lines that are operated on individually, but the entire input is treated as a single stream. Thus `tr` can replace the special "newline" characters that encode the end of each line with some other character.

On the command line, such newline characters may be represented as `\n`, so a file with the following three lines

29. Unlike other tools, `tr` can only read its input from stdin.

```
line 1
line 2
line 3
```

could alternatively be represented as `line 1\nline 2\nline 3\n` (most files end with a final newline character). Supposing this file was called `lines.txt`, we could replace all of the `\n` newlines with # characters.

```
oneils@atmosphere ~/apcb/intro/fasta_stats$ cat lines.txt | tr '\n' '#'
line 1#line 2#line 3#oneils@atmosphere ~/apcb/intro/fasta_stats$
```

Notice in the above that even the final newline has been replaced, and our command prompt printed on the same line as the output. Similarly, `tr` (and `sed`) can replace characters with newlines, so `tr '#' '\n'` would undo the above.

Using `tr` in combination with other utilities can be helpful, particularly for formats like FASTA, where a single "record" is split across multiple lines. Suppose we want to extract all sequences from **`pz_cDNAs.fasta`** with nReads greater than 5. The strategy would be something like:

1. Identify a character not present in the file, perhaps an `@` or tab character `\t` (and check with `grep` to ensure it is not present before proceeding).
2. Use `tr` to replace all newlines with that character, for example, `tr '\n' '@'`.
3. Because > characters are used to indicate the start of each record in FASTA files, use `sed` to replace record start > characters with newlines followed by those characters: `sed -r 's/>/\n>/g'`.

   At this point, the stream would look like so, where each line represents a single sequence record (with extraneous `@` characters inserted):

   ```
   >PZ7180000027934 nReads=5 cov=2.32231@TTTAATGATCAGTAAAGTTATAGTAGTTGTATGTACAATATT
   >PZ456916 nReads=1 cov=1@AAACTGTCTCTAATTAATTTATAAAATTTAATTTTTTAGTAAAAAAGCTAAAATA
   >PZ7180000037718 nReads=9 cov=6.26448@ACTTTTTTTTTAATTTATTTAATTATATTAACTAATAAATCC
   >PZ7180000000004_TY nReads=86 cov=36.4238@CCAAAGTCAACCATGGCGGCCGGGGTACTTTATACTTA
   ```

4. Use `grep`, `sed`, `awk`, and so on to select or modify just those lines of interest. (If needed, we could also use `sed` to remove the inserted `@` characters so that we can process on the sequence itself.) For our example, use `sed -r 's/=/ /1' | awk '{if($3 > 5) {print $0}}'` to print only lines where the nReads is greater than 5.
5. Reformat the file back to FASTA by replacing the `@` characters for newlines, with `tr` or `sed`.

6. The resulting stream will have extra blank lines as a result of the extra newlines inserted before each > character. These can be removed in a variety of ways, including `awk '{if(NF > 0) print $0}'`.

Joining Files on a Common Column (and Related Row/Column Tasks)

Often, the information we want to work with is stored in separate files that share a common column. Consider the result of using `blastx` to identify top HSPs against the yeast open reading frame set, for example.

```
oneils@atmosphere ~/apcb/intro/fasta_stats$ blastx -query pz_cDNAs.fasta \
> -subject ../blast/orf_trans.fasta \
> -evalue 1e-6 \
> -max_target_seqs 1 \
> -max_hsps 1 \
> -outfmt 6 \
> -out pz_blastx_yeast_top1.txt
```

The resulting file **pz_blastx_yeast_top1.txt** contains the standard BLAST information:

```
PZ7180000000004_TX      YPR181C 58.33   36      15      0       891     998
PZ7180000000004_TY      YKL081W 31.07   338     197     8       13      993
PZ7180000000067_AF      YMR226C 40.00   60      34      1       60      239
PZ7180000031592 YGL130W 58.33   36      14      1       478     374     225
PZ1082_AB       YHR104W 44.92   118     62      3       4       348     196
PZ11_FX YLR406C 53.01   83      38      1       290     42      25      106
...
```

Similarly, we can save a table of sequence information from the `fasta_stats` program with the comment lines removed as **pz_stats.table**.

```
oneils@atmosphere ~/apcb/intro/fasta_stats$ ./fasta_stats pz_cDNAs.fasta | \
> grep -v '#' > pz_stats.table
```

Viewing the file with `less -S`:

```
PZ832049        0.321   218     CTTAA   4       unit:CGT        6       trinucle
PZ21878_A       0.162   172     ATTAA   8       unit:ATT        6       trinucle
PZ439397        0.153   111     TTAAT   5       unit:GAAAT      10      pentanuc
PZ16108_A       0.157   191     ATTAA   7       unit:ATT        6       trinucle
PZ21537_A       0.158   82      TTATT   3       unit:ATT        6       trinucle
PZ535325        0.108   120     AATTA   6       unit:TA 6       dinucleotide
...
```

Given such data, we might wish to ask which sequences had a hit to a yeast open reading frame *and* a GC content of over 50%. We could easily find out with `awk`, but first we need to invoke `join`, which merges two row/column text files based on lines with similar values in a specified "key" column. By default, `join` only outputs rows where data are present in both files. Both input files are required to be similarly sorted (either ascending or descending) on the key columns: `join -1 <key column in file1> -2 <key column in file2> <file1> <file2>`.

Like most tools, `join` outputs its result to standard output, which can be redirected to a file or other tools like `less` and `awk`. Ideally, we'd like to say `join -1 1 -2 1 pz_stats.txt pz_blastx_yeast_top1.txt` to indicate that we wish to join these files by their common first column, but as of yet the files are not similarly sorted. So, we'll first create sorted versions.

```
oneils@atmosphere ~/apcb/intro/fasta_stats$ cat pz_stats.txt | \
> sort -k1,1d > pz_stats.sorted.txt
oneils@atmosphere ~/apcb/intro/fasta_stats$ cat pz_blastx_yeast_top1.txt | \
> sort -k1,1d > pz_blastx_yeast_top1.sorted.txt
```

Now we can run our `join -1 1 -2 1 pz_stats.sorted.txt pz_blastx_yeast_top1.sorted.txt`, piping the result into `less`. The output contains all of the columns for the first file, followed by all of the columns of the second file (without the key column), separated by single spaces.

```
PZ1028_K 0.409 403 TTCAT 4 unit:CTTCT 10 pentanucleotide YDR146C 36.07 61 36 2 1
PZ1082_AB 0.404 373 TTTGC 4 unit:GCA 6 trinucleotide YHR104W 44.92 118 62 3 4 34
PZ11_FX 0.435 400 CCCTT 4 unit:CAT 9 trinucleotide YLR406C 53.01 83 38 1 290 42
PZ3202_E 0.463 496 AGAGT 5 unit:TGGC 8 tetranucleotide YBR247C 64.00 25 9 0 344
PZ483608 0.391 462 AATGT 4 unit:CAAGA 10 pentanucleotide YMR100W 44.44 27 15 0 4
PZ488295 0.665 428 CGCGC 9 unit:GC 10 dinucleotide YIL106W 35.00 140 87 3 410 3
...
```

Instead of viewing the output with `less`, piping it into an `awk '{if($1 > 0.5) print $1}'` would quickly identify those sequences with BLAST matches and GC content over 50%.

One difficulty with the above output is that it is quite hard to read, at least for us humans. The same complaint could be made for most files that are separated by tab characters; because of the way tabs are formatted in `less` and similar tools, different-length entries can cause columns to be misaligned (only visually, of course). The `column` utility helps in some cases. It reformats whitespace-separated row/column input so that the output is human readable, by replacing one or more spaces and tabs by an appropriate number of spaces so that columns are visually aligned: `column -t <file>` or `... | column -t`.

By now, it should be no surprise that `column` writes its output to standard output. Here's the result of `join -1 1 -2 1 pz_stats.sorted.txt pz_blastx_yeast_top1.sorted.txt | column -t | less -S`, which contains the same data as above, but with spaces used to pad out the columns appropriately.

```
PZ1028_K             0.409  403   TTCAT  4   unit:CTTCT   10  pentanucleotide  Y
PZ1082_AB            0.404  373   TTTGC  4   unit:GCA     6   trinucleotide    Y
PZ11_FX              0.435  400   CCCTT  4   unit:CAT     9   trinucleotide    Y
PZ3202_E             0.463  496   AGAGT  5   unit:TGGC    8   tetranucleotide  Y
PZ483608             0.391  462   AATGT  4   unit:CAAGA   10  pentanucleotide  Y
PZ488295             0.665  428   CGCGC  9   unit:GC      10  dinucleotide     Y
...
```

Because most tools like `awk` and `sort` use any number of whitespace characters as column delimiters, they can also be used on data post-`column`.

There are several important caveats when using `join`. First, if any entries in the key columns are repeated, the output will contain a row for each matching *pair* of keys.

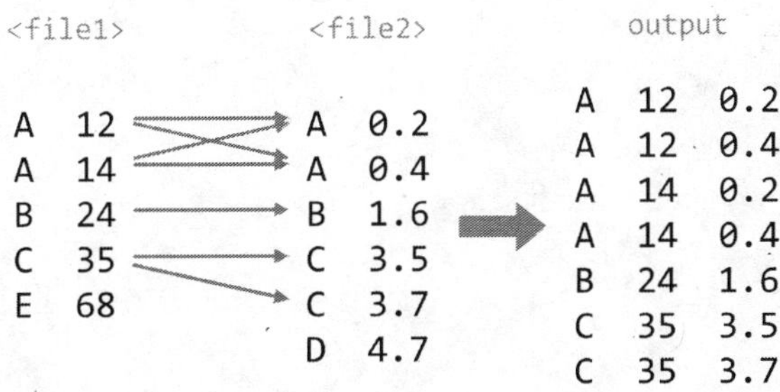

Second, the files must be similarly sorted—if they are not, `join` will at best produce a difficult-to-see warning. A useful trick when using `bash`-compatible shells is to make use of the features for "process substitution." Basically, any command that prints to standard output may be wrapped in

<(and) and used in place of a file name—the shell will automatically create a temporary file with the contents of the command's standard output, and replace the construct with the temporary file name. This example joins the two files as above, without separate sorting steps: `join -1 1 -2 1 <(sort -k1,1d pz_stats.txt) <(sort -k1,1d pz_blastx_yeast_top1.txt)`. Because the `pz_stats.txt` file was the result of redirecting standard output from `./fasta_stats pz_cDNAs.txt` through `grep -v '#'`, we could equivalently say `join -1 1 -2 1 <(./fasta_stats pz_cDNAs.fasta | grep -v '#' | sort -k1,1d) <(sort -k1,1d pz_blastx_yeast_top1.txt)`.

Finally, unless we are willing to supply an inordinate number of arguments, the default for `join` is to produce only lines where the key information is found in both files. More often, we might wish for all keys to be included, with "missing" values (e.g., `NA`) assumed for data present in only one file. In database parlance, these operations are known as an "inner join" and "full outer join" (or simply "outer join"), respectively.

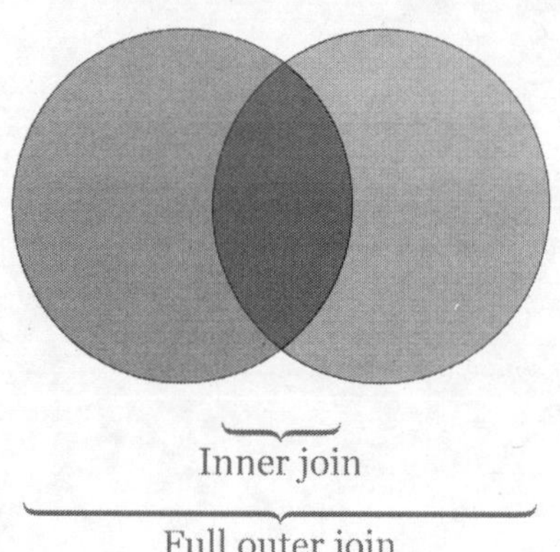

Where `join` does not easily produce outer joins, more sophisticated tools can do this and much more. For example, the Python and R programming languages (covered in later chapters) excel at manipulating and merging collections of tabular data. Other tools utilize a specialized language for database manipulation known as Structured Query Language, or SQL. Such databases are often stored in a binary format, and these are queried with software like MySQL and Postgres (both require administrator access to manage), or simpler engines like `sqlite3` (which can be installed and managed by normal users).[30]

30. While binary databases such as those used by `sqlite3` and Postgres have their place (especially when large tables need to be joined or searched), storing data in simple text files makes for easy access and manipulation. A discussion of SQL syntax and relational databases is beyond the scope of this book; see Jay Kreibich's *Using SQLite* (Sebastopol, CA: O'Reilly Media, Inc., 2010) for a friendly introduction to `sqlite3` and its syntax.

Counting Duplicate Lines

We saw that `sort` with the `-u` flag can be used to remove duplicates (defined by the key columns used). What about isolating duplicates, or otherwise counting or identifying them? Sadly, `sort` isn't up to the task, but a tool called `uniq` can help. It collapses consecutive, identical lines. If the `-c` flag is used, it prepends each line with the number of lines collapsed: `uniq <file>` or `... | uniq`.

Because `uniq` considers entire lines in its comparisons, it is somewhat more rigid than `sort -u`; there is no way to specify that only certain columns should be used in the comparison.[31] The `uniq` utility will also only collapse identical lines if they are consecutive, meaning the input should already be sorted (unless the goal really is to merge only already-consecutive duplicate lines). Thus, to identify duplicates, the strategy is usually:

1. Extract columns of interest using `awk`.
2. Sort the result using `sort`.
3. Use `uniq -c` to count duplicates in the resulting lines.

Let's again consider the output of `./fasta_stats pz_cDNAs.fasta`, where column 4 lists the most common 5-mer for each sequence. Using this extract/sort/uniq pattern, we can quickly identify how many times each 5-mer was listed.

```
oneils@atmosphere ~/apcb/intro/fasta_stats$ ./fasta_stats pz_cDNAs.fasta | \
> grep -v '#' | \
> awk '{print $4}' | \
> sort | \
> uniq -c | \
> less -S
```

The result lists the counts for each 5-mer. We could continue by sorting the output by the new first column to identify the 5-mers with the largest counts.

```
     44 AAAAA
      1 AAAAG
     12 AAAAT
     24 AAATA
      1 AAATG
     10 AAATT
...
```

31. This isn't quite true: the `-f <n>` flag for `uniq` removes the first `<n>` fields before performing the comparison.

It is often useful to run `uniq -c` on lists of counts produced by `uniq -c`. Running the result above through `awk '{print $1}' | sort -k1,1n | uniq -c` reveals that 90 5-mers are listed once, 18 are listed twice, and so on.

```
     90 1
     18 2
      4 3
      5 4
      5 5
      1 6
...
```

Counting items with `uniq -c` is a powerful technique for "sanity checking" data. If we wish to check that a given column or combination of columns has no duplicated entries, for example, we could apply the extract/sort/uniq strategy followed by `awk '{if($1 > 1) print $0}'`. Similarly, if we want to ensure that all rows of a table have the same number of columns, we could run the data through `awk '{print NF}'` to print the number of columns in each row and then apply extract/sort/uniq, expecting all column counts to be collapsed into a single entry.

Basic Plotting with gnuplot

Further illustrating the power of `awk`, `sed`, `sort`, and `uniq`, we can create a text-based "histogram" of coverages for the sequences in the `pz_cDNAs.fasta` file, which are stored in the third column of the header lines (e.g., `>PZ7180000000004_TX nReads=26 cov=9.436`). We'll start by isolating the coverage numbers with `grep` (to select only header lines), `sed` (to replace = characters with spaces), and `awk` (to extract the new column of coverage numbers), while simultaneously printing only the integer portion of the coverage column.

```
oneils@atmosphere ~/apcb/intro/fasta_stats$ cat pz_cDNAs.fasta | \
> grep '>' | \
> sed -r 's/=/ /g' | \
> awk '{print int($5)}'
9
2
2
...
```

The output is a simple column of integers, representing rounded-down coverages. Next, a `sort -k1,1n | uniq -c` will produce the counts for each coverage bin, revealing that most sequences (281) are at 1X coverage, a handful are at 2X and 3X, and higher coverages are increasingly rare.

```
    281 1
     48 2
     41 3
     26 4
     16 5
      8 6
...
```

Although languages like Python and R provide user-friendly data-plotting packages, it can sometimes be helpful to quickly plot a data set on the command line. The `gnuplot` program can produce not only image formats like PNG and PDF, but also text-based output directly to standard output. Here's the command to plot the above histogram (but with points, rather than the traditional boxes), along with the output.

```
oneils@atmosphere ~/apcb/intro/fasta_stats$ cat pz_cDNAs.fasta | \
> grep '>' | \
> sed -r 's/=/ /g' | \
> awk '{print int($5)}' | \
> sort -k1,1n | \
> uniq -c | \
> gnuplot -e 'set term dumb; plot "-" using 2:1 with points'


300 ++-----+------+------+------+------+-----+------+------+------+-----++
    A      +      +      +      +      +     +         "-" using 2:1 + A  +
    |                                                                     |
250 ++                                                                   ++
    |                                                                     |
    |                                                                     |
200 ++                                                                   ++
    |                                                                     |
    |                                                                     |
150 ++                                                                   ++
    |                                                                     |
    |                                                                     |
    |                                                                     |
100 ++                                                                   ++
    |                                                                     |
    |                                                                     |
 50 +A                                                                   ++
    |A                                                                    |
    + A    +      +      +      +      +     +      +      +      +       +
  0 ++-AAAAAAAAAAAA-----A+------+-----A+---A-+------+------+------+---A-A+
    0      20     40     60     80    100   120    140    160    180    200
```

It's a bit difficult to see, but the plotted points are represented by `A` characters. Breaking down the `gnuplot` command, `set term dumb` instructs `gnuplot` to produce text-based output, `plot "-"` indicates that we want to plot data from the standard input stream, `using 2:1` indicates that *X* values should be drawn from the second column and *Y* values from the first, and `with points` specifies that points—as opposed to lines or boxes—should be drawn. Because the counts of coverages drop off so rapidly, we may want to produce a log/log plot, and we can add a title, too: `gnuplot -e 'set term dumb; set logscale xy; plot "-" using 2:1 with points' title "Coverage Counts"`

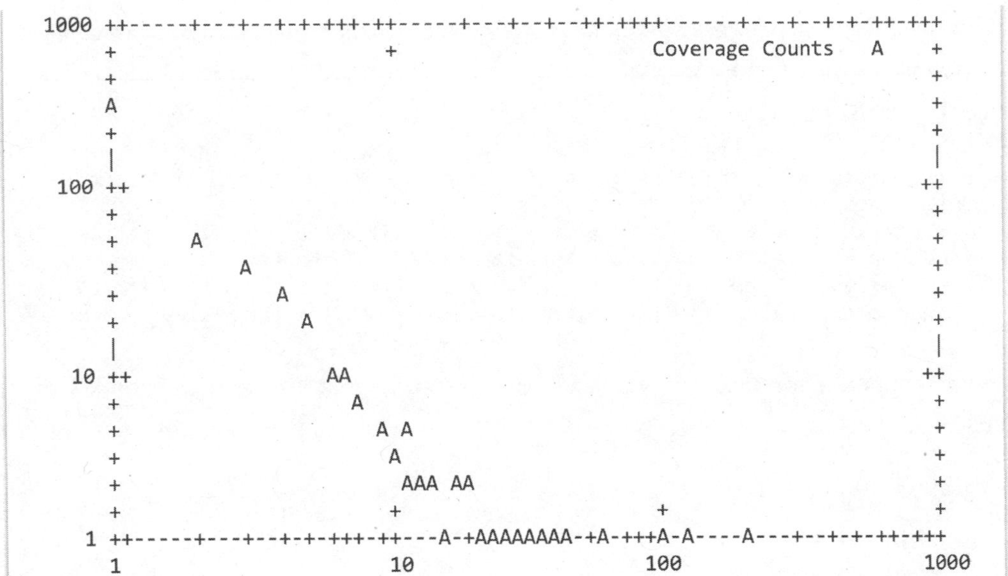

Although we've only shown the most basic usage of `gnuplot`, it is actually a sophisticated plotting package—the various commands are typically not specified on the command line but are rather placed in an executable script. For a demonstration of its capabilities, visit http://gnuplot.info.

For-Loops in bash

Sometimes we want to run the same command or similar commands as a set. For example, we may have a directory full of files ending in `.tmp`, but we wished they ended in `.txt`.

```
oneils@atmosphere ~/apcb/intro/temp$ ls
file10.tmp  file13.tmp  file16.tmp  file19.tmp  file2.tmp  file5.tmp  file8.tmp
file11.tmp  file14.tmp  file17.tmp  file1.tmp   file3.tmp  file6.tmp  file9.tmp
file12.tmp  file15.tmp  file18.tmp  file20.tmp  file4.tmp  file7.tmp
```

Because of the way command line wildcards work, we can't use a command like `mv *.tmp *.txt`; the `*.tmp` would expand into a list of all the files, and `*.txt` would expand into nothing (as it matches no existing file names).

Fortunately, bash provides a looping construct, where elements reported by commands (like ls *.tmp) are associated with a variable (like $i), and other commands (like mv $i $i.txt) are executed for each element.

```
oneils@atmosphere ~/apcb/intro/temp$ for i in $(ls *.tmp); do mv $i $i.txt; done
oneils@atmosphere ~/apcb/intro/temp$ ls
file10.tmp.txt  file14.tmp.txt  file18.tmp.txt  file2.tmp.txt  file6.tmp.txt
file11.tmp.txt  file15.tmp.txt  file19.tmp.txt  file3.tmp.txt  file7.tmp.txt
file12.tmp.txt  file16.tmp.txt  file1.tmp.txt   file4.tmp.txt  file8.tmp.txt
file13.tmp.txt  file17.tmp.txt  file20.tmp.txt  file5.tmp.txt  file9.tmp.txt
```

It's more common to see such loops in executable scripts, with the control structure broken over several lines.

```
#!/bin/bash

for i in $(ls *.tmp); do
    mv $i $i.txt;
done
```

This solution works, though often looping and similar programming techniques (like if-statements) in bash become cumbersome, and using a more robust language like Python may be the better choice. Nevertheless, bash does have one more interesting trick up its sleeve: the bash shell can read data on standard input, and when doing so attempts to execute each line. So, rather than using an explicit for-loop, we can use tools like awk and sed to "build" commands as lines. Let's remove the .tmp from the middle of the files by building mv commands on the basis of a starting input of ls -1 *.tmp* (which lists all the files matching *.tmp* in a single column). First, we'll build the structure of the commands.

```
oneils@atmosphere ~/apcb/intro/temp$ ls -1 *.tmp* | \
> awk '{print "mv "$1" "$1}'
mv file10.tmp.txt file10.tmp.txt
mv file11.tmp.txt file11.tmp.txt
mv file12.tmp.txt file12.tmp.txt
...
```

To this we will add a sed -r s/\.tmp//2 to replace the second instance of .tmp with nothing (remembering to escape the period in the regular expression), resulting in lines like

```
mv file10.tmp.txt file10.txt
mv file11.tmp.txt file11.txt
mv file12.tmp.txt file12.txt
...
```

After the `sed`, we'll pipe this list of commands to `bash`, and our goal is accomplished.

```
oneils@atmosphere ~/apcb/intro/temp$ ls -1 *.tmp.* | \
> awk '{print "mv "$1" "$1}' | \
> sed -r 's/\.tmp//2' | \
> bash
oneils@atmosphere ~/apcb/intro/temp$ ls
file10.txt  file13.txt  file16.txt  file19.txt  file2.txt  file5.txt  file8.txt
file11.txt  file14.txt  file17.txt  file1.txt   file3.txt  file6.txt  file9.txt
file12.txt  file15.txt  file18.txt  file20.txt  file4.txt  file7.txt
```

Version Control with git

In chapter 6, "Installing (Bioinformatics) Software," we worked on a rather sophisticated project, involving installing software (in our `$HOME/local/bin` directory) as well as downloading data files and writing executable scripts (which we did in our `$HOME/projects/p450s` directory). In particular, we initially created a script to automate running HMMER on some data files, called **`runhmmer.sh`**. Here are the contents of the project directory when last we saw it:

```
oneils@atmosphere ~/projects/p450s$ ls
dmel-all-translation-r6.02.fasta  p450s.fasta.aln      p450s_hmmsearch_dmel.txt
p450s.fasta                       p450s.fasta.aln.hmm  runhmmer.sh
```

It may be that as we continue working on the project, we will make adjustments to the `runhmmer.sh` script or other text files in this directory. Ideally, we would be able to access previous versions of these files—in case we need to refer back for provenance reasons or we want to undo later edits. One way to accomplish this task would be to frequently create backup copies of important files, perhaps with file names including the date of the backup. This would quickly become unwieldy, however.

An alternative is to use *version control*, which is a system for managing changes to files (especially programs and scripts) over time and across various contributors to a project. A version control system thus allows a user to log changes to files over time, and it even allows multiple users to log changes, providing the ability to examine differences between the various edits. There are a number of popular version control programs, like `svn` (subversion) and `cvs` (concurrent versioning system). Because the job of tracking changes in files across time and users is quite complex (especially when multiple users may be simultaneously making independent edits), using version control software can be a large skill set in itself.

One of the more recently developed version control systems is `git`, which has gained popularity for a number of reasons, including its use in managing popular software projects like the Linux kernel.[32] The `git` system (which is managed by a program called `git`) uses a number of vocabulary words we should define first.

Repository
: Also known as a "repo," a `git` repository is usually just a folder/directory.

Version
: A version is effectively a snapshot of a selected set of files or directories in a repository. Thus there may be multiple versions of a repository across time (or even independently created by different users).

Commit
: Committing is the action of storing a set of files (in a repository) to a version.

Diff
: Two different versions may be "diffed," which means to reveal the changes between them.

Stage
: Not all files need to be included in a version; staging a set of files marks them for inclusion in the version when the next commit happens.

The `git` system, like all version control systems, is fairly complex and provides many features. The basics, though, are: (1) there is a folder containing the project of interest; (2) changes to some files are made over time; (3) edited files can periodically be "staged"; and (4) a "commit" includes a

32. Linus Torvalds, who also started the Linux kernel project, developed the `git` system. Quoting Linus: "I'm an egotistical bastard, and I name all my projects after myself. First 'Linux,' now 'Git.'" (Here "git" refers to the British slang for a "pig-headed and argumentative" person.)

snapshot of all staged files and stores the information in a "version." (For the record, all of this information is stored in a hidden directory created within the project directory called `.git`, which is managed by the `git` program itself.)

To illustrate, let's create a repository for the `p450s` project, edit the `runhmmer.sh` script file as well as create a `README.txt` file, and commit those changes. First, to turn a directory into a `git` repository, we need to run `git init`:

```
oneils@atmosphere ~/projects/p450s$ git init
Initialized empty Git repository in /home/oneils/projects/p450s/.git/
```

This step creates the hidden directory `.git` containing the required files for tracking by the system. We don't usually need to work directly with this directory—the `git` software will do that for us. Next, we will create our first version by staging our first files, and running our first commit. We could keep tracked versions of all the files in this directory, but do we want to? The data files like `dmel-all-translation-r6.02.fasta` are large and unlikely to change, so logging them would be unnecessary. Similarly, because the output file `p450s_hmmsearch_dmel.txt` is generated programmatically and can always be regenerated (if we have a version of the program that created it), we won't track that, either. To "stage" files for the next commit, we use `git add`; to stage all files in the project directory, we would use `git add -A`, but here we want to stage only `runhmmer.sh`, so we'll run `git add runhmmer.sh`.

```
oneils@atmosphere ~/projects/p450s$ git add runhmmer.sh
```

No message has been printed, but at any time we can see the status of the `git` process by running `git status`.

```
oneils@atmosphere ~/projects/p450s$ git status
# On branch master
#
# Initial commit
#
# Changes to be committed:
#   (use "git rm --cached <file>..." to unstage)
#
#    new file:   runhmmer.sh
#
# Untracked files:
#   (use "git add <file>..." to include in what will be committed)
#
#    dmel-all-translation-r6.02.fasta
#    p450s.fasta
#    p450s.fasta.aln
#    p450s.fasta.aln.hmm
#    p450s_hmmsearch_dmel.txt
```

The status information shows that we have one new file to track, `runhmmer.sh`, and a number of untracked files (which we've left untracked for a reason). Now we can "commit" these staged files to a new version, which causes the updated staged files to be stored for later reference. When committing, we need to enter a commit message, which gives us a chance to summarize the changes that are being committed.

```
oneils@atmosphere ~/projects/p450s$ git commit -m 'Added runhmmer.sh'
[master (root-commit) ec46950] Added runhmmer.sh
 1 file changed, 20 insertions(+)
 create mode 100755 runhmmer.sh
```

At this point, a `git status` would merely inform us that we still have untracked files. Let's suppose we make some edits to `runhmmer.sh` (adding a new comment line, perhaps), as well as create a new `README.txt` file describing the project.

```
oneils@atmosphere ~/projects/p450s$ ls
dmel-all-translation-r6.02.fasta  p450s.fasta.aln.hmm       runhmmer.sh
p450s.fasta                       p450s_hmmsearch_dmel.txt
p450s.fasta.aln                   README.txt
```

Running `git status` at this point would report a new untracked file, `README.txt`, as well as a line reading `modified: runhmmer.sh` to indicate that this file has changed since the last commit. We could continue to edit files and work as needed; when we are ready to commit the changes, we just need to stage the appropriate files and run another commit.

```
oneils@atmosphere ~/projects/p450s$ git add runhmmer.sh
oneils@atmosphere ~/projects/p450s$ git add README.txt
oneils@atmosphere ~/projects/p450s$ git commit -m 'New readme, changed runhmmer'
[master 50c11fe] New readme, changed runhmmer
 2 files changed, 6 insertions(+), 1 deletion(-)
 create mode 100644 README.txt
```

Every version that we commit is saved, and we can easily see a quick log of the history of a project with `git log`.

```
oneils@atmosphere ~/projects/p450s$ git log
commit 50c11fe4fbe23c2615c3e56161cd122432f323df
Author: Shawn O'Neil <shawn.oneil@cgrb.oregonstate.edu>
Date:   Thu Feb 26 01:44:35 2015 -0700

    New readme, changed runhmmer

commit ec46950b36d3fe027b6ae50f0358f8ead02cb937
Author: Shawn O'Neil <shawn.oneil@cgrb.oregonstate.edu>
Date:   Thu Feb 26 01:38:50 2015 -0700

    Added runhmmer.sh
```

Notice that each commit is given a long serial number, such as `ec46950b36...`. To see the differences between two commits, we can run `git diff` with just the few characters of each serial number, as in `git diff 50c11fe ec4695`. The output format isn't remarkably readable by default.

```
diff --git a/README.txt b/README.txt
deleted file mode 100644
index 6bf2ae5..0000000
--- a/README.txt
+++ /dev/null
@@ -1,4 +0,0 @@
-This project aims at using HMMER to search
-for p450-1A1 genes against the D. melanogaster
-protein dataset.
-
diff --git a/runhmmer.sh b/runhmmer.sh
index 652b435..04b368c 100755
--- a/runhmmer.sh
+++ b/runhmmer.sh
@@ -1,7 +1,5 @@
 #!/bin/bash

...
```

Many other operations can be performed by `git`, such as viewing the contents of files from previous versions and "reverting" a project to a previous state (at least for those files that are tracked).

There are two other features worth mentioning. First, it's not uncommon to have many files that we'd like to leave untracked, but adding all of the rest one at a time with `git add` is tedious. Fortunately, `git add -A` looks at the contents of the file `.gitignore` (which may need to be created): any files or directories listed in `.gitignore` will not be staged by `git add -A`. (And the `.gitignore` file can be tracked.)

Second, the `git` system makes it relatively easy to share projects online with others, by creating repositories on sites like GitHub. After setting up an account at http://github.com (or on a similar site, such as http://bitbucket.com), you can "push" the current state of your project from the command line to the web. (Future commits can also be pushed as you create them.) Other users can then "clone" the project from the website by using the `git clone` command discussed briefly in chapter 6. GitHub and similar websites feature excellent tutorials for interfacing their products with your own command line repositories, should you desire to use them.

Exercises

1. In the file **`pz_cDNAs.fasta`**, sequence IDs are grouped according to common suffixes like `_TY`, `_ACT`, and the like. Which group has the largest number of sequences, and how many are in that group?
2. Using the various command line tools, extract all sequences composed of only one read (`nReads=1`) from **`pz_cDNAs.fasta`** to a FASTA formatted file called `pz_cDNAs_singles.fasta`.
3. In the annotation file **`PZ.annot.txt`**, each sequence ID may be associated with multiple gene ontology (GO) "numbers" (column 2) and a number of different "terms" (column 3). Many IDs are associated with multiple GO numbers, and there is nothing to stop a particular number or term from being associated with multiple IDs.

```
...
PZ7180000023260_APN     GO:0005515     btb poz domain containing protein
PZ7180000035568_APN     GO:0005515     btb poz domain containing protein
PZ7180000020052_APQ     GO:0055114     isocitrate dehydrogenase (nad+)
PZ7180000020052_APQ     GO:0006099     isocitrate dehydrogenase (nad+)
PZ7180000020052_APQ     GO:0004449     isocitrate dehydrogenase (nad+)
...
```

Which GO number is associated with largest number of unique IDs? How many different IDs is it associated with? Next, answer the same questions using the GO term instead of GO number. For the latter, beware that the column separators in this file are tab characters, `\t`, but `awk` by default uses any whitespace, including the spaces found in the terms column. In this file, though, `isocitrate` is not a term, but `isocitrate dehydrogenase (nad+)` is.

Part II: Programming in Python

Chapter 13

Hello, World

Before we begin with programming in Python, it is useful to consider how the language fits into the landscape and history of similar languages. Initially, computer programming was not far removed from the hardware on which it was being coded. This meant writing "bytecode"—or its human-readable equivalent, assembly code—that explicitly referenced memory (RAM) locations and copied data to and from the relatively small number of CPU registers (the storage units directly accessible by the CPU). Unfortunately, this meant that code had to be rewritten for each of the many types of CPUs. Later, more portable languages like C were developed. These languages still work close to the hardware in many ways; in particular, the programmer must tell the computer how it should allocate and de-allocate memory from RAM. On the other hand, some abstraction provides higher-level coding constructs that are not specific to CPU type. This code is then compiled into bytecode by a compiling program for each specific CPU (as discussed in previous chapters, we had to compile some software from source to install it). The result is a fast and often optimized program that frees the programmer from having to worry about the huge variety of CPU types.

Later, some programmers decided that they didn't want to have to worry about specifying how RAM space should be allocated and de-allocated. They also wanted more features built into their languages to help them quickly architect complicated programs. One of the languages meant to accomplish these goals is Python, a project started in 1988 by mathematician, computer scientist, and Monty Python fan Guido van Rossum.[33] "High-level" languages like Python and R (covered in later chapters) provide many built-in features to the programmer, and they are even more abstract than languages like C.

Unfortunately, because of the added abstractions, languages like Python can't easily be compiled (like C can) to be run directly on the CPU.[34] In fact, these languages are not run the same way compiled or assembly programs are: they are interpreted by another program that is written in a compiled language like C and runs on the CPU. So, a Python "program" is just a text file of commands that are interpreted by another program that is actually interacting with the CPU and RAM.

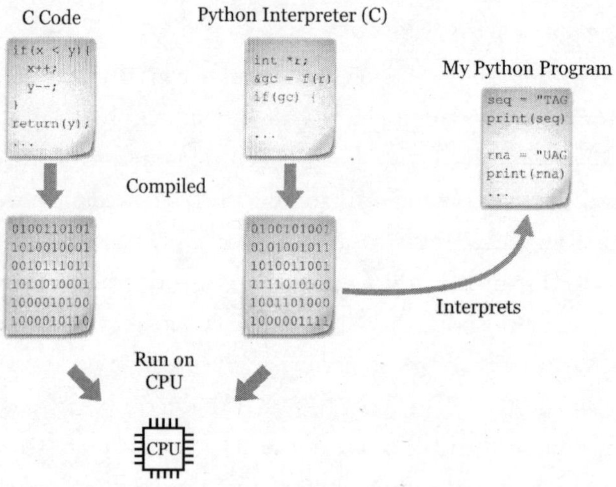

The added ease and flexibility of interpreted languages generally comes at a cost: because of the extra execution layer, they tend to be 2 to 100 times slower and use 2 to 20 times more memory than carefully constructed C programs, depending on what is being computed. These languages are significantly easier to use, however, and we can get our ideas into code far more quickly.

Work on the Python language and interpreters for it has progressed steadily since the 1990s, emphasizing a "one best way" approach. Rather than providing multiple versions of basic commands that do the same thing, Python provides as few commands as possible while attempting

33. The history of computing is full of twists, turns, and reinvention of wheels. LISP, for example, is a language that incorporates many of the same high-level features as Python, but it was first developed in 1958!

34. On the other hand, ambitious programmers are currently working on projects like Cython to do exactly this.

to not limit the programmer. Python also emphasizes code readability: most syntax is composed of English-like words, shortcuts and punctuation characters are kept to a minimum, and the visual structure of "blocks" of code are enforced with indentation.

For these reasons, Python use has grown significantly in recent years, especially for bioinformatics and computational biology. The emphasis on readability and "one best way" facilitates learning, particularly for those who are brand-new to programming.[35] Most importantly, Python allows us to focus on the concepts of programming without struggling through an abundance of choices and confusing syntax, and new programmers can frequently read and understand code written by others. Finally, Python incorporates a number of modern programming paradigms making it appropriate for both small tasks and larger software engineering projects—it's an official language at Google (along with C++ and Java), and it's taught in introductory courses at Johns Hopkins University, New York University, the Massachusetts Institute of Technology, and many others.

All of this isn't to say that any programming language is devoid of quirks and difficulties. We'll only be covering some of what Python offers—the parts that are most basic and likely to be found in many languages. Topics that are highly "Pythonic" will be highlighted as we go along.

Python Versions

In this book we will be working with Python version 2.7; that is, we're going to assume that the Python executable found in your `$PATH` variable is version 2.7 (perhaps 2.7.10, which is the last of the 2.7 series as of 2015). You can check this by running `python --version` on the command line. While newer versions are available (up to 3.4 and higher), they are not yet universally used. These newer versions change some syntax from 2.7. For many of the concepts introduced in this book, if you stick with the syntax as shown, your code should be compatible with these newer versions as well, but possibly not backward-compatible with older versions such as 2.5 or 2.6. This is an unfortunate artifact of Python's "one best way" philosophy: on occasion, the Python designers change their minds about what the best way is!

35. Many believe that the best way to learn programming is in a hardware-oriented fashion with assembly or a language like C. This is a legitimate philosophy, but for the intended audience of this book, we'll stick with the higher-level languages Python and R.

To give an example, the print function `print("hello there")` works in Python versions 2.6, 2.7, 3.0, 3.1, and so on, whereas the keyword version `print "hello there"` (notice the lack of parentheses) would only work in versions 2.6 and 2.7. In some cases where differences in behavior would occur in later versions, we'll note them in footnotes.

Hello, World

Because we're working in an interpreted language, in order to write a program, we'll need to create a file of Python code, and then supply it as input to the interpreting program. There are a few ways to do this: (1) use an interactive graphical environment like Jupyter notebook; (2) run the interpreter ourselves on the command line, giving the file name containing our code as an argument; or (3) making the code file an executable script in the command line environment using `#!` syntax.

Jupyter Notebook

For those wishing to program Python without working on the command line, a variety of graphical environments are available. A typical installation of Python from http://python.org includes the "Idle" code editor. One of the nicer alternatives to this default is known as Jupyter, which runs in a web browser allows the programmer to interleave sections of code and documentation.

Installing Jupyter requires that Python already be installed (from http://python.org), and then requires using the command line terminal in Linux, OS X, or Windows; see http://jupyter.org/install for details. Once installed, it can be started from the command line by running `jupyter notebook`:

```
Terminal — -bash — 66×5
[soneil@mbp ~]$ jupyter notebook
```

The Jupyter interface will open in the default desktop web browser, showing the list of folders and files in whatever directory the command was run from.

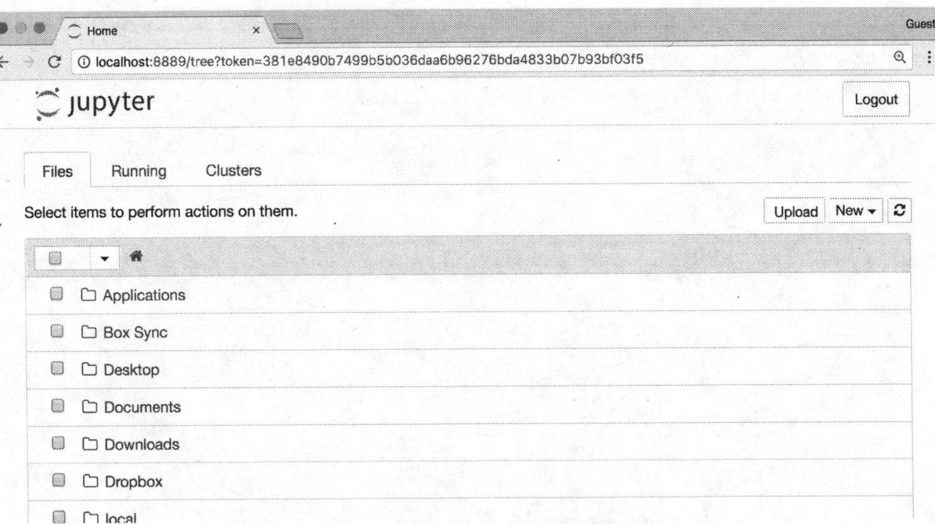

Clicking the "New" button, followed by "Python Notebook" will create a new notebook document composed of "cells." Cells in a notebook can contain human-readable text (as documentation) or lines of Python code. Whether the text in the cell is interpreted as code or text depends on the choice made in the "Cell" menu.

Each cell may be "executed" by clicking on the "Play" button; doing so causes text cells to change to a nicely formatted output, and executes lines of code for code cells. But beware: the output of a given cell often depends on what other cells have been executed and in which order (see the "Cell output depends on the order of execution" cell in the figure below).

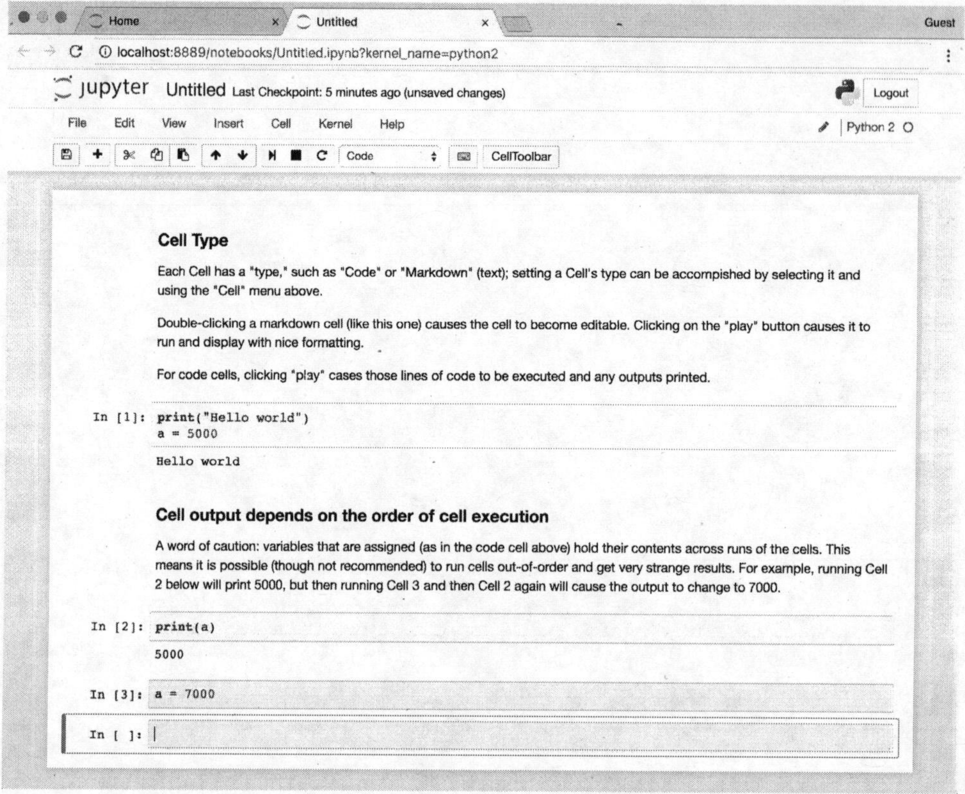

For this reason, I highly recommend making the assumption that all code cells will be executed in top-to-bottom order, which can be accomplished by selecting "Run All" from the "Cell" menu whenever you want to execute any code. Doing so causes all the cells to be re-executed each time it is run, it but has the advantage of ensuring the correctness of the overall notebook as changes are made to cells over time.

Specified Interpreter

As convenient as iPython notebook is, because the previous part of this book focused on the command line, and Python interfaces quite nicely with it, the examples here will be from the command line environment. Because Python programs are interpreted scripts, we can manually specify the interpreter on the command line each time we run such a script.

For this method, we first have to edit a code file that we'll call `helloworld.py` (`.py` is the traditional extension for Python programs). On the command line, we'll edit code files with our text editor `nano`, passing in a few extra parameters:

```
oneils@atmosphere ~/apcb/py$ nano -w -i -E -T 4 helloworld.py
```

The `-w` tells `nano` not to automatically wrap long lines (we're writing code after all, not an essay), `-i` says to automatically indent newlines to the current indentation level, `-T 4` says that tab-stops should be four spaces wide, and `-E` says that tabs should be converted to spaces (four of them). This usage of four spaces per indentation level is widely agreed upon by the Python community as being easy on the eyes. ("One best way," remember?) We'll put a simple call to the `print()` function in the file:

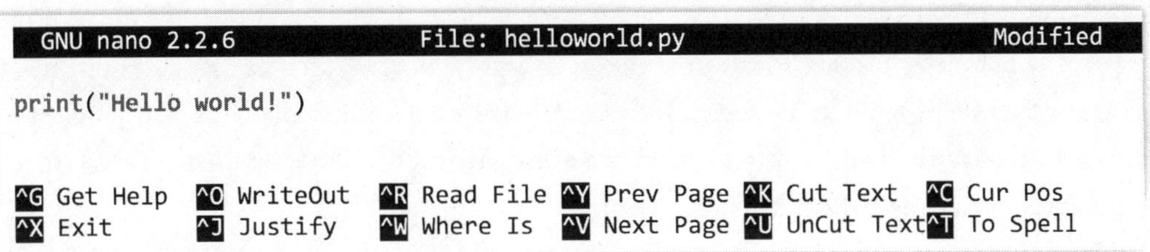

As usual, `Control-o` saves the file (press Enter if prompted about the file name) and `Control-x` exits `nano`. Next, to run it, all we need to do is call the Python interpreter on the file:

```
oneils@atmosphere ~/apcb/py$ python helloworld.py
Hello world!
oneils@atmosphere ~/apcb/py$
```

Success! We've written our first Python program!

Making the File Executable

An alternative method is to make the code file an executable script. First, we have to edit the code file to include a special first line:

```
 GNU nano 2.2.6              File: helloworld.py

#!/usr/bin/env python
print("Hello world!")

                             [ Wrote 2 lines ]
^G Get Help  ^O WriteOut  ^R Read File ^Y Prev Page ^K Cut Text  ^C Cur Pos
^X Exit      ^J Justify   ^W Where Is  ^V Next Page ^U UnCut Text^T To Spell
```

For this method to work, the first two characters of the file must be `#!` (in fact, the entire line needs to be replicated exactly); although `nano` is displaying what looks to be a blank line above our `#!` line, there isn't really one there.

In chapter 5, "Permissions and Executables," we discussed the `#!` line as containing the absolute path to the interpreter, as in `#!/usr/bin/bash` for `bash` scripts. In this case, we are specifying something slightly different: `#!/usr/bin/env python`. The `env` program, among other things, searches for the installed location of the given argument and executes that. A `#!` line like this will cause a Python program to be successfully executed, even if it is installed in a nonstandard location. (One may ask if `env` is ever installed in a nonstandard location. Fortunately, it is rare to find `env` located anywhere other than in `/usr/bin`.)

Next, we need to exit `nano` and make the file executable by using the `chmod` utility, and finally we can run it with `./helloworld.py`. This specifies that the program `helloworld.py` should be run and that it exists in the current directory (`./`).

```
oneils@atmosphere ~/apcb/py$ ls
helloworld.py
oneils@atmosphere ~/apcb/py$ chmod +x helloworld.py
oneils@atmosphere ~/apcb/py$ ./helloworld.py
Hello world!
oneils@atmosphere ~/apcb/py$
```

Configuring and Using nano

Generally, you won't want to type `nano -w -i -E -T 4 ...` every time you want to edit a Python code file. Fortunately, `nano` can be configured to automatically use these options if they are specified correctly in a file called `.nanorc` in your home directory. But this may not be the best choice, either: when editing files that are not Python code, you likely don't want to convert all your tab entries to spaces. Instead, you may want to define a shell alias called `nanopy` specifically for editing Python code. To have this shell alias preserved for each login session, the relevant code would need to be added to your `.bashrc` (assuming your shell is `bash`):

```
oneils@atmosphere ~$ echo "alias nanopy='nano -w -i -E -T 4 '" >> ~/.bashrc
```

If you are going to perform the above, double-check that the command is exactly as written. After logging out and back in, you can edit a Python code file with the alias using `nanopy helloworld.py`.

As evident from the code sample above, `nano` can also provide syntax highlighting (coloring of code for readability) if your `$HOME/.nanorc` and related files are configured properly, though it isn't necessary for programming.

Don't forget that it is often useful to have multiple terminal windows open simultaneously. You can use one for editing, one for running your program and testing, and perhaps a third running `top`, displaying how much CPU and RAM your program is using.

Although not as powerful as more sophisticated text editors such as `emacs` or `vim`, `nano` is easy to use and includes a number of features such as editing multiple files, cutting and pasting within and between files, regular-expression-based search and search/replace, spell check, and more. While editing, nano can also take you directly to a line number (`Control--`), which will come in handy when you need to go directly to a line that caused an error.

Whatever editor you choose to use, reading some documentation will help you be much more productive. For the rest of our code samples, we'll be showing screenshots from within `vim`, primarily because it provides prettier syntax highlighting.

Exercises

1. Create a file of Python code on the command line containing a simple `print("Hello!")` statement. Execute it by specifying the interpreter. If you are running iPython notebook, try to create a similarly simple notebook and execute the cell.

2. Create a file of Python code, make it an executable script, and execute it. Does that work? Why or why not?

3. Determine which version of Python you are running (perhaps by running `python --version`). Test to see which versions of print work for you: `print("Hello!")` or `print "Hello!"`. (The former is much preferred.)

Chapter 14

Elementary Data Types

Variables are vital to nearly all programming languages. In Python, *variables* are "names that refer to data." The most basic types of data that can be referred to are defined by how contemporary computers work and are shared by most languages.

Integers, Floats, and Booleans

Consider the integer 10, and the real number 5.64. It turns out that these two are represented differently in the computer's binary code, partly for reasons of efficiency (e.g., storing `10` vs. `10.0000000000`). Python and most other languages consider integers and real numbers to be two different "types": real numbers are called *floats* (short for "floating point numbers"), and integers are called *ints.* We assign data like floats and ints to variables using the = operator.

```
print("Hello world!")
```

```
exons = 10                          # an int holding 10
theta = 5.64                        # a float holding 5.64
```

While we're on the topic of variables, variable names in Python should always start with a lowercase letter and contain only letters, underscores, and numbers.

Note that the interpreter ignores # characters and anything after them on the line.[36] This allows us to put "comments" in our code. Blank lines don't matter, either, allowing the insertion of blank lines in the code for ease of reading.

We can convert an int type into a float type using the `float()` function, which takes one parameter inside the parentheses:

```
exons_as_float = float(exons)      # return 10.0 (a float)
```

Similarly, we can convert floats to ints using the `int()` function, which truncates the floating point value at the decimal point (so 5.64 will be truncated to the int type 5, while -4.67 would be truncated to the int type -4):

```
theta_as_int = int(theta)          # returns 5 (an int)
```

This information is useful because of a particular caveat when working with most programming languages, Python included: if we perform mathematical operations using only int types, the result will always be an int. Thus, if we want to have our operations return a floating point value, we need to convert at least one of the inputs on the right-hand side to a float type first. Fortunately, we can do this in-line:

```
a = 10                             # an int
b = 3                              # an int
answer_1 = a/b                     # int/int returns int: 3
answer_2 = a/float(b)              # int/float returns float: 3.333
c = 2.55                           # a float
answer_3 = (b + c)*a               # (int+float)*int returns (float)*int
                                   #     returns float: 55.5
```

In the last line above, we see that mathematical expressions can be grouped with parentheses in the usual way (to override the standard order of operations if needed), and if a subexpression returns a float, then it will travel up the chain to induce floating-point math for the rest of the expression.[37]

Another property of importance is that the right-hand side of an assignment is evaluated before the assignment happens.

36. The # symbols are ignored unless they occur within a pair of quotes. Technically, the interpreter also ignores the `#!` line, but it is needed to help the system find the interpreter in the execution process.

37. This isn't true in Python 3.0 and later; e.g., 10/3 will return the float 3.33333.

```
value = 7
value = value + 1
print(value)                          # prints 8
```

Aside from the usual addition and subtraction, other mathematical operators of note include `**` for exponential powers and `%` for modulus (to indicate a remainder after integer division, e.g., `7 % 3` is `1`, `8 % 3` is `2`, `9 % 3` is `0`, and so on).

```
a = 7
b = 3
c = a % b                             # modulus ("remainder"): 1
c = a ** b                            # exponent: 343
```

Notice that we've reused the variables `a`, `b`, and `c`; this is completely allowable, but we must remember that they now refer to different data. (Recall that a variable in Python is a name that refers to, or references, some data.) In general, execution of lines within the same file (or cell in an iPython notebook) happens in an orderly top-to-bottom fashion, though later we'll see "control structures" that alter this flow.

Booleans are simple data types that hold either the special value `True` or the special value `False`. Many functions return Booleans, as do comparisons:

```
sun_is_yellow = True                  # boolean True
test = 2 < -3                         # boolean False
```

For now, we won't use Boolean values much, but later on they'll be important for controlling the flow of our programs.

Strings

Strings, which hold sequences of letters, digits, and other characters, are the most interesting basic data type.[38] We can specify the contents using either single or double quotes, which can be useful if we want the string itself to contain a quote. Alternatively, we can *escape* odd characters like quotes if they would confuse the interpreter as it attempts to parse the file.

38. Unlike C and some other languages, Python does not have a "char" datatype specifically for storing a single character.

```
first = "Shawn"
last = "O'Neil"                    # string
last = 'O\'Neil'                   # same string
```

Strings can be added together with + to concatenate them, which results in a new string being returned so that it can be assigned to a variable. The `print()` function, in its simplest form, takes a single value such as a string as a parameter. This could be a variable referring to a piece of data, or the result of a computation that returns one:

```
full = first + last
print(full)                        # prints "ShawnO'Neil"
print(first + " " + last)          # prints "Shawn O'Neil"
```

We cannot concatenate strings to data types that aren't strings, however.

```
height = 5.5
sentence = full + " height is " + height   # Error on this line!
```

Running the above code would result in a `TypeError: cannot concatenate 'str' and 'float' objects`, and the offending line number would be reported. In general, the actual bug in your code might be before the line reporting the error. This particular error example wouldn't occur if we had specified `height = "5.5"` in the previous line, because two strings can be concatenated successfully.

Fortunately, most built-in data types in Python can be converted to a string (or a string describing them) using the `str()` function, which returns a string.

```
sentence = full + " height is " + str(height)
print(sentence)
## or
print(full + " height is " + str(height))
```

string float

string

string

As the above illustrates, we may choose in some cases to store the result of an expression in a variable for later use, or we may wish to use a more complex expression directly. The choice is a balance between verbose code and less verbose code, either of which can enhance readability. You should use whatever makes the most sense to you, the person most likely to have to read the code later!

Python makes it easy to extract a single-character string from a string using brackets, or "index" syntax. Remember that in Python, the first letter of a string occurs at index `0`.

```
seq = "ACTAG"
```



```
middle_base = seq[2]                # "T"
```

The use of brackets is also known as "slice" syntax, because we can them it to extract a slice (substring) of a string with the following syntax: `string_var[begin_index:end_index]`. Again, indexing of a string starts at `0`. Because things can't be too easy, the beginning index is inclusive, while the ending index is exclusive.

```
seq = "ACTAG"
subseq = seq[1:4]                   # "CTA"
```

A good way to remember this confusing bit of syntax is to think of indices as occurring between the letters.

If a string looks like it could be converted to an int or float type, it probably can be with the `float()` or `int()` conversion functions. If such a conversion doesn't make sense, however, Python will crash with an error. For example, the last line below will produce the error `ValueError: could not convert string to float: XY_2.7Q.`

```
exons_str = "10"
theta_str = "5.64"
evalue_str = "2.5e-4"

exons = int(exons_str)              # int 10
theta = float(theta_str)            # float 5.46
evalue = float(evalue_str)          # float 0.00025

test = float("XY_2.7Q")             # Error on this line!
```

To get the length of a string, we can use the `len()` function, which returns an int. We can use this in conjunction with `[]` syntax to get the last letter of a string, even if we don't know the length of the string before the program is run. We need to remember, though, that the index of the last character is one less than the length, based on the indexing rules.

```
seq = "ACTAG"
seq_len = len(seq)                  # int 5
last_letter = seq[seq_len - 1]      # string "G"
## or
last_letter = seq[len(seq) - 1]     # string "G"
```

Similarly, if we want a substring from position `2` to the end of the string, we need to remember the peculiarities of the `[]` slice notation, which is inclusive:exclusive.

```
end_seq = seq[2:len(seq)]           # indices 2 to 5, inclusive:exclusive
print(end_seq)                      # prints "TAG"
```

Immutability

In some languages it is possible to change the contents of a string after it's been created. In Python and some other languages, this is not the case, and strings are said to be immutable. Data are said to be *immutable* if they cannot be altered after their initial creation. The following line of code, for example, would cause an error like `TypeError: 'str' object does not support item assignment`:

```
seq = "ACTAG"
seq[2] = "C"                        # Error on this line!
```

Languages like Python and Java make strings immutable for a variety of reasons, including computational efficiency and as a safeguard to prevent certain common classes of programming bugs. For computational biology, where we often wish to modify strings representing biological sequences, this is an annoyance. We'll learn several strategies to work around this problem in future chapters.

In many cases, we can make it look like we are changing the contents of some string data by reusing the variable name. In the code below, we are defining strings `seqa` and `seqb`, as well as `seqc` as the concatenation of these, and then `seqd` as a substring of `seqc`. Finally, we reuse the `seqa` variable name to refer to different data (which gets copied from the original).

```
seqa = "ACTAG"                    # "ACTAG"
seqb = "GGAC"                     # "GGAC"
seqc = seqa + seqb                # "ACTAGGGAC"
seqd = seqc[2:7]                  # "TAGGG"

seqa = "ATG" + seqa               # redefine seqa, "ATGACTAG"
```

Here's how we might represent these variables and the data stored in memory, both before and after the reassignment of `seqa`.

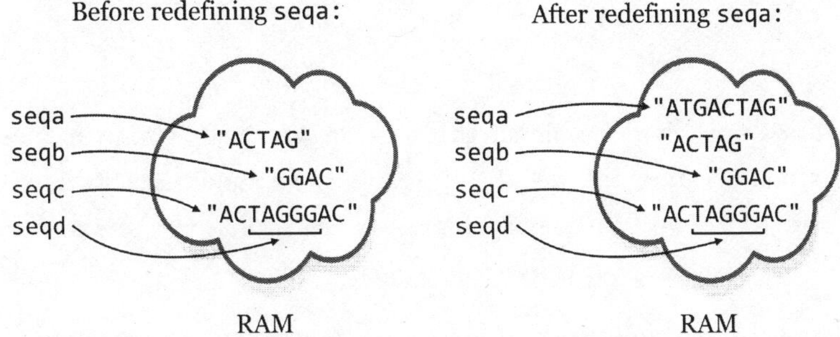

Because the string `"ACTAG"` is immutable, redefining `seqa` results in an entirely different piece of data being created. The original string `"ACTAG"` will still exist in memory (RAM) for a short time, but because it is not accessible via any variables, Python will eventually clean it out to make room in memory in a process known as *garbage collection*.[39] Garbage collection is an automatic, periodic process of de-allocating memory used by data that are no longer accessible (and hence no longer needed) by the program.

This immutability of strings could result in code that takes much longer to execute than expected, because concatenating strings results in copying of data (see the results of `seqc = seqa + seqb` above). For example, if we had a command that concatenated chromosome strings (millions of

39. Garbage collection is a common feature of high-level languages like Python, though some compiled languages support it as well. C does not: programmers are required to ensure that all unused data are erased during program execution. Failure to do so is known as a "memory leak" and causes many real-world software crashes.

letters each) to create a genome string, `genome = chr1 + chr2 + chr3 + chr4`, the result would be a copy of all four chromosomes being created in memory! On the other hand, in many cases, Python can use the immutability of strings to its advantage. Suppose, for example, that we wanted to get a large substring from a large string, `centromere_region = chr1[0:1500000]`. In this case, Python doesn't need to make a copy of the substring in memory. Because the original string can never change, all it needs is some bookkeeping behind the scenes to remember that the `centromere_region` variable is associated with part of string `chr1` references. This is why `seqd` in the figure above does not duplicate data from `seqc`.

None of this discussion is to imply that you should worry about the computational efficiency of these operations at this point. Rather, the concept of immutability and the definition of a variable (in Python) as a "name that refers to some data" are important enough to warrant formal discussion.

Exercises

1. Create and run a Python program that uses integers, floats, and strings, and converts between these types. Try using the `bool()` function to convert an integer, float, or string to Boolean and print the results. What kinds of integers, floats, and strings are converted to a Boolean `False`, and what kinds are converted to `True`?
2. We know that we can't use the `+` operator to concatenate a string type and integer type, but what happens when we multiply a string by an integer? (This is a feature that is fairly specific to Python.)
3. What happens when you attempt to use a float type as an index into a string using `[]` syntax? What happens when you use a negative integer?
4. Suppose you have a sequence string as a variable, like `seq = "ACTAGATGA"`. Using only the concepts from this chapter, write some code that uses the `seq` variable to create two new variables, `first_half` and `second_half` that contain the first half (rounded down) and the second half (rounded up) of `seq`. When printed, these two should print `"ACTA"` and `"GATGA"`, respectively.

```
#!/usr/bin/env python

seq = "ACTAGATGA"
## Write some code here:



print(first_half)           # should print "ACTA"
print(second_half)          # should print "GATGA"
```

Importantly, your code should work no matter the string `seq` refers to, without changing any other code, so long as the length of the string is at least two letters. For example, if `seq = "TACTTG"`, then the same code should result in `first_half` referring to `"TAC"` and `second_half` referring to `"TTG"`.

Chapter 15

Collections and Looping: Lists and for

A list, as its name implies, is a list of data (integers, floats, strings, Booleans, or even other lists or more complicated data types). Python lists are similar to arrays or vectors in other languages. Like letters in strings, elements of a list are indexed starting at `0` using `[]` syntax. We can also use brackets to create a list with a few elements of different types, though in practice we won't do this often.

```
a_list = [1, 2.4, "CYP6B", 724]          # list of 4 elements
an_el = a_list[1]                        # 2.4
```

Just like with strings, we can use `[]` notation to get a sublist "slice," and we can use the `len()` function to get the length of a list.

```
sublist = a_list[1:3]                    # list of two: [2.4, "CYP6B"]
list_len = len(a_list)                   # 4
last_el = a_list[list_len - 1]           # 725
```

Unlike strings, though, lists are *mutable*, meaning we can modify them after they've been created, for example, by replacing an element with another element. As mentioned above, lists can even contain other lists!

```
inner_list = [0.245, False]              # list of two elements
a_list[3] = inner_list                   # a_list now refers to
                                         # [2, 2.4, "CYP6B", [0.245, False]]
```

We will typically want our code to create an empty list, and then add data elements to it one element at a time. An empty list is returned by calling the `list()` function with no parameters. Given a variable which references a list object, we can append an element to the end using the `.append()` method, giving the method the element we want to append as a parameter.

```
new_list = list()                           # new_list refers to an empty list
new_list.append("A")                        # new_list now refers to ["A"]
new_list.append("G")                        # new_list now refers to ["A", "G"]
```

This syntax might seem a bit odd compared to what we've seen so far. Here `new_list.append("G")` is telling the list object the `new_list` variable refers to to run its `.append()` method, taking as a parameter the string `"G"`. We'll explore the concepts of objects and methods more formally in later chapters. For now, consider the list not just a collection of data, but a "smart" object with which we can interact using `.` methods.

Note that the `.append()` method asks the list to modify itself (which it can do, because lists are mutable), but this operation doesn't return anything of use.[40]

```
a_result = new_list.append("C")     # ask list to .append(), store returned value

print(a_result)                     # prints None
print(new_list)                     # prints ["A", "G", "C"]
```

Modifies list, returns None

This type of command opens up the possibility for some insidious bugs; for example, a line like `new_list = new_list.append("C")` looks innocent enough and causes no immediate error, but it is probably not what the programmer intended. The reason is that the `new_list.append("C")` call successfully asks the list to modify itself, but then the `None` value is returned, which would be assigned to the `new_list` variable with the assignment. At the end of the line, `new_list` will refer to `None`, and the list itself will no longer be accessible. (In fact, it will be garbage collected in due time.) In short, use `some_list.append(el)`, not `some_list = some_list.append(el)`.

We often want to sort lists, which we can do in two ways. First, we could use the `sorted()` function, which takes a list as a parameter and returns a new copy of the list in sorted order, leaving the original alone. Alternatively, we could call a lists `.sort()` method to ask a list to sort itself in place.

40. It returns a special data type known as `None`, which allows for a variable to exist but not reference any data. (Technically, `None` *is* a type of data, albeit a very simple one.) `None` can be used as a type of placeholder, and so in some situations isn't entirely useless.

```
c_list = ["T", "C", "A", "G"]
sorted_copy = sorted(c_list)                    # ["A", "C", "G", "T"]

print(c_list)                                   # prints ["T", "C", "A", "G"]
c_list.sort()                                   # asks c_list to sort itself
print(c_list)                                   # prints ["A", "C", "G", "T"]
```

As with the `.append()` method above, the `.sort()` method returns `None`, so the following would almost surely have resulted in a bug: `a_list = a_list.sort()`.

At this point, one would be forgiven for thinking that `.` methods always return `None` and so assignment based on the results isn't useful. But before we move on from lists, let's introduce a simple way to split a string up into a list of substrings, using the `.split()` method on a string data type. For example, let's split up a string wherever the subsequence `"TA"` occurs.

```
seq = "CGCGTAGTACAGA"

subs_list = seq.split("TA")
print(subs_list)                                # prints ["CGCG", "G", "CAGA"]
```

If the sequence was instead `"CGCGTATACAGA"`, the resulting list would have contained `["CGCG", "", "CAGA"]` (that is, one of the elements would be a zero-length empty string). This example illustrates that strings, like lists, are also "smart" objects with which we can interact using `.` methods. (In fact, so are integers, floats, and all other Python types that we'll cover.)

Tuples (Immutable Lists)

As noted above, lists are mutable, meaning they can be altered after their creation. In some special cases, it is helpful to create an immutable version of a list, called a "tuple" in Python. Like lists, tuples can be created in two ways: with the `tuple()` function (which returns an empty tuple) or directly.

```
ids = tuple()                                   # empty tuple
ids = ("CYP6B", "AGP4", "CATB")                 # a tuple of length 3
```

Tuples work much like lists—we can call `len()` on them and extract elements or slices with `[]` syntax. We can't change, remove, or insert elements.[41]

Looping with for

A for-loop in Python executes a block of code, once for each element of an *iterable* data type: one which can be accessed one element at a time, in order. As it turns out, both strings and lists are such iterable types in Python, though for now we'll explore only iterating over lists with for-loops.

A *block* is a set of lines of code that are grouped as a unit; in many cases they are executed as a unit as well, perhaps more than one time. Blocks in Python are indicated by being indented an additional level (usually with four spaces—remember to be consistent with this indentation practice).

When using a for-loop to iterate over a list, we need to specify a variable name that will reference each element of the list in turn.

```
gene_ids = ["CYP6B", "AGP4", "CATB"]

for gene_id in gene_ids:
    print("gene_id is " + gene_id)

print("Done.")
```

In the above, one line is indented an additional level just below the line defining the for-loop. In the for-loop, the `gene_id` variable is set to reference each element of the `gene_ids` list in turn. Here's the output of the loop:

```
gene_id is CYP6B
gene_id is AGP4
gene_id is CATB
Done.
```

41. Tuples are a cause of one of the more confusing parts of Python, because they are created by enclosing a list of elements inside of parentheses, but function calls also take parameters listed inside of parentheses, and mathematical expressions are grouped by parentheses, too! Consider the expression `(4 + 3) * 2`. Is `(4 + 3)` an integer, or a single-element tuple? Actually, it's an integer. By default, Python looks for a comma to determine whether a tuple should be created, so `(4 + 3)` is an integer, while `(4 + 3, 8)` is a two-element tuple and `(4 + 3,)` is a single-element tuple. Use parentheses deliberately in Python: either to group mathematical expressions, create tuples, or call functions—where the function name and opening parenthesis are neighboring, as in `print(a)` rather than `print (a)`. Needlessly adding parentheses (and thereby accidentally creating tuples) has been the cause of some difficult-to-find bugs.

Using for-loops in Python often confuses beginners, because a variable (e.g., `gene_id`) is being assigned without using the standard = assignment operator. If it helps, you can think of the first loop through the block as executing `gene_id = gene_ids[0]`, the next time around as executing `gene_id = gene_ids[1]`, and so on, until all elements of `gene_ids` have been used.

Blocks may contain multiple lines (including blank lines) so that multiple lines of code can work together. Here's a modified loop that keeps a `counter` variable, incrementing it by one each time.

```
gene_ids = ["CYP6B", "AGP4", "CATB"]
counter = 0

for gene_id in gene_ids:
    print("gene_id is " + gene_id)

    counter = counter + 1

print("Done.")
print(counter)
```

The output of this loop would be the same as the output above, with an additional line printing `3` (the contents of `counter` after the loop ends).

Some common errors when using block structures in Python include the following, many of which will result in an `IndentationError`.

1. Not using the same number of spaces for each indentation level, or mixing tab indentation with multiple-space indentation. (Most Python programmers prefer using four spaces per level.)
2. Forgetting the colon : that ends the line before the block.
3. Using something like a for-loop line that requires a block, but not indenting the next line.
4. Needlessly indenting (creating a block) without a corresponding for-loop definition line.

We often want to loop over a range of integers. Conveniently, the `range()` function returns a list of numbers.[42] It commonly takes two parameters: (1) the starting integer (inclusive) and (2) the ending integer (exclusive). Thus we could program our for-loop slightly differently by generating a list of integers to use as indices, and iterating over that:

```
gene_ids = ["CYP6B", "AGP4", "CATB"]

indices = range(0,3)
for index in indices:                              # for index in [0, 1, 2]
    gene_id = gene_ids[index]
    print("gene id " + str(index) + " is " + gene_id)

print("Done.")

## Or, more succinctly:
for index in range(0, len(gene_ids)):              # for index in [0, 1, 2]
    print("gene id " + str(index) + " is " + gene_ids[index])

print("Done.")
```

The output of one of the loops above:

```
gene_id 0 is CYP6B
gene_id 1 is AGP4
gene_id 2 is CATB
Done.
```

The second example above illustrates the rationale behind the inclusive/exclusive nature of the `range()` function: because indices start at zero and go to one less than the length of the list, we can use `range(0, len(ids))` (as opposed to needing to modify the ending index) to properly iterate over the indices of `ids` without first knowing the length of the list. Seasoned programmers generally find this intuitive, but those who are not used to counting from zero may need some practice. You should study these examples of looping carefully, and try them out. These concepts are often more difficult for beginners, but they are important to learn.

42. An alternative to `range()` in Python 2.7 is `xrange()`, which produces an iterable type that works much like a list of numbers but is more memory efficient. In more recent versions of Python (3.0 and above) the `range()` function works like `xrange()` and `xrange()` has been removed. Programmers using Python 2.7 may wish to use `xrange()` for efficiency, but we'll stick with `range()` so that our code works with the widest variety of Python versions, even if it sacrifices efficiency in some cases. There is one important difference between `range()` and `xrange()` in Python 2.7: `range()` returns a list, while `xrange()` returns an iterable type that lacks some features of true lists. For example, `nums = range(0, 4)` followed by `nums[3] = 1000` would result in `nums` referencing `[0, 1, 2, 1000]`, while `nums = xrange(0, 4)` followed by `nums[3] = 1000` would produce an error.

Loops and the blocks they control can be nested, to powerful effect:

```
total = 0

for i in range(0,4):
    total = total + i
    for j in range(0,3):
        total = total + j + i
        print("total is: " + str(total))
    sum = sum + i

print("Done! Final total is: " + str(total))
```

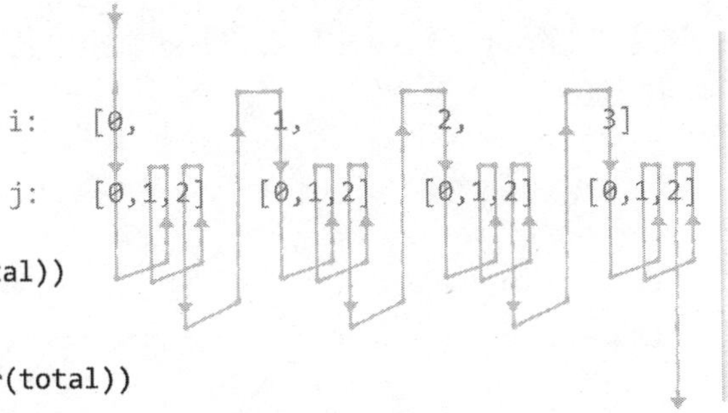

In the above, the outer for-loop controls a block of five lines; contained within is the inner for-loop controlling a block of only two lines. The outer block is principally concerned with the variable `i`, while the inner block is principally concerned with the variable `j`. We see that both blocks also make use of variables defined outside them; the inner block makes use of `sum`, `i`, and `j`, while lines specific to the outer block make use of `sum` and `i` (but not `j`). This is a common pattern we'll be seeing more often. Can you determine the value of `total` at the end without running the code?

List Comprehensions

Python and a few other languages include specialized shorthand syntax for creating lists from other lists known as *list comprehensions*. Effectively, this shorthand combines a for-loop syntax and list-creation syntax into a single line.

Here's a quick example: starting with a list of numbers `[1, 2, 3, 4, 5, 6]`, we generate a list of squares (`[1, 4, 9, 16, 25, 36]`):

```
nums = [1, 2, 3, 4, 5, 6]
squares = [num ** 2 for num in nums]   # [1, 4, 9, 16, 25, 36]
```

Here we're using a naming convention of `num in nums`, but like a for-loop, the looping variable can be named almost anything; for example, `squares = [x ** 2 for x in nums]` would accomplish the same task.

List comprehensions can be quite flexible and used in creative ways. Given a list of sequences, we can easily generate a list of lengths.

```
seqs = ["TAC", "TC", "CGAGG", "TAG", "A"]
lens = [len(seq) for seq in seqs]      # [3, 2, 5, 3, 1]
```

These structures support "conditional inclusion" as well, though we haven't yet covered operators like ==:

```
lens = [len(seq) for seq in seqs if seq[0] == "T"] # [3, 2, 3]
```

The next example generates a list of `1`s for each element where the first base is `"T"`, and then uses the `sum()` function to sum up the list, resulting in a count of sequences beginning with `"T"`.

```
count_start_ts = sum([1 for seq in seqs if seq[0] == "T"])
print(count_start_ts)                   # prints 3
```

Although many Python programmers often use list comprehensions, we won't use them much in this book. Partially, this is because they are a feature that many programming languages don't have, but also because they can become difficult to read and understand owing to their compactness. As an example, what do you think the following comprehension does? `[x for x in range(2, n) if x not in [j for i in range(2, sqrtn) for j in range (i*2, n, i)]]` (Suppose `n = 100` and `sqrtn = 10`. This example also makes use of the fact that `range()` can take a step argument, as in `range(start, stop, step)`.)

Exercises

1. What is the value of `total` at the end of each loop set below? First, see if you can compute the answers by hand, and then write and execute the code with some added `print()` statements to check your answers.

```
total = 0
for i in range(0,4):
    total = total + 1

print(total)                    ## What is printed?

total = 0
for i in range(0,4):
    for j in range(0,4):
        total = total + 1

print(total)                    ## What is printed?

total = 0
for i in range(0,4):
    for j in range(0,4):
        for k in range(0,4):
            total = total + 1

print(total)                    ## What is printed?
```

2. Suppose we say the first for-loop block above has "depth" 1 and "width" 4, and the second has depth 2 and width 4, and the third has depth 3 and width 4. Can you define an equation that indicates what the `total` would be for a nested for-loop block with depth *d* and width *w*? How does this equation relate to the number of times the interpreter has to execute the line `total = total + 1`?

3. Determine an equation that relates the final value of `total` below to the value of `x`.

```
total = 0
x = 10

for i in range(0, x):
    for j in range(0, i):
        total = total + 1

print(total)                    ## How does total relate to x?
```

4. Given a declaration of a sequence string, like `seq = "ATGATAGAGGGATACGGGATAG"`, and a subsequence of interest, like `subseq = "GATA"`, write some code that prints all of the locations of that substring in the sequence, one per line, using only the Python concepts we've covered so far (such as `len()`, for-loops, and `.split()`). For the above example, the output should be `3`, `11`, and `18`.

Your code should still work if the substring occurs at the start or end of the sequence, or if the subsequence occurs back to back (e.g., in `"GATACCGATAGATA"`, `"GATA"` occurs at positions 1, 7, and 11). As a hint, you may assume the subsequence is not self-overlapping (e.g., you needn't worry about locating `"GAGA"` in `"GAGAGAGAGA"`, which would occur at positions 1, 3, 5, and 7).

5. Suppose we have a matrix represented as a list of columns: `cols = [[10, 20, 30, 40], [5, 6, 7, 8], [0.9, 0.10, 0.11, 0.12]]`. Because each column is an internal list, this arrangement is said to be in "column-major order." Write some code that produces the same data in "row-major order"; for example, `rows` should contain `[[10, 5, 0.9], [20, 6, 0.10], [30, 7, 0.11], [40, 8, 0.12]]`. You can assume that all columns have the same number of elements and that the matrix is at least 2 by 2.

   This problem is a bit tricky, but it will help you organize your thoughts around loops and lists. You might start by first determining the number of rows in the data, and then building the "structure" of `rows` as a list of empty lists.

Chapter 16

File Input and Output

So far, all the data we've been working with have been "hard-coded" into our programs. In real life, though, we'll be seeking data from external files.

We can access data that exist in an external file (a text file, usually) using a *file handle*, a special kind of data type we can connect to using an external source. In its simplest form, a file handle for reading data works sort of like a pipe: the operating system puts data (usually strings) in one end, and we extract data out of the other end, usually one line at a time. File handles for writing to files move the data in the opposite direction.

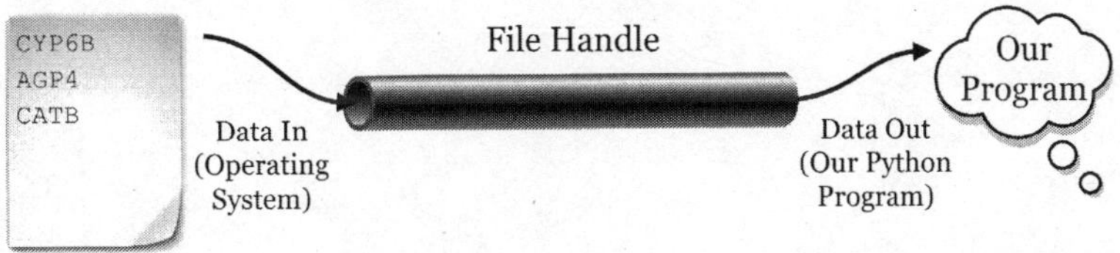

To use a file handle in Python, we have to tell the interpreter that we want to utilize some input/output functions by first putting the special command `import io` near the top of our program. This command imports the `io` module, making available extra functions that exist within it. These functions also use a dot syntax to specify the module they exist in, which is confusingly similar to the dot syntax discussed previously for methods. To associate a file handle called `fhandle` with the file "`filename.txt`", for example, the command is `fhandle = io.open("filename.txt", "rU")`, where `open()` is a function in the imported `io` module. The `r` indicates that we're creating a read-only file handle, and the `U` lets the interpreter know that we want a "universal" interpretation for newlines, to allow reading files created on Microsoft Windows systems or Unix-like systems.[43] The file handle data type has a method called `.readline()`, which returns the next available line from the pipe as a string; so, `line = fhandle.readline()` extracts a string (line) from the file handle and assigns it to `line`.

Let's create a small file with three lines each containing a gene ID, a tab, and a corresponding GC-content value.

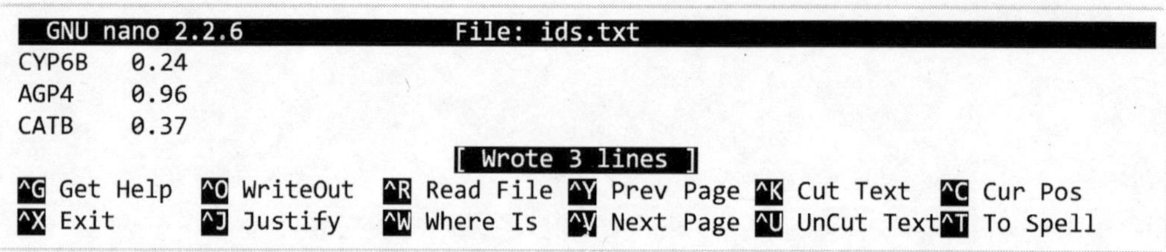

Now let's create a program that reads and prints the data. When we are finished reading data from a file, we need to run the file handle's `.close()` method. Using `.close()` causes the file handle to alert the operating system that we are done reading data from the handle. At this point the operating system is free to remove the back-end "pipe" and any data that might be in it.

```
import io

fhandle = io.open("ids.txt", "rU")

line1 = fhandle.readline()
line2 = fhandle.readline()
line3 = fhandle.readline()

print("Starting output:")
print(line1)
print(line2)
print(line3)
print("Done.")

fhandle.close()
```

43. Python includes a simple `open()` function that works largely the same way as the `io.open()` shown here. In the newest versions of Python (Python 3.0 and up), `open()` is actually a shortcut to `io.open()`, whereas in older versions the two operate slightly differently. For consistency, let's stick with an explicit `io.open()`, which works the same in all recent versions (Python 2.7 and up). With regard to the `"U"` parameter, Microsoft- and Unix-based systems utilize different encodings for newlines. The `"U"` parameter handles this transparently, so that all types of newlines can be considered as Unix based, `\n`, even if the file in question was created on a Microsoft system.

If the file doesn't exist or is not readable by the program, you will get an `IOError` indicating either `No such file or directory` or `Permission denied` with the file name you attempted to open.[44] Because our file does exist, we can successfully execute the code and inspect the results:

```
Starting output:
CYP6B   0.24

AGP4    0.96

CATB    0.37

Done.
```

There are a few things to note here. First, each time we call `fhandle.readline()`, the file handle returns a different string: the next string waiting to be popped out of the pipe. Second, we see that the output contains our three lines, but separated by blank lines: this is because there are "newline" characters already in the file. We don't see these newline characters, but we can represent them ourselves if we like, in a string with the control code `\n`. Similarly, tab characters have a control code like `\t`. The file is actually a single serial string, and the `.readline()` method asks the file handle to return everything up to and including the next `\n`.

44. In many languages, it is good practice to test whether a file handle was successfully opened, and if not, exit the program with an error. We could also do such tests in Python, but because Python by default quits with an informative error in this situation anyway, we won't worry as much about checking for errors here. Python also includes a special `with` keyword that handles both opening and closing the file:

```
with io.open("ids.txt", "rU") as fhandle:
    line1 = fhandle.readline()
    line2 = fhandle.readline()
    line3 = fhandle.readline()

    print(line1)
    print(line2)
    print(line3)

print("Done processing file; file handle is closed.")
```

Python documentation recommends using `with` in this way, as it is slightly safer in cases of program crashes that occur when a file handle is open. Still, we leave this idea to a footnote, as the open and close are explicit and straightforward.

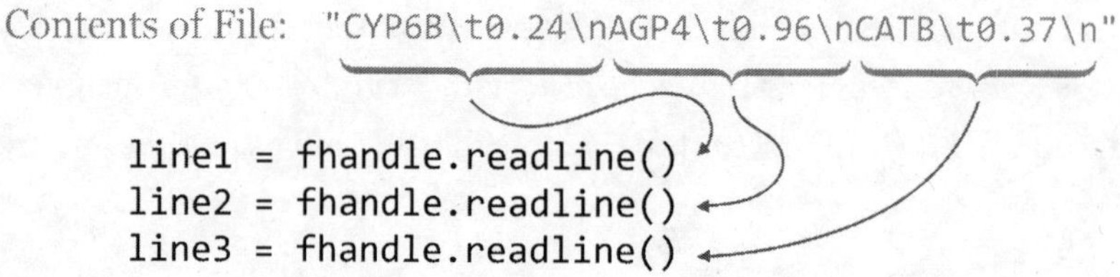

Contents of File: `"CYP6B\t0.24\nAGP4\t0.96\nCATB\t0.37\n"`

```
line1 = fhandle.readline()
line2 = fhandle.readline()
line3 = fhandle.readline()
```

In the end, the reason for the blank lines in our output is that the `print()` function, for our convenience, appends a `\n` to the end of any string that we print (otherwise, most of our output would all be on the same line). Thus each line read in from the file is being printed with two `\n` characters. We'll learn how to print without an additional newline later, when we learn how to write data to files. For now, we'll solve the problem by removing leading and trailing whitespace (spaces, tabs, and newlines) by asking a string to run its `.strip()` method.

```
import io

fhandle = io.open("ids.txt", "rU")

line1 = fhandle.readline()
line1stripped = line1.strip()
line2 = fhandle.readline()
line2stripped = line2.strip()
line3 = fhandle.readline()
line3stripped = line3.strip()

print("Starting output:")
print(line1stripped)
print(line2stripped)
print(line3stripped)
print("Done.")

fhandle.close()
```

Although we're calling `import io` again, it is just for clarity: a program only needs to import a module once (and usually these are collected near the top of the program). Here's the modified output:

```
Starting output:
CYP6B   0.24
AGP4    0.96
CATB    0.37
Done.
```

If you feel adventurous (and you should!), you can try *method chaining*, where the dot syntax for methods can be appended so long as the previous method returned the correct type.

```
line1stripped = fhandle.readline().strip()
                file handle
                    string "AGP4\n"
                        string "AGP4"
print(line2stripped)        # prints line without redundant newline
```

To tie this into earlier concepts, after we've extracted each line and stripped off the trailing whitespace, we can use `.split("\t")` to split each line into a list of strings. From there, we can use `float()` to convert the second element of each into a float type, and compute the mean GC content.

```
import io

fhandle = io.open("ids.txt", "rU")
line1 = fhandle.readline().strip()
line2 = fhandle.readline().strip()
line3 = fhandle.readline().strip()

fhandle.close()

line1_list = line1.split("\t")
line2_list = line2.split("\t")
line3_list = line3.split("\t")

gc1 = float(line1_list[1])
gc2 = float(line2_list[1])
gc3 = float(line3_list[1])

mean_gc = (gc1 + gc2 + gc3)/3.0
print(mean_gc)
```

The above prints the average GC content of the three lines as `0.523333333`. (Because `.strip()` also returns a string, we could have further chained methods, as in `line1_list = fhandle.readline().strip().split("\t")`.)

Because file handles work like a pipe, they don't allow for "random access"; we can get the next bit of data out of the end of the pipe, but that's it.[45] One might think that a command like `line5 = fhandle[4]` would work, but instead it would produce an error like `TypeError: '_io.BufferedReader' object has no attribute '__getitem__'`.

On the other hand, like lists, file handles are iterable, meaning we can use a for-loop to access each line in order. A simple program to read lines of a file and print them one at a time (without extra blank lines) might look like this:

```
import io

fhandle = io.open("ids.txt", "rU")

print("Starting output:")
for line in fhandle:
    linestripped = line.strip()
    print(linestripped)
print("Done")

fhandle.close()
```

Like `.readline()`, using a for-loop extracts lines from the pipe. So, if you call `.readline()` twice on a file handle attached to a file with 10 lines, and then run a for-loop on that file handle, the for-loop will iterate over the remaining 8 lines. This call could be useful if you want to remove a header line from a text table before processing the remaining lines with a loop, for example.

Writing Data

Writing data to a file works much the same way as reading data: we open a file handle (which again works like a pipe), and call a method on the handle called `.write()` to write strings to it. In this

45. In more sophisticated usage of file handles, our pipe analogy breaks down. It's possible to "rewind" or "fast-forward" the operating system's current position in the file being read, if we know how far the "read head" needs to move to get to the data we want. Perhaps a better analogy for file handles would be an old-fashioned magnetic cassette tape, perhaps a fitting analogy given that files were once stored on such tapes.

case, instead of using the "rU" parameter in calling io.open(), we'll use "w" to indicate that we want to write to the file. Be warned: when you open a file handle for writing in this manner, it overwrites any existing contents of the file. If you'd rather append to the file, you can instead use "a".[46]

Unlike the print() function, the .write() method of a file handle does not automatically include an additional newline character "\n". Thus, if you wish to write multiple lines to a file handle, you have to add them yourself. Here's an example that prints the numbers 0 through 9 to a file, one per line.

```
import io

fhandle = io.open("data/numbers.txt", "w")    # overwrites data/numbers.txt
for value in range(0, 10):
    fhandle.write(str(value) + "\n")

fhandle.close()
```

As mentioned above, a file handle opened for writing works like a pipe that is set up by the operating system. When writing, however, we put data into the pipe, and the operating system pulls the data out the other side, writing it to the file on disk.

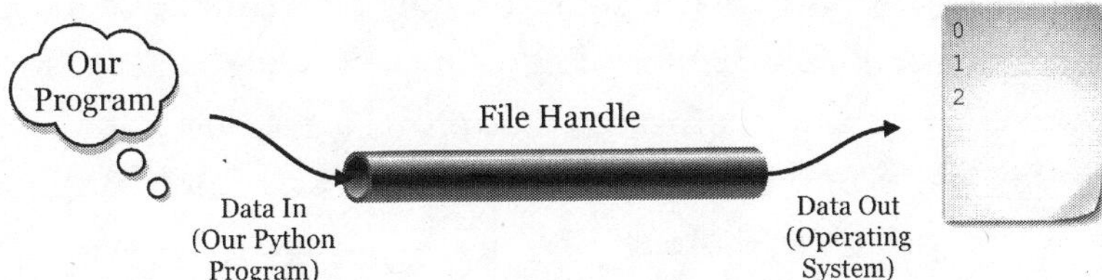

Because the operating system is busy handling similar requests for many programs on the system, it may not get around to pulling the data from the pipe and writing to the file on disk right away. So, it's important that we remember to call the .close() method on the file handle when we are done writing data; this tells the operating system that any information left in the pipe should be flushed out and saved to disk, and that the pipe structure can be cleaned up. If our program were to crash (or were to be killed) before the handle was properly closed, data in the pipe may never get written.

46. It's also possible to open a file handle for simultaneous reading and writing. We won't cover it here, as that functionality is less often needed.

Computing a Mean

Let's put many of the skills we've learned so far into practice to compute the mean of the *E* values from the output of BLAST, which is formatted as a simple text table with tabs separating the columns. Here's what the file, **`pz_blastx_yeast_top1.txt`**, looks like in `less -S`. Although the reader will have to take our word for it, the eleventh column of this file contains *E* values like `5e-14`, `2e-112`, and `1e-18`.

```
PZ7180000000004_TY      YKL081W 31.07   338     197     8       13      993
PZ1082_AB       YHR104W 44.92   118     62      3       4       348     196
PZ11_FX YLR406C 53.01   83      38      1       290     42      25      106
PZ7180000036154 YNL245C 36.27   102     60      3       105     395     1
PZ605962        YKR079C 29.57   115     66      4       429     121     479
PZ856513        YKL215C 48.39   155     73      3       3       452     109
...
```

When solving a problem like this, it's usually a good idea to first write, in simple English (or your nonprogramming language of choice), the *strategy* you intend to use to solve the problem. The strategy here will be as follows: the mean of a set of numbers is defined as their sum, divided by the count of them. We'll need to keep at least two important variables, `eval_sum` and `counter`. After opening the file with `io.open()`, we can loop through the lines and extract each *E* value. (This will require cleaning up the line with `.strip()`, splitting it into pieces using `.split("\t")`, and finally converting the *E* value to a float rather than using the string.) For each line that we see, we can add the *E* value extracted to the `eval_sum` variable, and we'll add `1` to the `counter` variable as well. In the end, we can simply report `eval_sum/counter`.

It often helps to convert this strategy into something partway between natural language and code, called *pseudocode*, which can help tremendously in organizing your thoughts, particularly for complex programs:

```
import io
open fhandle

counter = 0
sum_eval = 0.0

for each line in fhandle
    linestripped = line.strip()
    break into a list of strings with line_list = linestripped.split("\t")
    eval as a string is in line_list at index 10 (11th column)
    add float(eval) to sum_eval and save in sum_eval
    add 1 to count and save in count

mean = sum_eval divided by counter
print("mean is " + mean)
```

With the pseudocode sketched out, we can write the actual code for our program. When executed, it reliably prints `Mean is: 1.37212611293e-08`.

```
#!/usr/bin/env python
import io

blast_handle = io.open("pz_blastx_yeast_top1.txt", "rU")

counter = 0
eval_sum = 0.0

for line in blast_handle:
    line_stripped = line.strip()
    line_list = line_stripped.split("\t")
    eval_str = line_list[10]

    eval_sum = eval_sum + float(eval_str)
    counter = counter + 1

blast_handle.close()

mean = eval_sum/counter
print("Mean is: " + str(mean))
```

Note that the actual Python code (in **`blast_mean.py`**) ended up looking quite a lot like the pseudocode—this is one of the frequently cited selling points for Python. (For this reason, we'll also skip the pseudocode step for most examples in this book, though it can still be a valuable technique when programming in any language.)

This may seem like a fair amount of work to compute a simple mean, but it is a consequence of writing software "from scratch," and we have to start somewhere! Additionally, the benefit of learning these techniques will pay off over the long run when it comes to solving novel problems.

The Process of Programming

Although the process of designing a strategy (and pseudocode) seems tedious, it's highly recommended. As you progress in your abilities, your strategies and pseudocode will become more terse and higher level, but you should never skip the planning steps before coding. (On the other hand, there is also the danger of over-planning, but this more often effects teams of coders working on large projects.)

There's another thing worth noting at this point: programs like the above are almost never written top-to-bottom in their entirety, at least not without requiring significant debugging! You may forget that the result of splitting a string results in a list of strings, resulting in an error on a line like `eval_sum = eval_sum + eval_str` because `eval_sum` is a float, while `eval_str` is a string. After finding and fixing this error, you may find yet another error if you attempt to print with a line like `print("Mean is: " + mean)`, because again the types don't match and can't be concatenated. After all this, perhaps you find the resulting mean to unexpectedly be a large number, like **`131.18`**, only to find out it was because you accidently used `eval_str = line_list[11]`, forgetting that list indices start at **`0`**.

There are two strategies for avoiding long chains of annoying bugs like this, and harder-to-find (and thus more dangerous) bugs that result in incorrect output that isn't obviously incorrect. The first strategy is to only write a few lines at a time, and test newly added lines with `print()` statements that reveal whether they are doing what they are supposed to. In the example above, you might write a few lines and do some printing to ensure that you can successfully open the file handle and read from it (if the file is large, create a smaller version for testing). Then write a simple for-loop and do some printing to ensure you can successfully loop over the lines of the file and split them into lists of strings. Continue on by filling in some of the code in the for-loop, again printing and testing to ensure the code is doing what you think it is. And so on via iterative development.

The second strategy, to avoid the more dangerous bugs that aren't immediately obvious, is to test and develop your programs using small files for which you can calculate the answer by hand (or some other trusted process). When you know from testing that your program or even just parts of your program produces the intended output, hidden bugs are much less likely to slip through the cracks. Some people produce several small inputs that represent the variety of different input situations, run the tests, and compare the outputs to the expected results in an automated fashion. This is known as "unit testing," a common practice in professional software development.

Finally, remember that every programmer, no matter her experience level, gets frustrated from time to time! This is entirely normal, and a good strategy when this happens is to take a brief walk, or even take a break for the day, picking up again the next day. Some find they program efficiently in the morning, others late at night. Whatever you do, *avoid the temptation to program by trial and error*, mindlessly tweaking the code in the hopes that it will produce the output you want. Even if you succeed this way (which rarely happens), you're likely to produce code with insidious bugs, and it may well be unreadable even to yourself in a few days' time.

Exercises

1. Write a Python program to compute the sample standard deviation of the *E* values in the file `pz_blastx_yeast_top1.txt`. As a reminder, the sample standard deviation is defined as the square root of the sum of squared differences from the mean, divided by the number of values minus 1:

$$stdev = \sqrt{\frac{\sum_{eval_i \in evals}(eval_i - mean)^2}{n-1}}$$

   To accomplish this, you'll need to make two passes over the data: one as in the example to compute the mean, and another to compute the sum of squared differences. This means you'll need to access the *E* values twice. Rather than close and reopen the data file, you should create an initially empty list to which you can append each *E* value (in your first pass over the data) for later use.

To compute the square root of a float, you will need to import the `math` module by calling `import math` near the top of your program. Then the `math.sqrt()` function will return the square root of a float; for example, `math.sqrt(3.0)` will return the float `1.7320508`.

2. If `a_list` is a list, then `b_list = reversed(a_list)` creates a "`listreverseiterator`" allowing one to loop over the elements with a for-loop in reverse order. Using this information, write a program called `reverse_blast.py` that reads the contents of **`pz_blastx_yeast_top1.txt`** and writes the lines in reverse order to a file called `pz_blastx_yeast_top1_reversed.txt`.

3. A *quine* (after the logician and philosopher W. V. Quine) is a nonempty program that prints its own source code exactly. Use the `io` module and file handles to write a quine called `quine.py`. (Quine programs aren't technically allowed to open files. Can you write a program that prints its own source code without using the `io` module?)

Chapter 17

Conditional Control Flow

The phrase "control flow" refers to the fact that constructs like for-loops change the flow of program execution away from the simple top-to-bottom order. There are several other types of control flow we will cover, two of which are "conditional" in nature.

Using If-Statements

If-statements allow us to conditionally execute a block of code, depending on a variable referencing a Boolean `True` or `False`, or more commonly a condition that returns a Boolean `True` or `False`. The syntax is fairly simple, described here with an example.

```
seq = "ATGATGACGTAGACGAGTA"
num_bases = len(seq)

if num_bases < 10:                                   } if required
    print("seq is less than 10 bases.")
elif num_bases < 20:
    print("seq is at least 10 bases...")
    print("... and it's less than 20 bases.")        zero or more elif blocks
elif num_bases < 40:                                 (optional)
    print("seq is at least 20 bases...")
    print("... and it's less than 40 bases.")
else:                                                } zero or one else block
    print("seq is at least 40 bases.")                 (optional)

print("Done checking length of seq.")
```

All the lines from the starting `if` to the last line in an `elif:` or `else:` block are part of the same logical construct. Such a construct *must* have exactly one `if` conditional block, *may* have one or more `elif` blocks (they are optional), and *may* have exactly one catchall `else` block at the end (also optional). Each conditional is evaluated in order: the first one that evaluates to `True` will run, and the rest will be skipped. If an `else` block is present, it will run if none of the earlier `if` or `elif` blocks did as a "last resort."

Just like with for-loops, if-statements can be nested inside of other blocks, and other blocks can occur inside if-statement blocks. Also just like for-loops, Python uses indentation (standard practice is four spaces per indentation level) to indicate block structure, so you will get an error if you needlessly indent (without a corresponding control flow line like `for`, `if`, `elif`, or `else`) or forget to indent when an indentation is expected.[47]

```
seqs = ["ACGTAGAC", "CAGTAGAGC", "GACGA", "CGATAGG"]
count_short = 0
count_long = 0

for seq in seqs:
    num_bases = len(seq)
    if num_bases < 8:
        count_short = count_short + 1
    else:
        count_long = count_long + 1

print("Number short: " + str(count_short) + " number long: " + str(count_long))
```

The above code would print `Number short: 2 number long: 2`.

Using While-Loops

While-loops are less often used (depending on the nature of the programming being done), but they can be invaluable in certain situations and are a basic part of most programming languages. A while-loop executes a block of code so long as a condition remains `True`. Note that if the condition never becomes `False`, the block will execute over and over in an "infinite loop." If the condition is `False` to begin with, however, the block is skipped entirely.

47. Python is one of the only languages that require blocks to be delineated by indentation. Most other languages use pairs of curly brackets to delineate blocks. Many programmers feel this removes too much creativity in formatting of Python code, while others like that it enforces a common method for visually distinguishing blocks.

```
counter = 0

while counter < 4:
    print("Counter is now: " + str(counter))
    counter = counter + 1

print("Done. Counter ends with: " + str(counter))
```

The above will print `Counter is now: 0`, followed by `Counter is now: 1`, `Counter is now: 2`, `Counter is now: 3`, and finally `Done. Counter ends with: 4`. As with using a for-loop over a range of integers, we can also use a while-loop to access specific indices within a string or list.

```
seq = "ACAGGACGT"
num_bases = len(seq)

base_index = 0
while base_index < num_bases:
    base_i = seq[base_index]
    print("base is: " + base_i)

    base_index = base_index + 1

print("Done")
```

The above code will print `base is: A`, then `base is: C`, and so on, ending with `base is: T` before finally printing `Done`. While-loops can thus be used as a type of fine-grained for-loop, to iterate over elements of a string (or list), in turn using simple integer indexes and `[]` syntax. While the above example adds `1` to `base_index` on each iteration, it could just as easily add some other number. Adding `3` would cause it to print every third base, for example.

Boolean Operators and Connectives

We've already seen one type of Boolean comparison, `<`, which returns whether the value of its left-hand side is less than the value of its right-hand side. There are a number of others:

Operator	Meaning	Example (with `a = 7, b = 3`)
<	less than?	`a < b # False`
>	greater than?	`a > b # True`
<=	less than or equal to?	`a <= b # False`
>=	greater than or equal to?	`a >= b # True`
!=	not equal to?	`a != b # True`
==	equal to?	`a == b # False`

These comparisons work for floats, integers, and even strings and lists. Sorting on strings and lists is done in *lexicographic order:* an ordering wherein item A is less than item B if the first element of A is less than the first element of B; in the case of a tie, the second element is considered, and so on. If in this process we run out of elements for comparison, that shorter one is smaller. When applied to strings, lexicographic order corresponds to the familiar alphabetical order.

Let's print the sorted version of a Python list of strings, which does its sorting using the comparisons above. Note that numeric digits are considered to be "less than" alphabetic characters, and uppercase letters come before lowercase letters.

```
seqs = ["ACT", "AC", "T", "TAG", "1AC", "A"]
seqs.sort()
print(seqs)   # prints ['1AC', 'A', 'AC', 'ACT', 'T', 'TAG']
```

Boolean connectives let us combine conditionals that return `True` or `False` into more complex statements that also return Boolean types.

Connective	Meaning	Example (with `a = 7, b = 3`)
`and`	`True` if both are `True`	`a < 8 and b == 3 # True`
`or`	`True` if one or both are `True`	`a < 8 or b == 9 # True`
`not`	`True` if following is `False`	`not a < 3 # True`

These can be grouped with parentheses, and usually should be to avoid confusion, especially when more than one test follow a `not`.[48]

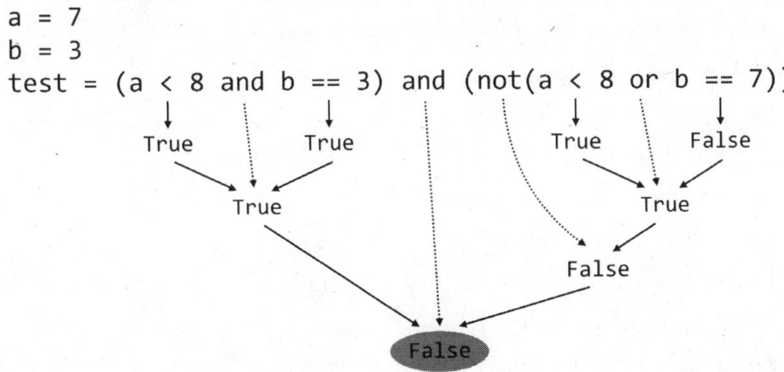

Finally, note that generally each side of an `and` or `or` should result in only `True` or `False`. The expression `a == 3 or a == 7` has the correct form, whereas `a == 3 or 7` does not. (In fact, `7` in the latter context will be taken to mean `True`, and so `a == 3 or 7` will always result in `True`.)

Logical Dangers

Notice the similarity between `=` and `==`, and yet they have dramatically different meanings: the former is the variable assignment operator, while the latter is an equality test. Accidentally using one where the other is meant is an easy way to produce erroneous code. Here `count == 1` won't initialize `count` to `1`; rather, it will return whether it already is `1` (or result in an error if `count` doesn't exist as a variable at that point). The reverse mistake is harder to make, as Python does not allow variable assignment in if-statement and while-loop definitions.

```
seq = "ACTAGGAC"
remainder = len(seq)%3
if remainder = 0:
    print("Number of bases is a multiple of 3")
    print("It's a candidate for an open reading frame.")
```

48. In the absence of parentheses for grouping, `and` takes precedence over `or`, much like multiplication takes precedence over addition in numeric algebra. Boolean logic, by the way, is named after the nineteenth-century mathematician and philosopher George Boole, who first formally described the "logical algebra" of combining truth values with connectives like "and," "or," and "not."

In the above, the intent is to determine whether the length of `seq` is a multiple of 3 (as determined by the result of `len(seq)%3` using the modulus operator), but the if-statement in this case should actually be `if remainder == 0:`. In many languages, the above would be a difficult-to-find bug (`remainder` would be assigned to `0`, and the result would be `True` anyway!). In Python, the result is an error: `SyntaxError: invalid syntax`.

Still, a certain class of dangerous comparison is common to nearly every language, Python included: the comparison of two float types for equality or inequality.

Although integers can be represented exactly in binary arithmetic (e.g., `751` in binary is represented exactly as `1011101111`), floating-point numbers can only be represented approximately. This shouldn't be an entirely unfamiliar concept; for example, we might decide to round fractions to four decimal places when doing calculations on pencil and paper, working with 1/3 as 0.3333. The trouble is that these rounding errors can compound in difficult-to-predict ways. If we decide to compute (1/3)*(1/3)/(1/3) as 0.3333*0.3333/0.3333, working left to right we'd start with 0.3333*0.3333 rounded to four digits as 0.1110. This is then divided by 0.3333 and rounded again to produce an answer of 0.3330. So, even though we know that (1/3)*(1/3)/(1/3) == 1/3, our calculation process would call them unequal because it ultimately tests 0.3330 against 0.3333!

Modern computers have many more digits of precision (about 15 decimal digits at a minimum, in most cases), but the problem remains the same. Worse, numbers that don't need rounding in our Base-10 arithmetic system do require rounding in the computer's Base-2 system. Consider 0.2, which in binary is 0.001100110011, and so on. Indeed, `0.2 * 0.2 / 0.2 == 0.2` results in `False`!

While comparing floats with `<`, `>`, `<=`, and `>=` is usually safe (within extremely small margins of error), comparison of floats with `==` and `!=` usually indicates a misunderstanding of how floating-point numbers work.[49] In practice, we'd determine if two floating-point values are sufficiently similar, within some defined margin of error.

```
a = 0.2
b = 0.2*0.2/0.2
epsilon = 0.00000001

if a + epsilon > b and a - epsilon < b:
    print("a and b are within " + str(epsilon))
```

49. You can see this for yourself: `print(0.2*0.2/0.2 == 0.2)` prints `False`! Some mathematically oriented languages are able to work entirely symbolically with such equations, bypassing the need to work with numbers at all. This requires a sophisticated parsing engine but enables such languages to evaluate even generic expressions like `x*x/x == x` as true.

Counting Stop Codons

As an example of using conditional control flow, we'll consider the file **`seq.txt`**, which contains a single DNA string on the first line. We wish to count the number of potential stop codons `"TAG"`, `"TAA"`, or `"TGA"` that occur in the sequence (on the forward strand only, for this example).

Our strategy will be as follows: First, we'll need to open the file and read the sequence from the first line. We'll need to keep a counter of the number of stop codons that we see; this counter will start at zero and we'll add one to it for each `"TAG"`, `"TAA"`, or `"TGA"` subsequence we see. To find these three possibilities, we can use a for-loop and string slicing to inspect every 3bp subsequence of the sequence; the 3bp sequence at index `0` of `seq` occurs at `seq[0:3]`, the one at position `1` occurs at `seq[1:4]`, and so on.

We must be careful not to attempt to read a subsequence that doesn't occur in the sequence. If `seq = "AGAGAT"`, there are only four possible 3bp sequences, and attempting to select the one starting at index 4, `seq[4:7]`, would result in an error. To make matters worse, string indexing starts at `0`, and there are also the peculiarities of the inclusive/exclusive nature of `[]` slicing and the `range()` function!

To help out, let's draw a picture of an example sequence, with various indices and 3bp subsequences we'd like to look at annotated.

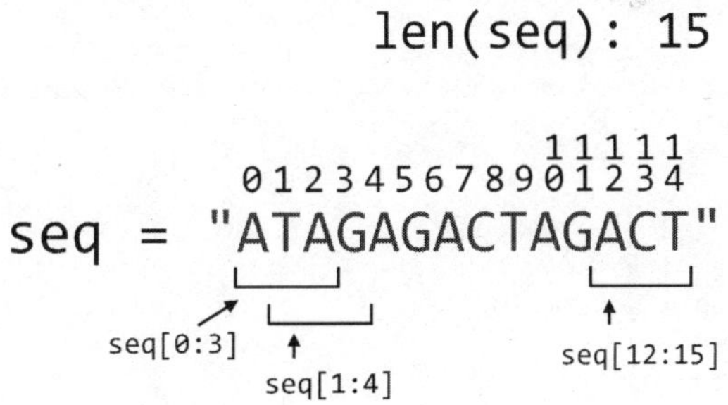

Given a starting index `index`, the 3bp subsequence is defined as `seq[index:index + 3]`. For the sequence above, `len(seq)` would return `15`. The first start index we are interested in is `0`, while the last start index we want to include is `12`, or `len(seq) - 3`. If we were to use the `range()` function to return a list of start sequences we are interested in, we would use `range(0, len(seq) - 3 + 1)`, where the `+ 1` accounts for the fact that `range()` includes the first index, but is exclusive in the last index.[50]

We should also remember to run `.strip()` on the read sequence, as we don't want the inclusion of any `\n` newline characters messing up the correct computation of the sequence length!

50. Yes, this sort of detailed logical thinking can be tedious, but it becomes easier with practice. Drawing pictures and considering small examples is also invaluable when working on programming problems, so keep a pencil and piece of paper handy.

Notice in the code below (which can be found in the file **stop_count_seq.py**) the commented-out line #print(codon).

```
#!/usr/bin/env python
import io

fhandle = io.open("seq.txt", "rU")
seq = fhandle.readline()
seq = seq.strip()

stop_counter = 0
for index in range(0, len(seq) - 3 + 1):
    codon = seq[index:index + 3]
    #print(codon)
    if codon == "TAG" or codon == "TAA" or codon == "TGA":
        stop_counter = stop_counter + 1

print(stop_counter)
```

While coding, we used this line to print each codon to be sure that 3bp subsequences were reliably being considered, especially the first and last in seq1.txt (ATA and AAT). This is an important part of the debugging process because it is easy to make small "off-by-one" errors with this type of code. When satisfied with the solution, we simply commented out the print statement.

For windowing tasks like this, it can occasionally be easier to access the indices with a while-loop.

```
import io

fhandle = io.open("seq.txt", "rU")
seq = fhandle.readline()
seq = seq.strip()

stop_counter = 0
index = 0
while index <= len(seq) - 3:
    codon = seq[index:index + 3]
    # print(codon)
    if codon == "TAG" or codon == "TAA" or codon == "TGA":
        stop_counter = stop_counter + 1

    index = index + 1

print(stop_counter)
```

If we wished to access nonoverlapping codons, we could use `index = index + 3` rather than `index = index + 1` without any other changes to the code. Similarly, if we wished to inspect 5bp windows, we could replace instances of `3` with `5` (or use a `windowsize` variable).

Exercises

1. The molecular weight of a single-stranded DNA string (in g/mol) is (count of `"A"`)*313.21 + (count of `"T"`)*304.2 + (count of `"C"`)*289.18 + (count of `"G"`)*329.21 – 61.96 (to account for removal of one phosphate and the addition of a hydroxyl on the single strand).

   Write code that prints the total molecular weight for the sequence in the file **`seq.txt`**. The result should be `21483.8`. Call your program `mol_weight_seq.py`.

2. The file **`seqs.txt`** contains a number of sequences, one sequence per line. Write a new Python program that prints the molecular weight of each on a new line. For example:

   ```
   21483.8
   19461.4
   32151.9
   ...
   ```

   You may wish to use substantial parts of the answer for question 1 inside of a loop of some kind. Call your program `mol_weight_seqs.py`.

3. The file **`ids_seqs.txt`** contains the same sequences as `seqs.txt`; however, this file also contains sequence IDs, with one ID per line followed by a tab character (`\t`) followed by the sequence. Modify your program from question 2 to print the same output, in a similar format: one ID per line, followed by a tab character, followed by the molecular weight. The output format should thus look like so (but the numbers will be different, to avoid giving away the answer):

   ```
   PZ7180000024555 402705.62
   PZ7180000000678_B       562981.52
   PZ7180000000003_KK      354193.41
   PZ7180000000005_NW      416120.04
   ```

   Call your program `mol_weight_ids_seqs.py`.

   Because the tab characters cause the output to align differently depending on the length of the ID string, you may wish to run the output through the command line tool `column` with a `-t` option, which automatically formats tab-separated input.

```
[oneils@atmosphere ~/apcb/py$ python mol_weight_ids_seqs.py | column -t
PZ7180000024555      402705.62
PZ7180000000678_B    562981.52
PZ7180000000003_KK   354193.41
PZ7180000000005_NW   416120.04
```

4. Create a modified version of the program in question 3 of chapter 15, "Collections and Looping, Part 1: Lists and `for`," so that it also identifies the locations of subsequences that are self-overlapping. For example, `"GAGA"` occurs in `"GAGAGAGAGATATGAGA"` at positions 1, 3, 5, 7, and 14.

Chapter 18

Python Functions

> The psychological profiling [of a programmer] is mostly the ability to shift levels of abstraction, from low level to high level. To see something in the small and to see something in the large.
>
> ~Donald Knuth

Functions (sometimes called "subroutines") are arguably the most important concept in programming. We've already seen their use in many contexts, for example, when using functions like `len()` and `float()`. Here's a bit of code that computes the GC content of a DNA sequence in a variable called `seq`:

```
base_counter = 0
seq_len = len(seq)

for index in range(0, seq_len):
    seq_base = seq[index]
    if seq_base == "G" or seq_base == "C":
        base_counter = base_counter + 1

gc = base_counter/float(seq_len)
```

What if we wanted to compute the GC content for multiple different variables in our code? Should we rewrite our GC-computing for-loop for each different variable? No way! First, we'd be at a much higher risk of bugs (it's a probabilistic fact that more typing leads to more typos). Second, the whole point of programming is for the computer to do all the work, not us.

Ideally, we'd like to encapsulate the functionality of computing GC content, just as the functionality of getting the length of a sequence is encapsulated by the `len()` function. We want to just be able to say `gc = gc_content(seq)`. Functions allow us to do exactly this: encapsulate a block of code for reuse whenever we need it. There are three important parts of a function:

1. The input (parameters given to the function).
2. The code block that is to be executed using those parameters. As usual, an additional level of indentation will define the block.
3. The output of the function, called the return value. This may be optional, if the function "does something" (like `print()`) rather than "returns something" (like `len()`).

Ignoring point 2, functions can actually represent a mathematical ideal: they relate inputs to outputs. They even have domains (the set of all valid inputs) and ranges (the set of all potential outputs).

We define functions in Python using the `def` keyword, and in Python functions must be defined before they can be executed. Here's an example function that computes a "base composition" (count of a character in a given string) given two parameters: (1) the sequence (a string) and (2) the base/character to look for (also a string).

```
## Given a DNA (A,C,T,G) string and a,
## 1-letter base string, returns the number of
## occurances of the base in the sequence.
def base_composition(seq, query_base):          } function definition, parameters
    base_counter = 0
    seq_len = len(seq)

    for index in range(0, seq_len):
        seq_base = seq[index]                              block
        if seq_base == query_base:
            base_counter = base_counter + 1

    return base_counter                         } return


seq1 = "ACTGATGCAT"
seq2 = "TTATCGAC"
seq1_ccomp = base_composition(seq1, "C")      # 2
seq2_tcomp = base_composition(seq2, "T")      # 3
```

The last two lines above call the function with different parameters—note that the parameter variable names (in this case `seq` and `query_base`) need not relate to the variable names of the data outside the function. This is an important point to which we'll return. When the interpreter reads the `def` line and corresponding block, the function is defined (available for use), but the lines are not run, or *called*, until the function is used in the last two lines.

One of the best things about functions is that they can call other functions, provided they've already been defined at the time they are called.

```
## Given a DNA (A,C,T,G) string and a,
## 1-letter base string, returns the number of
## occurances of the base in the sequence.
def base_composition(seq, query_base):
    base_counter = 0
    seq_len = len(seq)

    for index in range(0, seq_len):
        seq_base = seq[index]
        if seq_base == query_base:
            base_counter = base_counter + 1

    return base_counter
```

```
## Given a DNA (A,C,T,G) sequence string, returns the GC-content as float
def gc_content(seq):
    g_cont = base_composition(seq, "G")
    c_cont = base_composition(seq, "C")
    seq_len = len(seq)
    gc = (g_cont + c_cont)/float(seq_len)
    return gc

seq3 = "ACCCTAGACTG"
seq3_gc = gc_content(seq3)                          # 0.5454
```

Because functions need only to be defined before they are called, it is common to see collections of functions first in a program. Further, the order of definition need not correspond to their order of execution: either the `gc_content()` or the `base_composition()` function definition could occur first in this file and the computation would still work.

The idea of encapsulating small ideas into functions in this way is a powerful one, and from this point forward you should attempt to think mostly in terms of "what *function* am I writing/do I need," rather than "what *program* am I writing?"

Important Notes about Functions

In an effort to produce clean, readable, and reusable code, you should strive to follow these rules when writing functions.

1. Function blocks should only utilize variables that are either *created within the function block* (e.g., `g_cont` and `c_cont` within the `gc_content()` function above), or *passed in as parameters* (e.g., `seq`). Functions should be "worlds unto themselves" that only interact with the outside world through their parameters and return statement. The following redefinition of the `gc_content()` function would work, but it would be considered bad form because `seq4` is used but is not one of the two types listed.

```
## Assume a variable called seq4 holds a DNA sequence,
## returns the GC-content of it as float.
def gc_content():
    g_cont = base_composition(seq4, "G")
    c_cont = base_composition(seq4, "C")
    seq_len = len(seq4)
    gc = (g_cont + c_cont)/float(seq_len)
    return gc

seq4 = "ACCGTAGACTG"
seq4_gc = gc_content()                          # 0.5454
```

   Notice the difference: the function takes no parameters and so must use the same variable names as defined outside the function. This makes the function strongly tied to the context in which it is called. Because function definitions may be hundreds or thousands of lines removed from where they are called (or even present in a different file), this makes the task of coding much more difficult.

2. Document the use of each function with comments. What parameters are taken, and what types should they be? Do the parameters need to conform to any specification? For example: "this function works only for DNA strings, not RNA strings." What is returned? Although above we've commented on our functions by using simple comment lines, Python allows you to specify a special kind of function comment called a "docstring." These triple-quoted strings must occur at the top of the function block, and they may span multiple lines.

```
def gc_content(seq):
    """ Given a DNA (A,C,T,G) sequence string,
    returns the GC-content as float"""

    g_cont = base_composition(seq, "G")
    c_cont = base_composition(seq, "C")
    seq_len = len(seq)
    gc = (g_cont + c_cont)/float(seq_len)
    return gc
```

Later, when we discuss documentation for Python code, we'll learn how the Python interpreter can automatically collate these comments into nice, human-readable documents for others who might want to call your functions. Otherwise, we'll be using regular comments for function documentation.

3. Functions shouldn't be "too long." This is subjective and dependent on context, but most programmers are uncomfortable with functions that are more than one page long in their editor window.[51] The idea is that a function encapsulates a single, small, reusable idea. If you find yourself writing a function that is hard to read or understand, consider breaking it into two functions that need to be called in sequence, or into one short function that calls another short function.

4. Write lots of functions! Even if a section of code is only going to be called once, it is completely acceptable make a function out of it if it encapsulates some idea or well-separable block. After all, you never know if you might need to use it again, and just the act of encapsulating the code helps you ensure its correctness, allowing you to forget about it when working on the rest of your program.

GC Content over Sequences in a File

In chapter 17, "Conditional Control Flow," one of the exercises involved computing the molecular weight for each sequence in a tab-separated file of IDs and sequences, **`ids_seqs.txt`**. Here are the lines of this file viewed with `less -S`:

51. With the advent of large monitors, perhaps this is no longer a good rule of thumb! Personally, I find that few of my functions need to exceed 30 lines in length, depending only somewhat on the language I'm writing in.

```
PZ7180000000004_TX      AATGCGAATATTTTTATTTACAATCAATTACAATCAAGTCTTAAACTTATAGATTA
PZ7180000027934 TTTAATGATCAGTAAAGTTATAGTAGTTGTATGTACAATATTGTGCTAGCTTGAGACAGTCATG
PZ7180000000004_TY      CCAAAGTCAACCATGGCGGCCGGGGTACTTTATACTTATCCGACAAACTTCCGTGC
PZ7180000024036 GTAATACAAACGCTTTTATTAAATAAATACTTAAGTTACAATTAGTGTTATGAGAACATTAGAT
ids_seqs.txt (END)
```

If you completed the molecular weight exercise in chapter 17, you might have found it somewhat challenging. Without using functions, you likely needed to use a for-loop to iterate over each line of the file, and within there another for-loop to iterate over each base of the current sequence, and inside of that an if-statement to determine the molecular weight of each base.

Here we'll write a similar program, except this time we'll make use of a function that returns the GC content of its parameter. The entire solution is actually quite short and readable.

```
#!/usr/bin/env python
import io


## Given a DNA (A,C,T,G) string and a 1-letter base string,
## returns the number of occurances of the base in the sequence.
def base_composition(seq, query_base):
    base_counter = 0
    seq_len = len(seq)

    for index in range(0, seq_len):
        seq_base = seq[index]
        if seq_base == query_base:
            base_counter = base_counter + 1

    return base_counter

## Given a DNA (A,C,T,G) sequence string, returns the GC-content as float
def gc_content(seq):
    g_cont = base_composition(seq, "G")
    c_cont = base_composition(seq, "C")
    seq_len = len(seq)
    gc = (g_cont + c_cont)/float(seq_len)
    return gc

## Open file, and loop over lines
fhandle = io.open("ids_seqs.txt", "rU")

for line in fhandle:
    linestripped = line.strip()
    linelist = line.split("\t")
    id = linelist[0]
    sequence = linelist[1]
    seqgc = gc_content(sequence)
    print(id + "\t" + str(seqgc))

fhandle.close()
```

In the above code (**ids_seqs_gcs.py**), there's a clear "separation of concerns" that makes designing a solution easier. One function handles counting bases in a DNA sequence, and another the computation of the GC content. With those out of the way, we can turn our focus to parsing the input file and printing the output.

To simplify exposition, most of the small code samples in the next few chapters don't involve creating new functions. Most of the exercises will suggest writing functions as a means of breaking a problem down into its component parts, however.

Exercises

1. We have now covered four of the basic control-flow structures common to most programming languages. What are they, and what do they do?

2. We often want to look at some sort of windows of a sequence; perhaps we want to look at the codons of a sequence like `"ACTTAGAGC"`(`"ACT"`, `"TAG"`, and `"AGC"`), which we can think of as a 3bp "window size" and a 3bp "step size." Or perhaps we want to consider overlapping windows of 6-mers (like `"ACTTAG"`, `"CTTAGA"`, `"TTAGAG"`, and `"TAGAGC"`, window size 6, step size 1).

   Write a program with a function called `get_windows()` that takes three parameters: the sequence to extract windows from (string), the window size (int), and the step size (int). The function should return a list of strings, one per window, with no "partial windows" if the last window would run off the end of the sequence.

```
## Given a string, windowsize (int) and step size (int),
## returns a list of windows of the windowsize.
## E.g. "TACTGG", 3, 2 => ["TAC", "CTG"]
def get_windows(seq, windowsize, stepsize):
    ## your code here
    ## etc.
```

   (You might find a while-loop to be more helpful than a for-loop for this exercise.) Use this code to test your function:

```
seq = "ACGGTAGACCT"
print(seq)

windows1 = get_windows(seq, 3, 1)
print(windows1)

windows2 = get_windows(seq, 3, 3)
print(windows2)

windows3 = get_windows(seq, 3, 5)
print(windows3)

windows4 = get_windows(seq, 5, 2)
print(windows4)
```

   This code should output the following:

```
ACGGTAGACCT
['ACG', 'CGG', 'GGT', 'GTA', 'TAG', 'AGA', 'GAC', 'ACC', 'CCT']
['ACG', 'GTA', 'GAC']
['ACG', 'AGA']
['ACGGT', 'GGTAG', 'TAGAC', 'GACCT']
```

3. Although the `get_windows()` function is designed to run on strings, indexing in strings and in lists works the same in Python. What happens if you run `get_windows([1, 2, 3, 4, 5, 6, 7, 8], 3, 2)`?

4. Redo exercise 1 from chapter 17, but this time write and use a `molecular_weight()` function that takes a DNA string as a parameter called `seq`.

Chapter 19

Command Line Interfacing

While Python is a great language for stand-alone programming, it's also a first-class language for interfacing with the Unix/Linux command line. This chapter explores some of those concepts, allowing us to use Python as "glue" to connect other powerful command line concepts.

Standard Input is a File Handle

In previous chapters, we saw that simple Unix/Linux utilities such as `grep` and `sort` can be chained together for a variety of data manipulation tasks.

```
oneils@atmosphere ~/apcb/intro/fasta_stats$ grep '_L' pz_stats.table | \
> sort -k2,2g | \
> less -S
```

The above command uses `grep` to match lines in the file **`pz_stats.table`** against the pattern `_L`, which are printed to standard output. Using the | redirect, we send this output to `sort` on its standard input and sort the lines on the second column in general numeric order, before piping this output to `less -S`.

```
PZ7180000000117_L      0.336   1000    AAATA   12      unit:TCT        9
PZ7180000000157_L      0.372   836     ATTTA   7       unit:AGGAA      10
PZ7180000000106_L      0.39    876     ACAAT   6       unit:TAA        6
PZ7180000000146_L      0.419   630     AAGAG   4       unit:CAAA       8
PZ7180000000124_L      0.422   812     TAATA   5       unit:TAGCTG     12
```

Again, `sort` and `less` are not reading their input from a file, but rather from standard input streams. Similarly, they are not writing their output to a file, but just printing to the standard output streams. It's often useful to design our own small utilities as Python programs that operate in the same way.

The "file handle as a pipe" analogy that we've been using will serve us well when writing programs that read input from standard input. The file handle associated with standard input is easy to use in Python. We just have to import the `sys` module, and then `sys.stdin` is a variable that references the read-only file handle that works just like any other. We don't even have to use `io.open()`. Here's a program that prints each line passed into it on standard input.

```
#!/usr/bin/env python
import sys

counter = 1
for line in sys.stdin:
    linestripped = line.strip()
    print("line " + str(counter) + " is: " + linestripped)
    counter = counter + 1
```

In this case, we're using `sys.stdin` much like the read-only file handles we saw in previous examples. We can test our program on a simple file holding some gene IDs:

```
oneils@atmosphere ~/apcb/py$ cat ids.txt
CYP6B
AGP4
CATB
oneils@atmosphere ~/apcb/py$ cat ids.txt | ./stdin_ex.py
line 1 is: CYP6B
line 2 is: AGP4
line 3 is: CATB
```

But if we attempt to run our program without giving it any input on the standard input stream, it will sit and wait for data that will never come.

```
oneils@atmosphere ~/apcb/py$ ./stdin_ex.py
...
```

To kill the program, we can use the usual `Control-c`. The reason for this behavior is that the standard input stream can be used for input from standard out of another program (as we wish to do), or it can be used to build interactive programs. To see this, we can run the same program, except this time we'll type some input using the keyboard, and when we're done we'll send the control code `Control-d`, which is a way to tell Python we are done sending input.

```
oneils@atmosphere ~/apcb/py$ ./stdin_ex.py
Typing line one, then enter
Typing line two, then control-d
line 1 is: Typing line one, then enter
line 2 is: Typing line two, then control-d
```

This hints at some interesting possibilities for interactive programs,[52] but in this case it would likely be confusing for someone wanting to utilize our program by piping data to it. Fortunately, `sys.stdin` has a method called `.isatty()` that returns `True` if there are no data present on the input stream when it is called. (TTY is short for "TeleTYpewriter," a common name for a keyboard-connected input device of decades past.) So, we can fairly easily modify our program to quit with some helpful usage text if it detects that there are no data present on standard input.

```
#!/usr/bin/env python
import sys

if(sys.stdin.isatty()):
    print("Usage: cat <file> | ./stdin_ex.py")
    quit()

counter = 1
for line in sys.stdin:
    linestripped = line.strip()
    print("line " + str(counter) + " is: " + linestripped)
    counter = counter + 1
```

It's almost always a good idea for a program to check whether it is being called correctly, and to give some helpful usage information if not. It's also common to include, in comments near the top of the program, authoring information such as date of last update and contact information. You may also need or want to include some copyright and licensing text, either of your own choice or your institution's.

52. Try using a line of code like `value = input("What is your name? ")` followed by `print(value)`. Note that when running the program and prompted, you'll need to include quotes when typing your name (e.g., `"Shawn"`).

Note that we haven't called `.close()` on the standard input file handle. Although Python won't complain if you do, it is not required, as standard input is a special type of file handle that the operating system handles automatically.

Standard Output as a File Handle

Anything printed by a Python program is printed to the standard output stream, and so this output can also be sent to other programs for processing.

```
oneils@atmosphere ~/apcb/py$ cat ids.txt | ./stdin_ex.py | sort -k4,4d
line 2 is: AGP4
line 3 is: CATB
line 1 is: CYP6B
```

However, the `sys` module also provides access to the standard output file handle, `sys.stdout`. Just like other output-oriented file handles, we can call `.write()` on this file handle to print. The main difference between this method and calling `print()` is that while `print()` by default adds a `"\n"` newline character `sys.stdout.write()` does not.[53] This feature can be particularly useful when we wish to print a table in an element-by-element fashion because the data are not stored in strings representing individual rows.

53. In Python 3.0 and above, the `print()` function can also take some parameters to specify whether the `"\n"` should be printed or not, or even be replaced by some other character.

```
#!/usr/bin/env python
import sys

row_ids = ["AGP", "CYP", "CAT"]
columns = [[1856, 2262, 457], [0.26, 0.16, 0.73], ["T", "T", "F"]]

for row_index in range(0, len(row_ids)):
    row_id = row_ids[row_index]           # e.g. "AGP"
    sys.stdout.write(row_id + "\t")

    for col_index in range(0, len(columns)):
        column = columns[col_index]       # e.g. [1856, 2262, 457]
        value = column[row_index]         # e.g. 1856

        sys.stdout.write(str(value) + "\t")

    sys.stdout.write("\n")
```

Output from the above:

```
AGP    1856    0.26    T
CYP    2262    0.16    T
CAT    457     0.73    F
```

While `print()` is a simple way to print a line of text, `sys.stdout.write()` gives you more control over the formatting of printed output. As with `sys.stdin`, there is no need to call `.close()` on the file handle. Unlike other output-oriented file handles, data are regularly flushed from the pipe (either for printing or to another program's standard input).

Command Line Parameters

So far, even though we've been executing our programs from the command line, none of our Python programs have accepted input parameters. Consider the example from chapter 18, "Python Functions," where we computed the GC content for each sequence in a file. Rather than hard-coding the file name into the `io.open()` call, it would be preferable to supply the file name to work with on the command line, as in `./ids_seqs_gcs.py ids_seqs.txt`.

The sys module again comes to the rescue. After importing sys, the variable `sys.argv` references a list of strings that contain, starting at index 0, the name of the script itself, then each parameter. Because `sys.argv` is always a list of strings, if we want to input a float or integer argument, we'll need to convert the appropriate parameter using `int()` or `float()` before use.

```
#!/usr/bin/env python
import sys

if len(sys.argv) < 3:
    print("Sorry, expected at least two paramaters.")
    print("Usage: ./input_params_ex.py param1 param2")
    quit()

print("Here is sys.argv:")
print(sys.argv)
```

This code also determines whether the expected number of parameters has been given by the user by looking at `len(sys.argv)`, exiting if this isn't the case.

```
oneils@atmosphere ~/apcb/py$ chmod +x input_params.py
oneils@atmosphere ~/apcb/py$ ./input_params.py arg1 50
Here is sys.argv:
['./input_params.py', 'arg1', '50']
```

As with other programs run on the command line, if we wish to send a single parameter that contains spaces, we need to wrap it in single or double quotes.

```
oneils@atmosphere ~/apcb/py$ ./input_params.py 'argument 1' 50
Here is sys.argv:
['./input_params.py', 'argument 1', '50']
```

Although we won't cover it here, the `argparse` module makes writing scripts that require input parameters of different types relatively easy. The `argparse` module also automates the printing and formatting of help information that can be accessed by users who run the program with a `-h` or `--help` flag. A variety of tutorials for `argparse` can be found online. (There is also a simpler, but less featureful, module `optparse`.)

Executing Shell Commands within Python

It is sometimes useful to execute other programs from within our own programs. This could be because we want to execute an algorithmically generated series of commands (e.g., perhaps we want to run a genome assembly program on a variety of files in sequence), or perhaps we want to get the output of a program run on the command line.

Consider the command line task of listing all of the files in the current directory matching the pattern `cluster*.fasta`. For this we can use `ls -1 cluster*.fasta`, where `-1` tells `ls` to print its output as a single column of file names, and `cluster*.fasta` matches all file names matching the desired pattern (like **`cluster_AB.fasta`**, **`cluster_AC.fasta`**, **`cluster_AG.fasta`**, and **`cluster_D.fasta`**.).

```
oneils@atmosphere ~/apcb/py$ ls -1 cluster*.fasta
cluster_AB.fasta
cluster_AC.fasta
cluster_AG.fasta
cluster_D.fasta
```

There are a few ways to get this information in a Python program, but one of the easiest is to run the `ls -1 cluster*.fasta` command from within our program, capturing the result as a string.[54] The `subprocess` module allows us to do exactly this by using the `subprocess.check_output()` function. This function takes a number of potential parameters, but we'll run it with just two: (1) the command we wish to execute (as a string) and (2) `shell = True`, which indicates that the Python interpreter should run the command as though we had run it on the command line.

The result of the function call is a single string containing whatever the command printed to standard output. Because it's a single string, if we want it represented as a list of lines, we need to first `.strip()` off any newline characters at the end, and then split it on the `\n` characters that separate the lines.

54. An alternative method to get a list of files is to use the `os` module and the `os.listdir()` function, which will return a list of file names. For listing and working with file names and paths, the functions in the `os` module are are preferred, because they will work whether the program is run on a Windows system or a Unix-like system.For running shell commands more generally, the `subprocess` module is useful. Here we're illustrating that on the Unix/Linux command line, it can do both jobs by calling out to system utilities.

```
#!/usr/bin/env python
import subprocess

cmd = "ls -1 cluster*.fasta"
result = subprocess.check_output(cmd, shell = True)
# eg "cluster_AB.fasta\ncluster_AC.fasta\n"

result_stripped = result.strip()
result_lines = result_stripped.split("\n")

print(result_lines)
```

Running the program, we see that the result is a list of strings, one per file:

```
oneils@atmosphere ~/apcb/py$ chmod +x list_cluster_fasta.py
oneils@atmosphere ~/apcb/py$ ./list_cluster_fasta.py
['cluster_AB.fasta', 'cluster_AC.fasta',
'cluster_AG.fasta', 'cluster_D.fasta']
```

With this list of file names in hand, we might wish to run a series of assemblies with the program `runAssembly` (produced by 454 Life Sciences for assembling transcript sequences). This program requires a `-o` parameter to specify where outputs should be written as well as a file name input; we generate these with string operations to "build" a command to execute. Once we are satisfied with the commands, we can uncomment the actual call to `subprocess.check_output()`.

```
#!/usr/bin/env python
import subprocess

cmd = "ls -1 cluster*.fasta"
result = subprocess.check_output(cmd, shell = True)
# eg "cluster_AB.fasta\ncluster_AC.fasta\n"

result_stripped = result.strip()
result_lines = result_stripped.split("\n")

for filename in result_lines:
    name_list = filename.split(".")  # eg ["cluster_AB", "fasta"]
    cmd = "runAssembly " + "-o " + name_list[0] + ".out " + filename
    # eg "runAssembly -o cluster_AB.out cluster_AB.fasta"

    print("Running: " + cmd)
    #subprocess.check_output(cmd, shell = True)
```

This version of the program reports the commands that will be run.[55]

```
oneils@atmosphere ~/apcb/py$ ./list_cluster_fasta.py
Running: runAssembly -o cluster_AB.out cluster_AB.fasta
Running: runAssembly -o cluster_AC.out cluster_AC.fasta
Running: runAssembly -o cluster_AG.out cluster_AG.fasta
Running: runAssembly -o cluster_D.out cluster_D.fasta
```

While a command is running via the `subprocess.check_output()` function in this way, the program will wait before continuing to the next line (though you'll likely only notice if the command takes a long time to execute). If the command crashes or is killed, the Python program will crash with a `CalledProcessError` too (usually a good thing!).

Other functions in the `subprocess` module allow for running commands in the background—so the script doesn't wait for them to complete before continuing in its own execution—and to work with multiple such processes while communicating with them. Other Python modules even allow for advanced work with parallel execution and threading, though these topics are beyond the scope of this chapter.

Don't forget that you are responsible for the execution of your programs! If you write a program that executes many other programs, you can quickly utilize more computational resources than you realize. If you are working on a system that you don't mind taking some risks with, you could try to run the following program, which executes a copy of itself (which in turn executes a copy of itself, which in turn executes a copy of itself, and on and on):

```
#!/usr/bin/env python
import subprocess

## This file is called runs_itself.py, and this program runs itself.
## Maybe not a good idea?
cmd = "./runs_itself.py"
result = subprocess.check_output(cmd, shell = True)
```

Exercises

1. The table-printing example used `sys.stdout` to nicely print a table in column-major order.

55. Here we are splitting filenames into a [name, extension] list with `filename.split(".")`. This method won't work if there are multiple `.`s in the file. A more robust alternative is found in the `os` module: `os.path.splitext("this.file.txt")` would return `["this.file", "txt"]`.

Write a function called `print_row_major()` that prints a table stored in row-major order, so a call like `print_row_major([["A", "B"], ["C", "D"], ["E", "F"]])` results in output looking like:

```
A    B
C    D
E    F
```

2. Write a program called `stdin_eval_mean.py` that reads a file in the format of **`pz_blastx_yeast_top1.txt`** on `sys.stdin`, computes the mean of the *E*-value column (column 11, with values like `1e-32`), and prints the mean to standard output. If there are no data on standard input, the program should produce a "usage" message.

   Next, try running the program with `cat pz_blastx_yeast_top1.txt | ./stdin_eval_mean.py`. Also, try prefiltering the lines before the computation with something like `cat pz_blastx_yeast_top1.txt | grep '_L' | ./stdin_eval_mean.py`.

   Now copy this program to one called `stdin_eval_sd.py` that reads in **`pz_blastx_yeast_top1.txt`** on `sys.stdin` and computes and prints the standard deviation of the *E* values. Again, try running the program with and without prefiltering with `grep`. (At this point you should be comfortable with writing `mean()` and `sd()` functions that take lists of floats and return answers. Your `sd()` function might well call the `mean()` function.)

3. Modify the `stdin_eval_mean.py` program above (call the new version `stdin_eval_mean_threshold.py`) to accept an *E*-value upper threshold as a first parameter. Thus `cat pz_blastx_yeast_top1.txt | ./stdin_eval_mean.py 1e-6` should compute and print the mean of the *E*-value column, but only for those lines whose *E* value is less than 1e-6. (You'll need to use `sys.argv` and be sure to convert the entry therein to a float.)

   Your program should print usage information and exit if either there are no data provided on standard input, or if there is no threshold argument.

Chapter 20

Dictionaries

Dictionaries (often called "hash tables" in other languages) are an efficient and incredibly useful way to associate data with other data. Consider a list, which associates each data element in the list with an integer starting at zero:

```
a_list = [0.65, 0.24, 0.73]        Values
          ↑     ↑     ↑
          0     1     2            Indices
```

A dictionary works in much the same way, except instead of indices, dictionaries use "keys," which may be integers or strings.[56] We'll usually draw dictionaries longwise, and include the keys within the illustrating brackets, because they are as much a part of the dictionary structure as the values:

```
[ "CYP6B" → 0.56 ]
[  "AGP2" → 0.24 ]
[  "CATB" → 0.73 ]
    Keys    Values
```

Let's create a dictionary holding these keys and values with some code, calling it `ids_to_gcs`. Note that this name encodes both what the keys and values represent, which can be handy when keeping track of the contents of these structures. To create an empty dictionary, we call the `dict()` function with no parameters,

56. Dictionary keys may be any immutable type, which includes integers and strings, but they also include a number of other more exotic types, like tuples (immutable lists). Although floats are also immutable, because keys are looked up on the basis of equality and rounding errors can compound, it is generally not recommended to use them as dictionary keys.

which returns the empty dictionary. Then, we can assign and retrieve values much like we do with lists, except (1) we'll be using strings as keys instead of integer indices, and (2) we can assign values to keys even if the key is not already present.

```
ids_to_gcs = dict()

ids_to_gcs["CYP6B"] = 0.56

key = "CATB"
val = 0.73
ids_to_gcs[key] = val

ids_to_gcs["AGP2"] = 0.24
```

We can then access individual values just like with a list:

```
val_catb = ids_to_gcs["CATB"]
print(val_catb)                       # prints 0.73

key_agp2 = "AGP2"
val_agp2 = ids_to_gcs[key_agp2]       # holds 0.24
```

However, we cannot access a value for a key that doesn't exist. This will result in a `KeyError`, and the program will halt.

```
test = ids_to_gcs["TL34_X"]           # Error!
```

Dictionaries go one way only: given the key, we can look up the value, but given a value, we can't easily find the corresponding key. Also, the name "dictionary" is a bit misleading, because although real dictionaries are sorted in alphabetical order, Python dictionaries have no intrinsic order. Unlike lists, which are ordered and have a first and last element, Python makes no guarantees about how the key/value pairs are stored internally. Also, each unique key can only be present once in the dictionary and associated with one value, though that value might be something complex like a list, or even another dictionary. Perhaps a better analogy would be a set of labeled shelves, where each label can only be used once.

There are a variety of functions and methods that we can use to operate on dictionaries in Python. For example, the `len()` function will return the number of key/value pairs in a dictionary. For the dictionary above, `len(ids_to_gcs)` will return `3`. If we want, we can get a list of all the keys in a dictionary using its `.keys()` method, though this list may be in a random order because dictionaries are unordered. We could always sort the list, and loop over that:

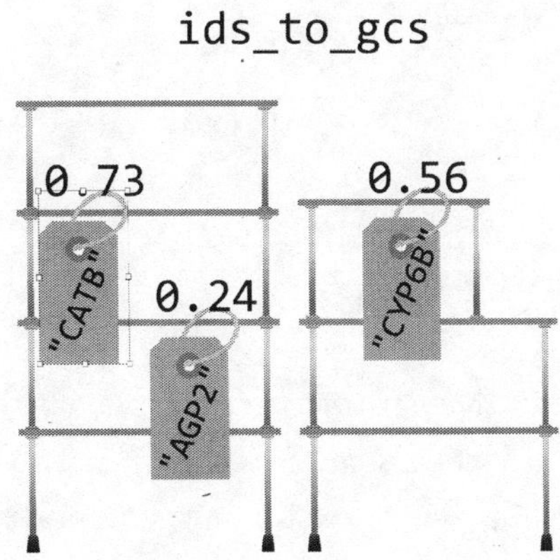


```
ids = ids_to_gcs.keys()        # list of keys
ids_sorted = sorted(ids)       # sorted version
for idkey in ids_sorted:       # for each key
    gc = ids_to_gcs[idkey]     # get the value
    print("id " + idkey + " has GC content " + str(gc))
```

Similarly, we can get a list of all the values using `.values()`, again in no particular order. So, `ids_to_gcs.values()` will return a list of three floats.[57]

If we try to get a value for a key that isn't present in the dictionary, we'll get a `KeyError`. So, we will usually want to test whether a key is present before attempting to read its value. We can do this with the dictionary's `.has_key()` method, which returns `True` if the key is present in the dictionary.[58]

Using the `in` keyword for dictionaries is generally preferred by the Python community, and in fact `.has_key()` has been removed in Python 3.0 and later. For the purposes of this book, we'll use `.has_key()` when working with dictionaries so it is clear exactly what is happening.

57. In Python 3.0 and later, what is returned by `.keys()` and `.values()` is not technically a list but a "view," which operates similarly to a list without using any additional memory. The code shown still works, but `ids.sort()` (the sort-in-place version) would not, as views do not have a `.sort()` method.

58. There is also a more Python-specific method of querying for a key in a dictionary: `"TL34_X" in ids_to_gcs` is equivalent to `ids_to_gcs.has_keys("TL34_X")`. Interestingly, the `in` keyword also works for lists: if `ids = ["CYP6B", "CATB", "AGP4"]`, then `"TL34_X" in ids` will return `False`. However, in this case every entry in `ids` must be compared to `"TL34_X"` to make this determination, and so `in` is much slower for lists than for dictionaries.

```
if ids_to_gcs.has_key("TL34_X"):
    tl34x_gc = ids_to_gcs["TL34_X"]
    print("TL34_X has GC " + str(tl34x_gc))
else:
    print("TL34_X not found in dictionary.")
```

Counting Gene Ontology Terms

To illustrate the usage of a dictionary in practice, consider the file **PZ.annot.txt**, the result of annotating a set of assembled transcripts with gene ontology (GO) terms and numbers. Each tab-separated line gives a gene ID, an annotation with a GO number, and a corresponding human-readable term associated with that number.

```
PZ7180000020811_DVU      GO:0003824      GJ12748 [Drosophila virilis]
PZ7180000020752_DVU      GO:0003824      GI16375 [Drosophila mojavensis]
PZ7180000034678_DWY      GO:0003824      hypothetical protein YpF1991016_1335 [Ye
PZ7180000024883_EZN      GO:0006548      sjchgc01974 protein
PZ7180000024883_EZN      GO:0004252      sjchgc01974 protein
PZ7180000024883_EZN      GO:0004500      sjchgc01974 protein
PZ7180000024883_EZN      GO:0006508      sjchgc01974 protein
PZ7180000023260_APN      GO:0005515      btb poz domain containing protein
...
```

In this file, each gene may be associated with multiple GO numbers, and each GO number may be associated with multiple genes. Further, each GO term may be associated with multiple different GO numbers. How many times is each ID found in this file? Ideally, we'd like to produce tab-separated output that looks like so:

```
1 PZ7180000020811_DVU
1 PZ7180000020752_DVU
1 PZ7180000034678_DWY
4 PZ7180000024883_EZN
1 PZ7180000023260_APN
...
```

Our strategy: To practice some of the command line interaction concepts, we'll have this program read the file on standard input and write its output to standard output (as discussed in chapter 19, "Command Line Interfacing"). We'll need to keep a dictionary, where the keys are the gene IDs and the values are the counts. A for-loop will do to read in each line, stripping off the ending newline and splitting the result into a list on the tab character, `\t`. If the ID is in the dictionary, we'll add one to the value. Because the dictionary will start empty, we will frequently run into IDs that aren't already present in the dictionary; in these cases we can set the value to `1`. Once we have processed the entire input, we can loop over the dictionary printing each count and ID.

In the code below (**`go_id_count.py`**), when the dictionary has the **`seqid`** key, we're both reading from the dictionary (on the right-hand side) and writing to the value (on the left-hand side).

```
#!/usr/bin/env python
import sys

if(sys.stdin.isatty()):
    print("Usage: cat <annotation file> | ./countannots.py")
    quit()

ids_to_counts = dict()

# Parse input
for line in sys.stdin:
    line_list = line.strip().split("\t")
    seqid = line_list[0]
    if ids_to_counts.has_key(seqid):
        ids_to_counts[seqid] = ids_to_counts[seqid] + 1
    else:
        ids_to_counts[seqid] = 1

# Print dict contents
ids_list = ids_to_counts.keys()
for seqid in ids_list:
    count = ids_to_counts[seqid]
    print(str(count) + "\t" + seqid)
```

But when the key is not present, we're simply writing to the value. In the loop that prints the dictionary contents, we are not checking for the presence of each `id` before reading it to print, because the list of `ids_list` is guaranteed to contain exactly those keys that are in the dictionary, as it is the result of `ids_to_counts.keys()`.

```
oneils@atmosphere ~/apcb/py$ cat PZ.annot.txt | ./go_id_count.py
2    PZ736262
4    PZ7180000000004_OO
5    PZ7180000000004_OM
3    PZ7180000000004_OJ
2    PZ7180000000021_AX
...
```

What's the advantage of organizing our Python program to read rows and columns on standard input and write rows and columns to standard output? Well, if we know the built-in command line tools well enough, we can utilize them along with our program for other analyses. For example, we can first filter the data with `grep` to select those lines that match the term `transcriptase`:

```
oneils@atmosphere ~/apcb/py$ cat PZ.annot.txt | grep 'transcriptase'
```

The result is only lines containing the word "transcriptase":

```
PZ7180000000003_PI      GO:0003723      reverse transcriptase
PZ7180000000003_PI      GO:0003964      reverse transcriptase
PZ7180000000003_PI      GO:0031072      reverse transcriptase
PZ7180000000003_PI      GO:0006278      reverse transcriptase
PZ840833_BZS    GO:0005488      reverse transcriptase
PZ858982_CAA    GO:0005488      reverse transcriptase
PZ858982_CAA    GO:0044464      reverse transcriptase
PZ7180000029134_AHQ     GO:0003723      reverse transcriptase
PZ7180000029134_AHQ     GO:0003964      reverse transcriptase
...
```

If we then feed those results through our program (`cat PZ.annot.txt | grep 'transcriptase' | ./go_id_count.py`), we see only counts for IDs among those lines.

```
4       PZ924_N
4       PZ7180000000089_N
1       PZ840833_BZS
1       PZCAP37180000034572_A
3       PZ7180000029134_AHQ
1       PZ492962
```

Finally, we could pipe the results through `wc` to count these lines and determine how many IDs were annotated at least once with that term (21). If we wanted to instead see which eight genes had the most annotations matching "transcriptase," we could do that, too, by sorting on the counts and using `head` to print the top eight lines (here we're breaking up the long command with backslashes, which allow us to continue typing on the next line in the terminal).[59]

59. For simple problems like this, if we know the command line tools well enough, we don't even need to use Python. This version of the problem can be solved with a pipeline like `cat PZ.annot.txt | grep 'transcriptase' | awk '{print $1}' | sort | uniq -c | sort -k1,1nr | head`.

```
oneils@atmosphere ~/apcb/py$ cat PZ.annot.txt | \
> grep 'transcriptase' | \
> ./go_id_count.py | \
> sort -k1,1nr | \
> head -n 8
7       PZ32722_B
5       PZ7180000000012_DC
4       PZ59_HO
4       PZ7180000000003_PI
4       PZ7180000000089_N
4       PZ924_N
3       PZ578878
3       PZ7180000000012_IL
```

It appears gene `PZ32722_B` has been annotated as a transcriptase seven times. This example illustrates that, as we work and build tools, if we consider how they might interact with other tools (even other pieces of code, like functions), we can increase our efficiency remarkably.

Extracting All Lines Matching a Set of IDs

Another useful property of dictionaries is that the `.has_key()` method is very efficient. Suppose we had an unordered list of strings, and we wanted to determine whether a particular string occurred in the list. This can be done, but it would require looking at each element (in a for-loop, perhaps) to see if it equaled the one we are searching for. If we instead stored the strings as keys in a dictionary (storing `"present"`, or the number `1`, or anything else in the value), we could use the `.has_key()` method, which takes a single time step (effectively, on average) no matter how many keys are in the dictionary.[60]

Returning to the GO/ID list from the last example, suppose that we had the following problem: we wish to first identify all those genes (rows in the table) that were labeled with `GO:0001539` (which we can do easily with `grep` on the command line), and then we wish to extract all rows from the table matching those IDs to get an idea of what other annotations those genes might have.

In essence, we want to print all entries of a file:

60. In computer science terms, we say that searching an unordered list runs in time $O(n)$, or "order n," where n is taken to be the size of the list. The `.has_key()` method runs in $O(1)$, which is to say that the time taken is independent of the size of the dictionary. If we need to do such a search many times, these differences can add up significantly. More information on run-time considerations is covered in chapter 25, "Algorithms and Data Structures."

```
PZ7180000020811_DVU     GO:0003824      GJ12748 [Drosophila virilis]
PZ7180000020752_DVU     GO:0003824      GI16375 [Drosophila mojavensis]
PZ7180000034678_DWY     GO:0003824      hypothetical protein YpF1991016_1335 [Ye
PZ7180000024883_EZN     GO:0006548      sjchgc01974 protein
PZ7180000024883_EZN     GO:0004252      sjchgc01974 protein
PZ7180000024883_EZN     GO:0004500      sjchgc01974 protein
PZ7180000024883_EZN     GO:0006508      sjchgc01974 protein
PZ7180000023260_APN     GO:0005515      btb poz domain containing protein
...
```

Where the first column matches any ID in the first column of another input:

```
oneils@atmosphere ~/apcb/py$ cat PZ.annot.txt | grep 'GO:0001539'
PZ7180000000028_AP     GO:0001539     troponin c 25d
PZ7180000000030_AP     GO:0001539     troponin c
PZ7180000000043_AP     GO:0001539     troponin c 25d
PZ7180000000044_AP     GO:0001539     troponin c 25d
PZ7180000000045_AP     GO:0001539     troponin c 25d
PZ7180000000046_AP     GO:0001539     troponin c 25d
...
...
```

As it turns out, the above problem is common in data analysis (subsetting lines on the basis of an input "query" set), so we'll be careful to design a program that is not specific to this data set, except that the IDs in question are found in the first column.[61]

We'll write a program called `match_1st_cols.py` that takes two inputs: on standard input, it will read a number of lines that have the query IDs we wish to extract, and it will also take a parameter that specifies the file from which matching lines should be printed. For this instance, we would like to be able to execute our program as follows:

```
oneils@atmosphere ~/apcb/py$ cat PZ.annot.txt | grep 'GO:0001539' | \
> ./match_1st_cols.py PZ.annot.txt
```

In terms of code, the program can first read the input from standard input and create a dictionary that has keys corresponding to each ID that we wish to extract (the values can be anything). Next, the program will loop over the lines of the input file (specified in `sys.argv[1]`), and for each ID check it with `.has_key()` against the dictionary created previously; if it's found, the line is printed.

61. The `grep` utility can perform a similar operation; `grep -f query_patterns.txt subject_file.txt` will print all lines in `subject_file.txt` that are matched by any of the patterns in `query_patterns.txt`. But this requires that all patterns are compared to all lines (even if the patterns are simple), and so our custom Python solution is much faster when the number of queries is large (because the `.has_key()` method is so fast).

```
#!/usr/bin/env python
import sys
import io

if sys.stdin.isatty() or len(sys.argv) != 2:
    print("Usage: cat <id_list> | ./match_1st_cols.py <search_file>")
    print("This script extracts lines from <search_file> where any entry")
    print(" in the the first column of <id_list> matches the first column")
    print(" of <search_file>")
    quit()

## Build IDs wanted dictionary from standard input
ids_wanted = dict()
for line in sys.stdin:
    line_stripped = line.strip()
    line_list = line_stripped.split("\t")
    id = line_list[0]
    ids_wanted[id] = "wanted"

## Loop over the file, print the lines that are wanted
fhandle = io.open(sys.argv[1], "rU")
for line in fhandle:
    line_stripped = line.strip()
    line_list = line_stripped.split("\t")
    id = line_list[0]
    # Is the ID one of the ones we want?
    if ids_wanted.has_key(id):
        print(line_stripped)

fhandle.close()
```

Making the program (**`match_1st_cols.py`**) executable and running it reveals all annotations for those IDs that are annotated with `GO:0001539`.

```
oneils@atmosphere ~/apcb/py$ grep 'GO:0001539' PZ.annot.txt | \
> ./match_1st_cols.py PZ.annot.txt
PZ7180000000028_AP    GO:0001539    troponin c 25d
PZ7180000000028_AP    GO:0009288    troponin c 25d
PZ7180000000028_AP    GO:0005509    troponin c 25d
PZ7180000000030_AP    GO:0001539    troponin c
PZ7180000000030_AP    GO:0009288    troponin c
PZ7180000000030_AP    GO:0005509    troponin c
PZ7180000000043_AP    GO:0001539    troponin c 25d
PZ7180000000043_AP    GO:0009288    troponin c 25d
```

As before, we can use this strategy to easily extract all the lines matching a variety of criteria, just by modifying one or both inputs. Given any list of gene IDs of interest from a collaborator, for example, we could use that on the standard input and extract the corresponding lines from the GO file.

Exercises

1. Dictionaries are often used for simple lookups. For example, a dictionary might have keys for all three base-pair DNA sequences (`"TGG"`, `"GCC"`, `"TAG"`, and so on) whose values correspond to amino acid codes (correspondingly, `"W"`, `"A"`, `"*"` for "stop," and so on). The full table can be found on the web by searching for "amino acid codon table."

   Write a function called `codon_to_aa()` that takes in a single three-base-pair string and returns a one-character string with the corresponding amino acid code. You may need to define all 64 possibilities, so be careful not to make any typos! If the input is not a valid three-base-pair DNA string, the function should return `"X"` to signify "unknown." Test your function with a few calls like `print(codon_to_aa("TGG"))`, `print(codon_to_aa("TAA"))`, and `print(codon_to_aa("BOB"))`.

2. Combine the result of the `codon_to_aa()` function above with the `get_windows()` function from the exercises in chapter 18, "Python Functions," to produce a `dna_to_aa()` function. Given a string like `"AAACTGTCTCTA"`, the function should return its translation as `"KLSL"`.

3. Use the `get_windows()` function to write a `count_kmers()` function; it should take two parameters (a DNA sequence and an integer) and return a dictionary of k-mers to count for those k-mers. For example, `count_kmers("AAACTGTCTCTA", 3)` should return a dictionary with keys `"AAA"`, `"AAC"`, `"ACT"`, `"CTG"`, `"TGT"`, `"GTC"`, `"TCT"`, `"CTC"`, `"CTA"` and corresponding values `1`, `1`, `1`, `1`, `1`, `1`, `2`, `1`, `1`. (K-mer counting is an important step in many bioinformatics algorithms, including genome assembly.)

4. Create a function `union_dictionaries()` that takes two dictionaries as parameters returns their "union" as a dictionary—when a key is found in both, the larger value should be used in the output. If dictionary `dict_a` maps `"A"`, `"B"`, `"C"` to `3`, `2`, `6`, and `dict_b` maps `"B"`, `"C"`, `"D"` to `7`, `4`, `1`, for example, the output should map `"A"`, `"B"`, `"C"`, `"D"` to `3`, `7`, `6`, `1`.

Chapter 21

Bioinformatics Knick-knacks and Regular Expressions

Chapter 14, "Elementary Data Types," mentioned that strings in Python are immutable, meaning they can't be modified after they have been created. For computational biology, this characteristic is often a hindrance. To get around this limitation, we frequently convert strings into lists of single-character strings, modify the lists as necessary, and convert the lists of characters back to strings when needed. Python comes built-in with a number of functions that operate on lists and strings useful for computational biology.

Converting a String into a List of Single-Character Strings

Remember that because Python strings are immutable, we can't run something like

```
seq = "ACTAG"
seq[2] = "U"                # Error!
```

To convert a string into a list of single-character strings, we can pass the string as a parameter to the `list()` function (which, when called with no parameters, creates an empty list):

```
seq = "ACTAG"
base_list = list(seq)
base_list[2] = "U"
print(base_list)            # prints ['A', 'C', 'U', 'A', 'G']
```

Converting a List of Strings into a String

If we happen to have a list of strings (such as the list of single-character strings above), we can join them into a single string using the `.join()` method of whatever string we wish to use as a separator for the output string.

```
rejoined = ":".join(base_list)       # 'A:C:U:A:G'
rejoined2 = "".join(base_list)       # 'ACUAG'
```

The syntax seems a bit backward, `"separator".join(list_variable)`, but it makes sense when considering what the method `.`-syntax says: it asks the string "`separator`" to produce a string given the input list `list_variable`. Because the output is of type string, it makes sense that we should ask a string to do this production.

Reverse a List

Sometimes we may want to reverse the order of a list. We can do this with the list's `.reverse()` method which asks the list to reverse itself in place (but returns `None`).

```
base_list.reverse()              # base_list now: ['G', 'A', 'U', 'C', 'C']
```

As with the `.sort()` and `.append()` methods from chapter 15, "Collections and Looping, Part 1: Lists and `for`," `.reverse()` returns `None`, so a line like `base_list = base_list.reverse()` would almost surely be a bug.

Reverse a String

There are two ways to reverse a string in Python: (1) first, convert it into a list of single-character strings, then reverse the list, then join it back into a string, or (2) use "extended slice syntax." The former solution uses the methods described above, whereas the second method extends the `[]` slice syntax so that the brackets can contain values for `[start:end:step]`, where `start` and `end` can be empty (to indicate the first and last indices of the string), and `step` can be `-1` (to indicate stepping backward in the string).

```
seq = "ACTAG"
seq_list = list(seq)
seq_list.reverse()
seq_rev = "".join(seq_list_rev)  # seq_rev: "GATCA"
# OR
seq_rev = seq[::-1]              # seq_rev: "GATCA"
```

Although the extended slice syntax is quite a bit faster and requires less typing, it is a bit more difficult to read. A search online for "python string reverse" would likely reveal this solution first.

Simple Find and Replace in a String

We've already seen how to split a string into a list of strings with the `.split()` method. A similar method, `.replace()`, allows us to replace all matching substrings of a string with another string. Because strings are immutable, this method returns a modified copy of the original string.

```
seq = "ACGTATATATGG"
replaced = seq.replace("TAT", "X")  # replaced: "ACGXAXGG"
rna = seq.replace("U", "T")         # rna: "ACGUAUAUAUGG"
                                    # seq is still: "ACGTATATATATGG"
```

Also because strings are immutable, the original data are left unchanged. Because a variable is simply a name we use to refer to some data, however, we can make it look like we've modified the original string by reassigning the `seq` variable to whatever is returned by the method.

```
seq = seq.replace("U", "T")         # seq now: "ACGUAUAUAUGG"
```

Commands versus Queries

Why do some operations (like a list's `.sort()` method) change the data in place but return `None`, while others (like a string's `.split()` method or the `len()` function) return something but leave the original data alone? Why is it so rare to see operations that do both? The reason is that the designers of Python (usually) try to follow what is known as the *principle of command-query separation*, a philosophical principle in the design of programming languages that states that single operations should either modify data or return answers to queries, but not both.

The idea behind this principle is that upon reading code, it should be immediately obvious what it does, a feature that is easily achieved if each operation only has one thing it *can* do. When operations both change data and return an answer, there is a temptation to "code by side effect," that is, to make use of simultaneous effects to minimize typing at the cost of clarity. Compared to many other languages, Python makes a stronger attempt to follow this principle.

Regular Expressions

Regular expressions, common to many programming languages and even command line tools like `sed`, are syntax for matching sophisticated patterns in strings. The simplest patterns are just simple strings; for example, `"ATG"` is the pattern for a start codon. Because Python treats backslashes as special in strings (e.g., `"\t"` is not actually `"\"` followed by a `"t"`, but rather a tab character), patterns for regular expressions in Python are usually expressed as "raw strings," indicated by prefixing them with an `r`. So, `r"ATG"` is also the pattern for a start codon, but `r"\t"` is the regular expression for `"\"` and `"t"` rather than the tab character.

```
print(len("\t"))        # prints 1 (tab character)
print(len(r"\t"))       # prints 2 (\ character and t character)
```

Python regular expression functionality is imported with the `re` module by using `import re` near the top of the script. There are many functions in the `re` module for working with strings and regular expression patterns, but we'll cover just the three most important ones: (1) searching for a pattern in a string, (2) replacing a pattern in a string, and (3) splitting a string into a list of strings based on a pattern.

Searching for a Pattern in a String

Searching for a pattern is accomplished with the `re.search()` function. Most functions in the `re` module return a special `SRE_Match` data type, or a list of them (with their own set of methods), or `None` if no match was found. The if- statement (and while-loops) in Python treats `None` as false, so we can easily use `re.search()` in an if-statement. The result of a successful match can tell us the location of the match in the query using the `.start()` method:

```
import re

seq = "CCGATGCATGCC"
if re.search(r"ATG", seq):
    print("seq has a start codon!")
    result = re.search(r"ATG", seq)
    print("the start codon is at position:")
    print(result.start())
```

Replace a Pattern in a String

The `re.subn()` function can be used to search and replace a pattern within a string. It takes at least four important arguments (and some optional ones we won't discuss here): (1) the pattern to search for, (2) the replacement string to replace matches with, (3) the string to look in, and (4) the maximum number of replacements to make (`0` to replace all matches). This function returns a list[62] containing two elements: at index `0`, the modified copy of the string, and at index `1`, the number of replacements made.

```
seq = "CCGATGCATGCC"
pattern = r"ATG"
replacement = "*"
result = re.subn(pattern, replacement, seq, 0)
print(result[0])                                # prints "CCG*C*CC"
print(result[1])                                # prints 2
```

Split a String into a List of Strings Based on a Pattern

We've already seen that we can split a string based on simple strings with `.split()`. Being able to split on complex regular expressions is often useful as well, and the `re.split()` function provides this functionality.

```
seq = "CCGATGCATGCC"
between_starts = re.split(r"ATG", seq)
print(between_starts)                           # prints ['CCG', 'C', 'CC']
```

62. Actually, it returns a tuple, which works much like a list but is immutable. Tuples are briefly covered in chapter 15.

We'll leave it as an exercise for the reader to determine what would be output for a sequence where the matches are back to back, or where the matches overlap, as in `re.split(r"ATA", "GCATATAGG")`.

The Language of Regular Expressions, in Python

In chapter 11, "Patterns (Regular Expressions)," we covered regular expressions in some detail when discussing the command line regular expression engine `sed`. Regular expressions in Python work similarly: simple strings like `r"ATG"` match their expressions, dots match any single character (`r"CC."` matches any `P` codon), and brackets can be used for matching one of a set of characters (`r"[ACTG]"` matches one of a DNA sequence, and `r"[A-Za-z0-9_]"` is shorthand for "any alphanumeric character and the underscore"). Parentheses group patterns, and the plus sign modifies the previous group (or implied group) to match one or more times (* for zero or more), and vertical pipes `|` work as an "or," as in `r"([ACTG])+(TAG|TAA|TGA)"`, which will match any sequence of DNA terminated by a stop codon (`TAG`, `TAA`, or `TGA`). Rounding out the discussion from previous chapters, Python regular expressions also support curly brackets (e.g., `r"(AT){10,100}"` matches an `"AT"` repeated 10 to 100 times) and standard notation for the start and end of the string. (Note that `r"^([ACTG])+$"` matches a string of DNA and only DNA. For more detailed examples of these regular expression constructs, refer to chapter 11.)

The regular expression syntax of Python, however, does differ from the POSIX-extended syntax we discussed for `sed`, and in fact provides an extra-extended syntax known as "Perl-style" regular expressions. These support a number of sophisticated features, two of which are detailed here.

First, operators that specify that a match should be repeated—such as plus signs, curly brackets, and asterisks—are by default "greedy." The same is true in the POSIX-extended syntax used by `sed`. In Python, however, we have the option of making the operator non-greedy, or more accurately, reluctant. Consider the pattern `r"([ACTG])+(TAG|TAA|TGA)"`, which matches a DNA sequence terminated by a stop codon. The greedy part, `([ACTG])+`, will consume all but the last stop codon, leaving as little of the remaining string as possible to make the rest of the pattern match.

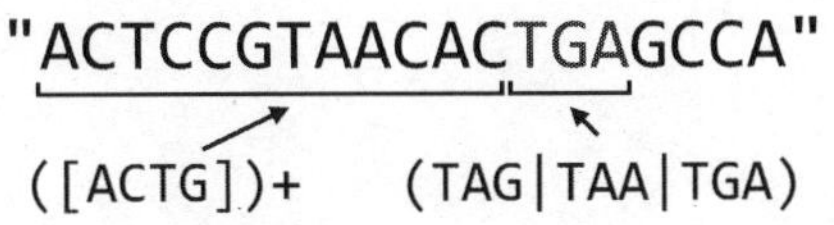

```
"ACTCCGTAACACTGAGCCA"
([ACTG])+    (TAG|TAA|TGA)
```

In Python, if we want to make the plus sign reluctant, we can follow it with a question mark, which causes the match to leave as much as possible for later parts of the pattern.

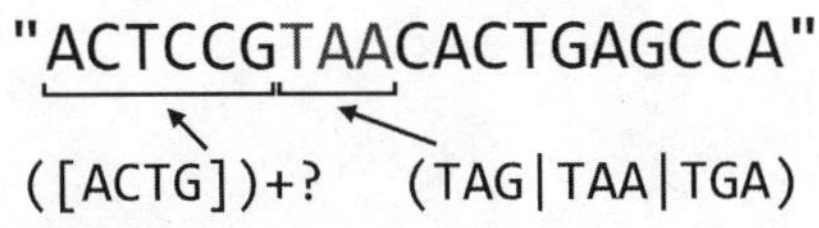

```
"ACTCCGTAACACTGAGCCA"
([ACTG])+?   (TAG|TAA|TGA)
```

The reluctance operator can also follow the asterisk and curly brackets, but beware: when used without following one of these three repetition operators, the question mark operator works as an "optional" operator (e.g., `r"C(ATG)?C"` matches `"CC"` and `"CATGC"`).

The second major difference between Python (Perl-style) regular expressions and POSIX-extended regular expressions is in how they specify character classes. As mentioned above, a pattern like `r"[A-Za-z0-9_]"` is shorthand for "any alphanumeric and the underscore," so a series of alphanumerics can be matched with `r"[A-Za-z0-9_]+"`. In POSIX regular expressions, the character class `A-Za-z0-9_` can be specified with `[:alnum:]`, and the pattern would then need to be used like `[[:alnum:]]+` in `sed`.

The Perl-style regular expression syntax used by Python introduced a number of shorthand codes for character classes. For example, `\w` is the shorthand for "any alphanumeric and the underscore," so `r"\w+"` is the pattern for a series of these and is equivalent to `r"[A-Za-z0-9_]+"`. The pattern `\W` matches any character that is not alphanumeric or the underscore, so `r"\W+"` matches a series of these (equivalent to `r"[^A-Za-z0-9_]+"`). Perhaps the most important shorthand class is `\s`, which matches a single whitespace character: a tab, space, or newline character (`\S` matches any non-whitespace character). This is most useful when parsing input, where it cannot be guaranteed that words are separated by tabs, spaces, or some combination thereof.

```
#!/usr/bin/env python
import re
import sys

## Parse input in two columns
for line in sys.stdin:
    line_stripped = line.strip()
    #line_list = line.split("\t")
    line_list = re.split(r"\s+", line_stripped)

    # ...
```

In the code above, we've replaced the split on `"\t"` that we're used to with a `re.split()` on `r"\s+"`, which will ensure that we correctly parse the pieces of the line even if they are separated with multiple tabs, spaces, or some combination thereof.

Counting Promoter Elements

Consider the file **`grape_promoters.txt`**, which contains on each line 1000bp upstream of gene regions in the *Vitis vinifera* genome:

```
GSVIVT01034325001_1     GATTTCAAAAGCATTCTGTTGTTCTTTGAGGTCAGCAACCTGACCAATAAAAACT
GSVIVT01034326001_2     TACGCTTGACAAGACGTCTCCATGTCCTTTCAAGCGACTTGCTACGCTATGCACC
GSVIVT01034329001_3     AAAATTGGATGCATAAAACAAAATAAATGTAAATACTAAAATAATGATCATATTC
GSVIVT01034331001_4     AAAAAATAAAGTGTTTTAAAATAAAATCATTTAATTATTTTCACCTATTTTTTAA
GSVIVT01034332001_5     CAAAGCCATAGCAAAAATTGCATTCTCGAACAACCAAATAAAATCGAAACTTGTA
GSVIVT01034334001_6     GATGTTAGGAATAGTGGTTAATGGTTGTTGTCCACGTGTATAGCTTTGTTAGAAT
GSVIVT01034337001_7     AAGATATTATAATTAAAAAATATTTAATATAATTTTTTTAAAATATTGCATTTTG
GSVIVT01034340001_8     AAATTTTGGAAACTTATTAATAAAGGAATGACTCACATATTGTTCTTATTCAAAA
GSVIVT01034341001_9     GTAGTGGTAAACCGTTGAGGTGGATGTCATGCTGTTGTCGACGGCGATTGGTGGT
GSVIVT01034344001_10    TTGTATCGCATATTCAACCAAATATAAGATATGATAAGTGATGATATATATTATT
GSVIVT01034346001_11    AACCCATTTCTTATCTCTTTCTTTCTTTCTTTCTTTCTTTCTTCTTCTTCTTCTT
```

These columns are not separated by a tab character, but rather by a variable number of spaces.

Promoter motifs are small DNA patterns nearby gene sequences to which the cellular machinery binds in order to help initiate the gene-transcription process. For example, the ABF protein binds to the DNA pattern `"CACGTGGC"` if it is near a gene in some plants. Some motifs are flexible and can be described by regular expressions; the GATA protein binds to any short DNA sequence matching the pattern `"[AT]GATA[GA]"`. We wish to analyze the *V. vinifera* upstream regions above, and count for each the number of occurrences of the GATA motif. Our output should look like so:

```
GSVIVT01034325001_1     3
GSVIVT01034326001_2     2
GSVIVT01034329001_3     2
GSVIVT01034331001_4     2
GSVIVT01034332001_5     1
GSVIVT01034334001_6     2
...
```

After importing the `io` and `re` modules, we'll first write a function `count_motifs()` that takes two parameters: first a sequence in which to count motif matches, and second a motif for which to search. There are a variety of ways we could code this function, but a simple solution will be to use `re.split()` to split the sequence on the motif regular expression—the number of motifs will then be the length of the result minus one (because if no motifs are found, no split will be done; if one is found, one split will be done, and so on).

With this function written and tested, we can open the file using `io.open()` and loop over each line, calling the `count_motifs()` function on each sequence with the motif `r"[AT]GATA[GA]"`. Because the columns are separated with a variable number of spaces instead of single tab or space characters, we'll use `re.split()` to split each line into pieces.

First, we'll write and test the function that counts motifs and offers an example usage, where the number of matches should be two.

```
#!/usr/bin/env python
import re
import io

def count_motifs(seq, motif):
    pieces = re.split(motif, seq)
    return len(pieces) - 1

seq = "AAAAAAATGATAGAAAAAGATAAAAAA"
print(count_motifs(seq, r"[AT]GATA[GA]"))    # prints 2
```

Next we can finish the program, using `re.split()` to process each line.

```
#!/usr/bin/env python
import re
import io

def count_motifs(seq, motif):
    pieces = re.split(motif, seq)
    return(len(pieces) - 1)


fhandle = io.open("grape_promoters.txt", "rU")

for line in fhandle:
    linestripped = line.strip()
    line_list = re.split(r"\s+", linestripped)
    gid = line_list[0]
    seq = line_list[1]

    num_motifs = count_motifs(seq, r"[AT]GATA[GA]")
    print(gid + "\t" + str(num_motifs))

fhandle.close()
```

When run, this simple program (**`grape_count_gata.py`**) produces the output desired above. Because the output is sent to standard output, we can further filter the results through the `sort` and `head` command line utilities to identify which promoters are most likely to be found by GATA: `./grape_count_gata.py | sort -k2,2nr | head -n 10`.

Exercises

1. Write a function called `reverse_complement()` that takes a DNA sequence parameter and returns its reverse complement (i.e., reverses the string, switches A's and T's, and switches G's and C's).

2. DNA sequences that are destined to be turned into proteins are "read" by the cellular machinery in one of six "reading frames" with lengths that are multiples of three. The first three are derived from the sequence itself, starting at index `0`, index `1`, and index `2`; the first three reading frames of `"ACTAGACG"` are `"ACTAGA"`, `"CTAGAC"`, and `"TAGACG"`.

   To derive reading frames three, four, and five, we first reverse-complement the sequence (`"CGTCTAGT"`) and then produce frames from indices `0`, `1`, and `2`, resulting in `"CGTCTA"`, `"GTCTAG"`, and `"TCTAGT"`.

Using the `reverse_complement()` function from above (and potentially the `get_windows()` function from chapter 18, "Python Functions"), write a `seq_to_six_frames()` function that takes a DNA sequence as a parameter and returns a list of the six reading frames (as strings).

3. Write a function called `longest_non_stop()` that takes a DNA sequence as a parameter and returns the longest amino acid sequence that can be produced from it. This is done by generating the six reading frames from the sequence, converting each to an amino acid sequence (probably using the `dna_to_aa()` function from chapter 20, "Dictionaries"), and then trimming the resulting sequences down to the first "stop" codon (`"*"`) if there are any.

   In the DNA sequence `seq = "AGCTACTAGGAAGATAGACGATTAGAC"`, for example, the six translations are `SY*EDRRLD` (Frame 1), `ATRKIDD*` (Frame 2), `LLGR*TIR` (Frame 3), `V*SSIFLVA` (Frame 4), `SNRLSS**` (Frame 5), and `LIVYLPSS` (Frame 6). In order of lengths, the possibilities are thus `"V"`, `"SY"`, `"LLGR"`, `"SNRLSS"`, `"ATRKIDD"`, and `"LIVYLPSS"`. As a result, `longest_non_stop(seq)` should return `"LIVYLPSS"`.

4. Modify the `grape_count_gata.py` program, calling it `motifs_count.py` so that it can read the file name and motifs to process (the latter as a comma-separated list) on `sys.argv`. When run as `./motifs_count.py grape_promoters.txt [AT]GATA[GA],[CGT]ACGTG[GT][AC],TTGAC`, the output should look like:

```
GSVIVT01034325001_1     3       [AT]GATA[GA]
GSVIVT01034325001_1     0       [CGT]ACGTG[GT][AC]
GSVIVT01034325001_1     2       TTGAC
GSVIVT01034326001_2     2       [AT]GATA[GA]
GSVIVT01034326001_2     0       [CGT]ACGTG[GT][AC]
GSVIVT01034326001_2     4       TTGAC
GSVIVT01034329001_3     2       [AT]GATA[GA]
GSVIVT01034329001_3     0       [CGT]ACGTG[GT][AC]
```

5. Genotyping by sequencing (GBS) is a method for identifying genetic variants in a DNA sample by first splitting chromosomes in pseudorandom locations through the application of restriction enzymes and then sequencing short reads from the ends of the resulting fragments:

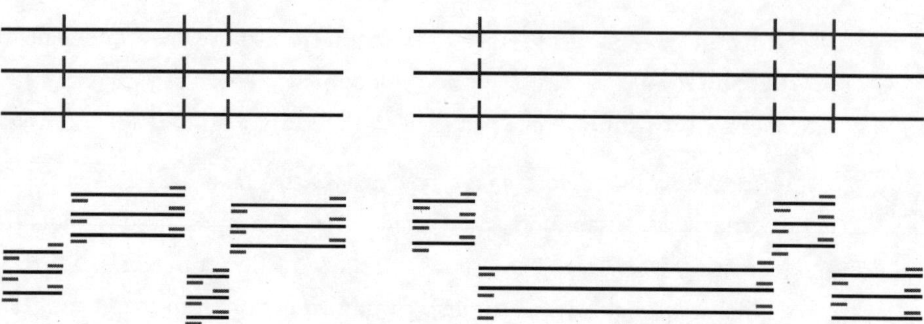

In the above, black lines represent input DNA sequence, red lines are cut sites, and green lines represent output from the sequencing machine. (Some simplifications have been made to the figure above for the sake of clarity.) The result is much higher depth of sequencing at fewer locations, which can be useful in a variety of contexts, including looking for genetic variants between different versions of the same chromosome (which requires many reads from the same location to statistically confirm those variants).

The restriction enzymes cut on the basis of patterns; for example, the ApeKI enzyme cuts DNA molecules at the pattern `"GC[AT]GC"`. Based on the recognized patterns, some enzymes are "frequent cutters" and some are not; more frequent cutters sample more locations in a genome but sacrifice depth of sequencing. For this reason, researchers often want to know, given an enzyme pattern and genome sequence, the distribution of fragment lengths that will result.

Write a function `gbs_cut()` that takes a DNA sequence, a regular expression pattern, and a "bin size." It should return a dictionary mapping sequence lengths, rounded down to the nearest bin size, to the number of fragments produced by the cutter in that bin. As an example, `"AAAAGCAGCAAAAAAGCTGCAAGCAGCAAAAA"` when processed with `"GC[AT]GC"` produces fragment sizes of 4, 6, 2, and 5. If grouped in a bin size of 3, the dictionary would have keys `0`, `3`, and `6`, and values `1`, `1`, and `2`.

Chapter 22

Variables and Scope

In chapter 18, "Python Functions," we learned that a function should use only variables that have either been defined within the block of the function, or passed in explicitly as parameters (Rule 1). This rule works closely with our definition of a variable in Python, as a name that refers to some data.

This is a powerful concept, allowing us to pass parameters to functions and get returned values. In Python, *a single piece of data may be referenced by multiple variables.* Parameter variables are new names for the same data. We can explore this concept with a simple example.

```
nums = [1, 2, 3, 4, 5]
numsb = nums
numsb[0] = 1000
print(nums)           # prints [1000, 2, 3, 4, 5]
print(numsb)          # prints [1000, 2, 3, 4, 5]
```

In this code sample, we have created a single list, but we are associating two variable names with it, `nums` and `numsb`. When we modify the list through one name, the change is reflected through the other (we can only change the data because lists are mutable, but even immutable data may be referred to by multiple names).

The same thing happens when we use a parameter in a function. Let's revisit the `gc_content()` function, which makes use of a `base_composition()` function that we wrote (but leave out here for clarity). In this listing, all of the variable names belonging to the function are highlighted in red, whereas other "global" variable names are highlighted in purple and underlined with a dashed line. According to Rule 1, we shouldn't see any purple/underlined variables in the function definition.

```
## Given a DNA (A,C,T,G) sequence string, returns the GC-content as float
def gc_content(seq):
    g_cont = base_composition(seq, "G")
    c_cont = base_composition(seq, "C")
    seq_len = len(seq)
    gc = (g_cont + c_cont)/float(seq_len)
    return gc

seq3 = "ACCCTAGACTG"
seq3_gc = gc_content(seq3)                          # 0.5454
```

Outside of the function, the string data "ACCCTAGACTG" was assigned to the variable seq3, but inside the function, it was assigned to the variable seq. Inside the function, the data value 0.54 was referenced by the variable gc, and outside it was referenced by seq3_gc.[63]

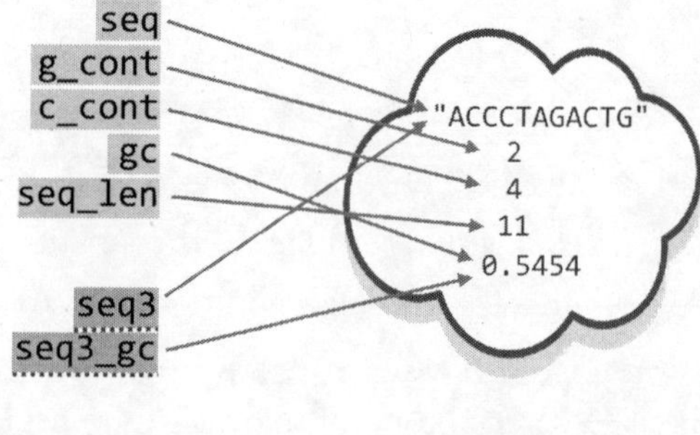

We know that we *could* access the variable seq3 from inside the function, even if we shouldn't. But what about the other way around? Could we access the gc variable, which was defined within the function, outside the function?

```
## Given a DNA (A,C,T,G) sequence string, returns the GC-content as float
def gc_content(seq):
    g_cont = base_composition(seq, "G")
    c_cont = base_composition(seq, "C")
    seq_len = len(seq)
    gc = (g_cont + c_cont)/float(seq_len)
    return gc

seq3 = "ACCCTAGACTG"
seq3_gc = gc_content(seq3)                          # 0.5454
print(gc)                # Error!
```

63. Many languages work like Python and allow different variables to refer to the same data, but in some languages this is impossible or less common. In some languages, all data are immutable, making the point moot. R is an example where most variables end up referring to unique data; see the box in chapter 29, "R Functions," for a more detailed comparison.

In fact, we cannot. The above `print(gc)` line, even after the computation of `seq3_gc`, would produce a `NameError: name 'gc' is not defined`. This error occurs because variables have *scope*, and the `gc` variable's scope is limited, from the time it is defined to the end of the function block in which it is defined (all of the red-highlighted variables share this feature). Variables with limited scope are called *local variables*. A variable's *scope* is the context in which can be used. It defines "how long the variable lives," or "where the variable exists." A variable with limited scope, as in the execution of a function, is called a *local variable*.

After the function call ends, the picture would look more like so:

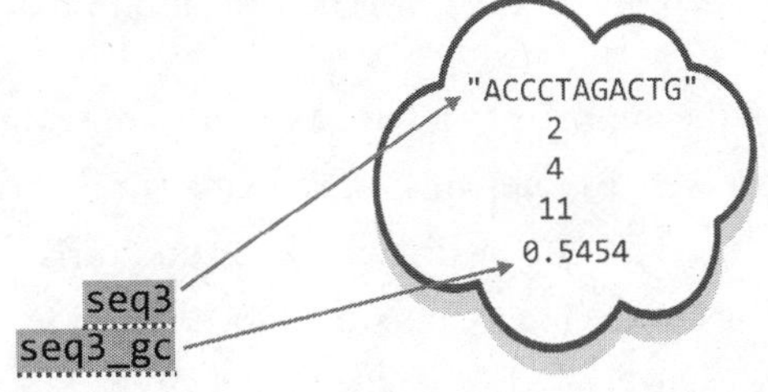

Eventually, the garbage collector will notice that no variables refer to the data `2`, `4`, and `11`, and it will clear those out of RAM to make room for new data.

Because scope talks about variables, it has little to do with the data to which a variable refers—except to say that data are garbage collected when there are no variables referring to that data in any current scope. Nevertheless, because the `gc` variable is not in scope after the function ends, we can't use it outside of the function block.

One of the particularly interesting and powerful features of local variables is that they exist independently of any other variables that might happen to exist at the time with the same name (sometimes they are said to "shadow" other variables with the same name). Consider the following code:

```
## Given a DNA (A,C,T,G) sequence string, returns the GC-content as float
def gc_content(seq):
    g_cont = base_composition(seq, "G")
    c_cont = base_composition(seq, "C")
    seq_len = len(seq)
    gc = (g_cont + c_cont)/float(seq_len)
    return gc

gc = "Measure of G and C bases in a sequence"
seq3 = "ACCCTAGACTG"
seq3_gc = gc_content(seq3)                          # 0.5454
print(gc)
```

What do you think the last line, `print(gc)`, will print? If you guessed `"Measure of G and C bases in a sequence"`, you are right! Even though the `gc_content()` function defines its own `gc` variable, because this variable is local, it does not "step on" the outside `gc` variable. In summary:

- Variables defined within a function (and parameters to a function) are local to the function call.
- These local variables "shadow" any other variables that might exist during the execution of the function.
- When the function call ends, the local variables are forgotten.

One of the important lessons of scope is knowing when it might lead to trouble. This is easiest to see if we were to compare how scoping works with another language, C++ in this case. In C++, variables are declared with their types, and variables declared within an if-statement block (or any other block) are also local to that block. This means that they *go out of scope* when the block ends. If we were to try and use them outside of the block in which they were declared, an error would occur.

In Python, on the other hand, variables declared in if-statements, for-loop blocks, and while-loop blocks are not local variables, and stay in scope outside of the block. Thus we say that C++ has "block-level" scoping, while Python uses only "function-level" scoping. The brackets in the figure below illustrate the scope for some variables in equivalent C++ and Python code. Because C++ uses block-level scoping, and `x` and `y` are declared inside of blocks, we cannot access their contents outside of those blocks.

```
/* C++ Example Code */
int x = 5;
if(x < 10) {
    int y = 20;
    if(x < 30) {
        int z = y*2;
    }
}

/* Print Contents of Variables */
cout << x << endl;  /* Ok, prints 5 */
cout << y << endl;  /* Error! */
cout << z << endl;  /* Error! */
```



```
# Python Example Code
x = 5
if x < 10:
    y = 20
    if x < 30:
        z = y * 2

# Print Contents of Variables
print(x)  # Ok, prints 5
print(y)  # Ok, prints 20
print(z)  # Ok, prints 40
```


This might seem like an advantage for Python, but in this case we think that that C++ may have it right. After all, if we *needed* for the `y` and `z` variables to be accessible after the if-blocks in the C++

example, we could have declared them along with `x`. We might even have given them a default value to hold using declarations like `int y = -1` and `int z = -1`; if these variables still hold `-1` after the if-blocks are run, then we know that those internal lines didn't execute. This would be especially helpful if the value of `x` wasn't set to five as it is here, but rather is dependent on the input data.

In Python, on the other hand, code like this might cause a problem. If the `x` variable is dependent on the input data, then the `y` and `z` variables might never be set at all, leading to an eventual `NameError` when they are later accessed. We certainly shouldn't consider any program that sometimes produces an error, depending on the input data, to be very good!

Probably the best way around this sort of problem is to pretend that Python uses block-level scoping, declaring variables in the widest block/level in which they will be used. `None` is a good default value that we can later check for if we want to conditionally execute later code on the basis of whether variables were actually set to something useful:

```
# Python example code
x = 5
y = None
z = None
if x < 10:
    y = 20
    if x < 30:
     z = y*2

# print contents of variables
print(x)
if x != None:
    print(y)
if z != None:
    print(z)
```

This code will not crash, no matter the value of x. Setting initial values for these variables to `None` before the if-blocks also provides a reminder and visual clue that these variables are intended to be used outside of the nested if-blocks, and that we should check their contents before using them later.

C++ and other block-level-scoped languages thus encourage "short-lived" variables, which is a good programming mantra. Defining variables only when needed and using them for as short a time as possible helps to produce clearer, more modular, and even more efficient code (because short-lived variables allow the garbage collector to clean out more data). In a way, breaking Rule 1 of functions is a similar type of abuse of variable scope.

Beginning programmers often find it easier to avoid these conventions by setting a large number of variables near the beginning of a program and then accessing and setting them at various points throughout the program. This is sometimes colloquially called "spaghetti code," because the long-distance connections created between the different areas of code resemble spaghetti. Rarely does this strategy pay off, however.

As discussed in chapter 25, "Algorithms and Data Structures," local variables and similar concepts allow us to solve complex problems in fascinating and elegant ways.

Exercises

1. Create a RAM/variable diagram that illustrates the following segment of code, and use it to explain the printed output.

```
def append_min(input_list):
    smallest = min(input_list)
    input_list.append(smallest)

    return input_list

nums = [4, 1, 5, 6, 7]
numsb = append_min(nums)
print(nums)
print(numsb)
```

2. Create a RAM/variable diagram that illustrates the following segment of code, and use it to explain the printed output.

```
def append_min(input_list):
    smallest = min(input_list)
    new_list = list()
    for el in input_list:
        new_list.append(el)

    new_list.append(smallest)

    return new_list

nums = [4, 1, 5, 6, 7]
numsb = append_min(nums)
print(nums)
print(numsb)
```

3. Step through the following code, and create a RAM/variable diagram for each time a "#point" is reached. (Because #point 1 is reached twice, that will mean four diagrams.) You can assume that data are never garbage collected, but that local variables do disappear.

```
def base_composition(seq, query_base):
    answer = 0
    seq_len = len(seq)

    for index in range(0, seq_len):
        seq_base = seq[index]
        if seq_base == query_base:
            answer = answer + 1

    # point 1
    return answer

def gc_content(seq):
    g_cont = base_composition(seq, "G")
    c_cont = base_composition(seq, "C")
    seq_len = len(seq)

    # point 2
    answer = (g_cont + c_cont)/float(seq_len)
    return answer


answer = "Computing answer..."
seq = "ACCCTAGACTG"
seq_gc = gc_content(seq)

# point 3
print(answer)
```

4. What is returned by a function in Python if no return statement is specified?

Chapter 23

Objects and Classes

This book, when initially introducing Python, mentioned some of the features of the language, such as its emphasis on "one best way" and readable code. Python also provides quite a bit of built-in functionality through importable modules such as `sys`, `re`, and `subprocess`.

Python provides another advantage: it is naturally "object oriented," even though we haven't discussed this point yet. Although there are competing paradigms for the best way to architect programs, the object-oriented paradigm is commonly used for software engineering and large-project management.[64] Even for small programs, though, the basic concepts of object orientation can make the task of programming easier.

An *object*, practically speaking, is a segment of memory (RAM) that references both data (referred to by *instance variables*) of various types and associated functions that can operate on the data.[65] Functions belonging to objects are called methods. Said another way, a *method* is a function that is associated with an object, and an *instance variable* is a variable that belongs to an object.

In Python, objects are the data that we have been associating with variables. What the methods are, how they work, and what the data are (e.g., a list of numbers, dictionary of strings, etc.) are defined by a *class*: the collection of code that serves as the "blueprint" for objects of that type and how they work.

64. Another prominent methodology for programming is the "functional" paradigm, wherein functions are the main focus and data are usually immutable. While Python also supports functional programming, we won't focus on this topic. On the other hand, the R programming language emphasizes functional programming, so we'll explore this paradigm in more detail in later chapters.

65. This definition is not precise, and in fact intentionally misrepresents how objects are stored in RAM. (In reality, all objects of the same type share a single set of functions/methods.) But this definition will serve us well conceptually.

Thus the class (much like the blueprint for a house) defines the structure of objects, but each object's instance variables may refer to different data elements so long as they conform to the defined structure (much like how different families may live in houses built from the same blueprint). In Python, each piece of data we routinely encounter constitutes an object. Each data type we've dealt with so far (lists, strings, dictionaries, and so on) has a class definition—a blueprint—that defines it. For example, lists have data (numbers, strings, or any other type) and methods such as `.sort()` and `.append()`.

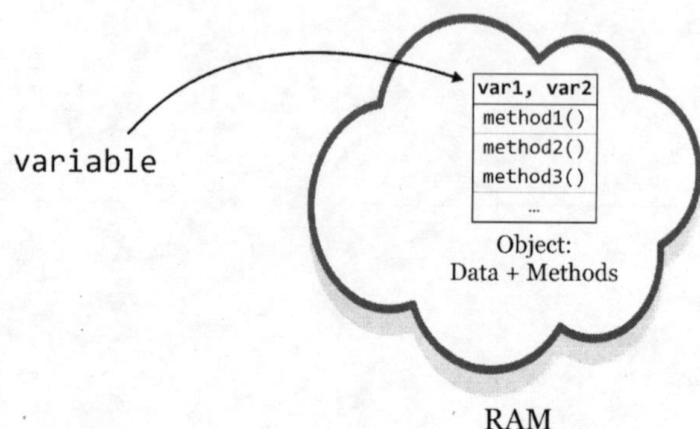

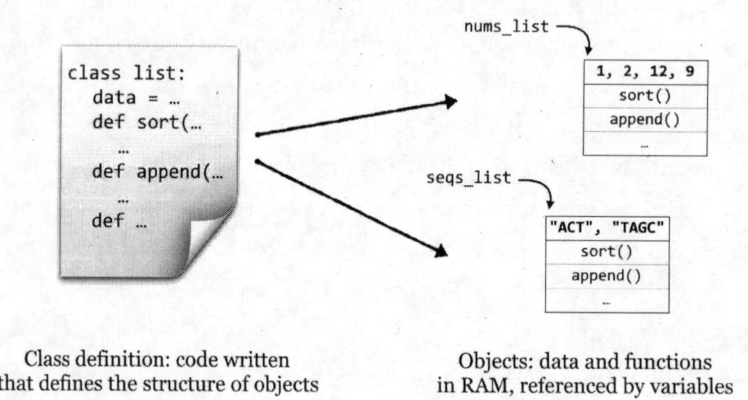

In a sense, calling object methods makes a request of the object: `nums_list.sort()` might be interpreted as "object referred to by `nums_list`, please run your `sort()` method." Upon receiving this message, the object will reorder its data.[66]

66. Although some have criticized the anthropomorphization of objects this way, it's perfectly fine—so long as we always say "please!"

Creating New Classes

Definitions for Python classes are just blocks of code, indicated by an additional level of indentation (like function blocks, if-statement blocks, and loop blocks). Each class definition requires three things, two of which we are already familiar with:

1. Methods (functions) that belong to objects of the class.
2. Instance variables referring to data.
3. A special method called a *constructor*. This method will be called automatically whenever a new object of the class is created, and must have the name `__init__`.

One peculiarity of Python is that each method of an object must take as its first argument a parameter called `self`,[67] which we use to access the instance variables. Let's start by defining a class, `Gene` (class names traditionally begin with a capital letter): each `Gene` object will have (1) an `id` (string) and (2) a `sequence` (also a string). When creating a `Gene` object, we should define its `id` and `sequence` by passing them as parameters to the `__init__` method.

Outside of the block defining the class, we can make use of it to create and interact with `Gene` objects.

67. This first parameter doesn't technically need to be named `self`, but it is a widely accepted standard.

```
#!/usr/bin/env python

class Gene:
    def __init__(self, creationid, creationseq):
        print("I'm a new Gene object!")
        print("My constructor got a param: " + str(creationid))
        print("Assigning that param to my id instance variable...")
        self.id = creationid
        print("Similarly, assigning to my sequence instance variable...")
        self.sequence = creationseq

    def print_id(self):
        print("My id is: " + str(self.id))

    def print_len(self):
        print("My sequence len is: " + str(len(self.sequence)))


print("\n***   Creating geneA:")
geneA = Gene("AY342", "CATTGAC")

print("\n***   Creating geneB:")
geneB = Gene("G54B", "TTACTAGA")

print("\n***   Asking geneA to print_id():")
geneA.print_id()

print("\n***   Asking geneB to print_id():")
geneB.print_id()

print("\n***   Asking geneA to print_len():")
geneA.print_len()
```

(Normally we don't include `print()` calls in the constructor; we're doing so here just to clarify the object creation process.) Executing the above:

```
***   Creating geneA:
I'm a new Gene object!
My constructor got a param: AY342
Assigning that param to my id instance variable...
Similarly, assigning to my sequence instance variable...

***   Creating geneB:
I'm a new Gene object!
My constructor got a param: G54B
Assigning that param to my id instance variable...
Similarly, assigning to my sequence instance variable...

***   Asking geneA to print_id():
My id is: AY342

***   Asking geneB to print_id():
My id is: G54B

***   Asking geneA to print_len():
My sequence len is: 7
```

Note that even though each method (including the constructor) takes as its first parameter `self`, we don't specify this parameter when calling methods for the objects. (For example, `.print_id()` takes a `self` parameter that we don't specify when calling it.) It's quite common to forget to include this "implicit" `self` parameter; if you do, you'll get an error like `TypeError: print_id() takes no arguments (1 given)`, because the number of parameters taken by the method doesn't match the number given when called. Also, any parameters sent to the creation function (`Gene("AY342", "CATTGAC")`) are passed on to the constructor (`__init__(self, creationid, creationseq)`).

What is `self`? The `self` parameter is a variable that is given to the method so that the object can refer to "itself." Much like other people might refer to you by your name, you might refer to yourself as "self," as in "self: remember to resubmit that manuscript tomorrow."

Interestingly, in some sense, the methods defined for classes are breaking the first rule of functions: they are accessing variables that aren't passed in as parameters! This is actually all right. The entire point of objects is that they hold functions *and* data that the functions can always be assumed to have direct access to.

Let's continue our example by adding a method that computes the GC content of the `self.sequence` instance variable. This method needs to be included in the block defining the class; notice that a method belonging to an object can call another method belonging to itself, so we can compute GC content as a pair of methods, much like we did with simple functions:

```
    # ... (inside class Gene:)

    def print_len(self):
        print("My sequence len is: " + str(len(self.sequence)))

    def base_composition(self, base):
        base_count = 0
        for index in range(0, len(self.sequence)):
            base_i = self.sequence[index]
            if base_i == base:
                base_count = base_count + 1
        return base_count

    def gc_content(self):
        g_count = self.base_composition("G")
        c_count = self.base_composition("C")
        return (g_count + c_count)/float(len(self.sequence))

print("\n***   Creating geneA:")
geneA = Gene("AY342", "CATTGAC")

# ...

print("\n***   Asking geneA to return its T content:")
geneA_t = geneA.base_composition("T")
print(geneA_t)


print("\n***   Asking geneA to return its GC content:")
geneA_gc = geneA.gc_content()
print(geneA_gc)
```

Resulting in the output:

```
***   Asking geneA to return its T content:
2

***   Asking geneA to return its GC content:
0.428571428571
```

It can also be useful to write methods that let us get and set the instance variables of an object. We might add to our class definition methods to get and set the sequence, for example, by having the methods refer to the `self.seq` instance variable.

```
    # ... (inside class Gene:)

    def gc_content(self):
        g_count = self.base_composition("G")
        c_count = self.base_composition("C")
        return (g_count + c_count)/float(len(self.sequence))

    def get_seq(self):
        return self.sequence

    def set_seq(self, newseq):
        self.sequence = newseq

print("***   Creating geneA:")
geneA = Gene("AY342", "CATTGAC")

# ...
```

We could make use of this added functionality later in our code with a line like `print("gene A's sequence is " + geneA.get_seq())` or `geneA.set_seq("ACTAGGGG")`.

Although methods can return values (as with `.base_composition()` and `.gc_content()`) and perform some action that modifies the object (as with `.set_seq()`), the principle of command-query separation states that they shouldn't do both unless it is absolutely necessary.

Is it possible for us to modify the instance variables of an object directly? It makes sense that we can; because the gene object's name for itself is `self` and sets its sequence via `self.sequence`, we should be able to set the gene object's sequence using our name for it, `geneA`. In fact, `geneA.sequence = "ACTAGGGG"` would have the same result as calling `geneA.set_seq("ACTAGGGG")`, as defined above.

So why might we want to use "getter" and "setter" methods as opposed to directly modifying or reading an object's instance variables? The difference is related a bit to politeness—if not to the object itself, then to whomever wrote the code for the class. By using methods, we are *requesting* that the object change its sequence data, whereas directly setting instance variables just reaches in and changes it—which is a bit like performing open-heart surgery without the patient's permission!

This is a subtle distinction, but it's considered serious business to many programmers. To see why, suppose that there are many methods that won't work at all on RNA sequences, so we must make sure that the `sequence` instance variable never has any `U` characters in it. In this case, we could have the `.set_seq()` method decide whether or not to accept the sequence:

```
def set_seq(self, newseq):
    if newseq.base_composition("U") != 0:
        print("Sorry, no RNA allowed.")
    else:
        self.sequence = newseq
```

Python has an `assert` statement for this sort of error checking. Like a function, it takes two parameters, but unlike a function, parentheses are not allowed.

```
def set_seq(self, newseq):
    assert newseq.base_composition("U") == 0, "Sorry, no RNA allowed."
    self.sequence = newseq
```

When using an assert, if the check doesn't evaluate to `True`, then the program will stop and report the specified error. The complete code for this example can be found in the file **`gene_class.py`**.

Using methods when working with objects is about encapsulation and letting the objects do as much work as possible. That way, they can ensure correct results so that you (or whomever you are sharing code with, which might be "future you") don't have to. Objects can have any number of instance variables, and the methods may access and modify them, but it's a good idea to ensure that all instance variables are left in a coherent state for a given object. For example, if a `Gene` object has an instance variable for the sequence, and another holding its GC content, then the GC content should be updated whenever the sequence is. Even better is to compute such quantities as needed, like we did above.[68]

The steps for writing a class definition are as follows:

1. Decide what concept or entity the objects of that class will represent, as well as what data (instance variables) and methods (functions) they will have.
2. Create a constructor method and have it initialize all of the instance variables, either with parameters passed into the constructor, or as empty (e.g., something like `self.id = ""` or `self.go_terms = list()`). Although instance variables can be created by any method, having them all initialized in the constructor provides a quick visual reference to refer to when coding.
3. Write methods that set or get the instance variables, compute calculations, call other methods

68. One might ask whether a `Gene` object would ever need to allow its sequence to change at all. One possible reason would be if we were simulating mutations over time; a method like `.mutate(0.05)` might ask a gene to randomly change 5% of its bases.

or functions, and so on. Don't forget the `self` parameter!

Exercises

1. Create a program `objects_test.py` that defines and uses a class. The class can be anything you want, but it should have at least two methods (other than the constructor) and at least two instance variables. One of the methods should be an "action" that you can ask an object of that class to perform, and the other should return an "answer."

   Instantiate your class into at least two objects, and try your methods on them.

2. Once defined, classes (and objects of those types) can be used anywhere, including other class definitions. Write two class definitions, one of which contains multiple instances of the other. For example, instance variables in a `House` object could refer to several different `Room` objects. (For a more biologically inspired example, a `Gene` object could have a `self.exons` that contains a list of `Exon` objects.)

   The example below illustrates this more thoroughly, but having some practice first will be beneficial.

3. If classes implement some special methods, then we can compare objects of those types with `==`, `<`, and the other comparison operators.

   When comparing two `Gene` objects, for example, we might say that they are equal if their sequences are equal, and `geneA` is less than `geneB` if `geneA.seq < geneB.seq`. Thus we can add a special method `__eq__()`, which, given the usual `self` and a reference to another object of the same type called `other`, returns `True` if we'd consider the two equal and `False` otherwise:

   ```
   def __eq__(self, other):
       if self.seq == other.get_seq():
           return True
       return False
   ```

   We can also implement an `__lt__()` method for "less than":

   ```
   def __lt__(self, other):
       if self.seq < other.get_seq():
           return True
       return False
   ```

With these, Python can work out how to compare `Gene` objects with < and ==. The other comparisons can be enabled by defining `__le__()` (for <=), `__gt__()` (for >), `__ge__()` (for >=) and `__ne__()` (for !=).

```
geneA = Gene("XY6B", "CATGATA")
geneB = Gene("LQQ5", "CATGATA")

print(geneA.__lt__(geneB))  # False
# same as:
print(geneA < geneB)        # False

print(geneA.__eq__(geneB))  # True
# same as:
print(geneA == geneB)       # True
```

Finally, if we have a list of `Gene` objects `genes_list` which define these comparators, then Python can sort according to our comparison criteria with `genes_list.sort()` and `sorted(genes_list)`.

Explore these concepts by defining your own ordered data type, implementing `__eq__()`, `__lt__()`, and the other comparison methods. Compare two objects of those types with the standard comparison operators, and sort a list of them. You might also try implementing a `__repr__()` method, which should return a string representing the object, enabling `print()` (as in `print(geneA)`).

Counting SNPs

As it turns out, multiple classes can be defined that interact with each other: instance variables of a custom class may refer to custom object types. Consider the file **`trio.subset.vcf`**, a VCF (variant call format) file for describing single-nucleotide polymorphisms (SNPs, pronounced "snips") across individuals in a group or population. In this case, the file represents a random sampling of SNPs from three people—a mother, a father, and their daughter—compared to the reference human genome.[69]

69. This file was obtained from the 1000 Human Genomes Project at ftp://ftp.1000genomes.ebi.ac.uk/vol1/ftp/pilot_data/release/2010_07/trio/snps/, in the file `CEU.trio.2010_03.genotypes.vcf.gz`. After decompressing, we sampled 5% of the SNPs at random with `awk '{if($1 ~ "##" || rand() < 0.05) {print $0}}' CEU.trio.2010_03.genotypes.vcf > trio.sample.vcf`; see chapter 10, "Rows and Columns" for details.

```
##fileformat=VCFv4.0
##INFO=<ID=AA,Number=1,Type=String,Description="Ancestral Allele, ftp://ftp.1000
##INFO=<ID=DP,Number=1,Type=Integer,Description="Total Depth">
##INFO=<ID=HM2,Number=0,Type=Flag,Description="HapMap2 membership">
##INFO=<ID=HM3,Number=0,Type=Flag,Description="HapMap3 membership">
##reference=human_b36_both.fasta
##FORMAT=<ID=GT,Number=1,Type=String,Description="Genotype">
##FORMAT=<ID=GQ,Number=1,Type=Integer,Description="Genotype Quality">
##FORMAT=<ID=DP,Number=1,Type=Integer,Description="Read Depth">
#CHROM  POS     ID          REF     ALT     QUAL    FILTER  INFO    FORMAT  NA12891
1       799739  rs57181708      A       G       .       PASS    AA=-;DP=141
1       805678  .       A       T       .       PASS    AA=a;DP=185     GT:GQ:DP
1       842827  rs4970461       T       G       .       PASS    AA=G;DP=114
1       847591  rs6689107       T       G       .       PASS    AA=G;DP=99
1       858267  rs13302914      C       T       .       PASS    AA=.;DP=84
1       877161  .       C       T       .       PASS    AA=.;DP=89      GT:GQ:DP
1       892860  rs7524174       G       A       .       PASS    AA=G;DP=105
1       917172  rs2341362       T       C       .       PASS    AA=t;DP=133;HM3
...
```

This file contains a variety of information, including header lines starting with # describing some of the coding found in the file. Columns 1, 2, 3, 4, and 5 represent the chromosome number of the SNP, the SNP's position on the chromosome, the ID of the SNP (if it has previously been described in human populations), the base present in the reference at that position, and an alternative base found in one of the three family members, respectively. Other columns describe various information; this file follows the "VCF 4.0" format, which is described in more detail at http://www.1000genomes.org/node/101. Some columns contain a `.` entry, which indicates that the information isn't present; in the case of the ID column, these represent novel polymorphisms identified in this trio.

For this example, we are interested in the first five columns, and the main questions are:

- How many transitions (A vs. G or C vs. T) are there within the data for each chromosome?
- How many transversions (anything else) are there within the data for each chromosome?

We may in the future have other questions about transitions and transversions on a per-chromosome basis. To answer the questions above, and to prepare for future ones, we'll start by defining some classes to represent these various entities. This example will prove to be a bit longer than others we've studied, partially because it allows us to illustrate answering multiple questions using the same codebase if we do some extra work up front, but also because object-oriented designs tend to result in significantly more code (a common criticism of using classes and objects).

SNP Class

A SNP object will hold relevant information about a single nonheader line in the VCF file. Instance variables would include the reference allele (a one-character string, e.g., `"A"`), the alternative allele (a one-character string, e.g., `"G"`), the name of the chromosome on which it exists (a string, e.g., `"1"`), the reference position (an integer, e.g., `799739`), and the ID of the SNP (e.g., `"rs57181708"` or `"."`). Because we'll be parsing lines one at a time, all of this information can be provided in the constructor.

SNP objects should be able to answer questions: `.is_transition()` should return `True` if the SNP is a transition and `False` otherwise by looking at the two allele instance variables. Similarly, `.is_transversion()` should return `True` if the SNP is a transversion and `False` otherwise.

Chromosome Class

A `Chromosome` object will hold data for an individual chromosome, including the chromosome name (a string, e.g., `"1"`), and all of the SNP objects that are located on that chromosome. We could store the SNP objects in a list, but we could also consider storing them in a dictionary, which maps SNP locations (integers) to the SNP objects. Then we can not only gain access to the list of SNPs (using the dictionary's `.values()` method) or the list of locations (using the dictionary's `.keys()` method), but also, given any location, we can get access to the SNP at that location. (We can even use `.has_key()` to determine whether a SNP exists at a given location.)

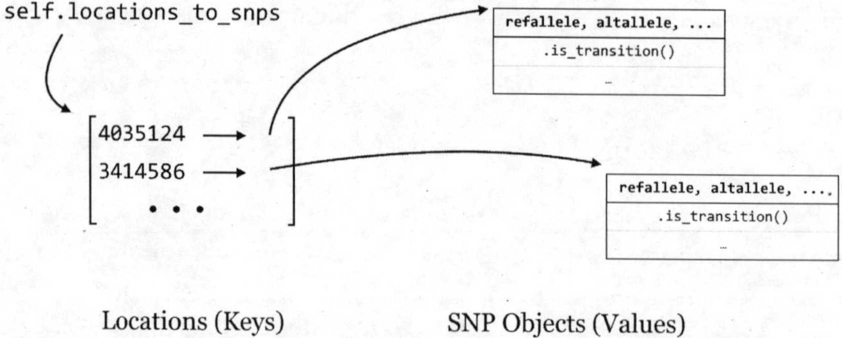

The chromosome constructor will initialize the name of the chromosome as `self.chrname`, but the `snps` dictionary will start as empty.

A `Chromosome` object should be able to answer questions as well: `.count_transitions()` should tell us the number of transition SNPs, and `.count_transversions()` should return the number of transversion SNPs. We're also going to need some way to *add* a SNP object to a chromosome's SNP dictionary because it starts empty. We'll accomplish this with an `.add_snp()` method, which will take all of the information for a SNP, create the new SNP object, and add it to the dictionary. If a SNP already exists at that location, an error should occur, because our program shouldn't accept VCF files that have multiple rows with the same position for the same chromosome.

For overall strategy, once we have our classes defined (and debugged), the "executable" portion of our program will be fairly simple: we'll need to keep a collection of `Chromosome` objects that we can interact with to add SNPs, and at the end we'll just loop through these and ask each how many transitions and transversions it has. It makes sense to keep these `Chromosome` objects in a dictionary as well, with the keys being the chromosome names (strings) and the values being the `Chromosome` objects. We'll call this dictionary `chrnames_to_chrs`.

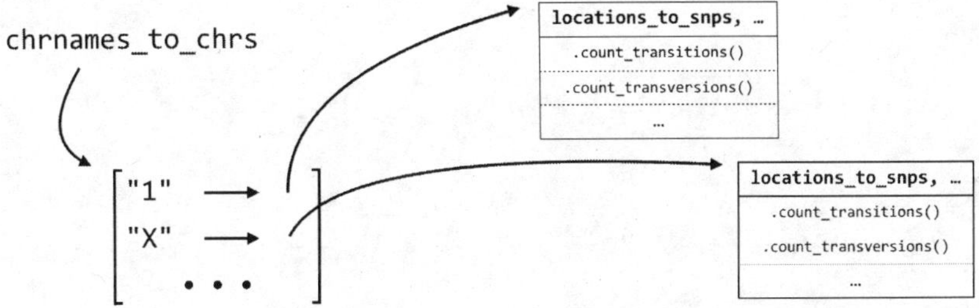

Chromosome Names (Keys) Chromosome Objects (Values)

As we loop through each line of input (reading a file name given in `sys.argv[1]`), we'll split it apart and check whether the chromosome name is in the dictionary with `.has_key()`. If so, we'll ask the object in that slot to add a SNP with `.add_snp()`. If not, then we'll need to first create a new `Chromosome` object, ask it to `.add_snp()`, and finally add it to the dictionary. Of course, all of this should happen only for nonheader lines.

We'll start with the SNP class, and then the `Chromosome` class. Although it is difficult to show here, it's a good idea to work on and debug each method in turn (with occasional `print()`

statements), starting with the constructors. Because a SNP is only a SNP if the reference and alternative allele differ, we'll `assert` this condition in the constructor so an error is produced if we ever try to create a nonpolymorphic SNP (which wouldn't actually be a SNP at all).

```
#!/usr/bin/env python
## Imports we are likely to need:
import io
import sys
import re

## A class representing simple SNPs
class SNP:
    def __init__(self, chrname, pos, snpid, refallele, altallele):
        assert refallele != altallele, "Error: ref == alt at pos " + str(pos)
        self.chrname = chrname
        self.pos = pos
        self.snpid = snpid
        self.refallele = refallele
        self.altallele = altallele


    ## Returns True if refallele/altallele is A/G, G/A, C/T, or T/C
    def is_transition(self):
        if self.refallele == "G" or self.refallele == "A":
            if self.altallele == "G" or self.altallele == "A":
                return True

        if self.refallele == "C" or self.refallele == "T":
            if self.altallele == "C" or self.altallele == "T":
                return True

        return False

    ## Returns True if the snp is a transversion (ie, not a transition)
    def is_transversion(self):
        if self.is_transition():
            return False
        return True
```

Note the shortcut that we took in the above code for the `.is_transversion()` method, which calls the `.is_transition()` method and returns the opposite answer. This sort of "shortcut" coding has its benefits and downsides. One benefit is that we can reuse methods rather than copying and pasting to produce many similar chunks of code, reducing the potential surface area for bugs to occur. A downside is that we have to be more careful—in this case, we've had to ensure that the alleles differ (via the `assert` in the constructor), so a SNP must be either a transition or transversion. (Is this actually true? What if someone were to attempt to create a SNP object with non-DNA characters? It's always wise to consider ways code could be inadvertently misused.)

The above shows the start of the script and the SNP class; code like this could be tested with just a few lines:

```
## transition test; should not result in "Failed Test"
snp1 = SNP("1", 12351, "rs11345", "C", "T")
assert snp1.is_transition() == True, "Failed Test"      ## Does not error

## transversion test; should not result in "Failed Test"
snp2 = SNP("1", 36642, "rs22541", "A", "T")
assert snp2.is_transversion() == True, "Failed Test"    ## Does not error

## error test; should result in "Error: ref == alt at pos 69835"
snp3 = SNP("1", 69835, "rs53461", "A", "A")                ## Results in error
```

Although we won't ultimately leave these testing lines in, they provide a good sanity check for the code. If these checks were wrapped in a function that could be called whenever we make changes to the code, we would have what is known as a *unit test*, or a collection of code (often one or more functions), with the specific purpose of testing functionality of other code for correctness.[70] These can be especially useful as code changes over time.

Let's continue on with the `Chromosome` class. Note that the `.add_snp()` method contains assertions that the SNP location is not a duplicate and that the chromosome name for the new SNP matches the chromosome's `self.chrname`.

70. Unit tests are often automated; in fact, Python includes a module for building unit tests called `unittest`.

```
# ...

## A class representing a chromosome, which has a collection of SNPs
class Chromosome:
    def __init__(self, chrname):
        self.chrname = chrname
        self.locations_to_snps = dict()

    ## Returns the chromosome name
    def get_name(self):
        return self.name


    ## Given all necessary information to add a new SNP, create
    ## a new SNP object and add it to the SNPs dictionary. If
    ## a SNP already exists at that location, or
    ## the given chrname doesn't match self.chrname, an error is reported.
    def add_snp(self, chrname, pos, snpid, refallele, altallele):
        ## If there is already an entry for that SNP, throw an error
        open_location = not(self.locations_to_snps.has_key(pos))
        assert open_location, "Duplicate SNP: " + self.chrname + ":" + str(pos)

        ## If the chrname doesn't match self.chrname, throw an error
        assert chrname == self.chrname, "Chr name mismatch!"

        ## Otherwise, create the SNP object and add it to the dictionary
        newsnp = SNP(chrname, pos, snpid, refallele, altallele)
        self.locations_to_snps[pos] = newsnp
```

Now we can write the methods for `.count_transitions()` and `.count_transversions()`. Because we've ensured that each SNP object is either a transition or a transversion, and no locations are duplicated within a chromosome, the `.count_transversions()` method can make direct use of the `.count_transitions()` method and the total number of SNPs stored via `len(self.locations_to_snps)`. (Alternatively, we could make a `count_transversions()` that operates similarly to `count_transitions()` by looping over all the SNP objects.)

```
# ... (inside class Chromosome:)

## Returns the number of transition snps stored in this chromosome
def count_transitions(self):
    count = 0

    locations = self.locations_to_snps.keys()
    for location in locations:
        snp = self.locations_to_snps[location]
        if snp.is_transition():
            count = count + 1

    return count

## Returns the number of transversion snps stored in this chromosome
def count_transversions(self):
    total_snps = len(self.locations_to_snps)
    return total_snps - self.count_transitions()
```

The corresponding test code is below. Here we are using `assert` statements, but we could also use lines like `print(chr1.count_transitions())` and ensure the output is as expected.

```
## A test chromosome
chr1 = Chromosome("testChr")
chr1.add_snp("testChr", 24524, "rs15926", "G", "T")
chr1.add_snp("testChr", 62464, "rs61532", "C", "T")

## These should not fail:
assert chr1.count_transitions() == 1, "Failed Test"
assert chr1.count_transversions() == 1, "Failed Test"

## This should fail with a "Duplicate SNP" error:
chr1.add_snp("testChr", 24524, "rs88664", "A", "C")
```

With the class definitions created and debugged, we can write the "executable" part of the program, concerned with parsing the input file (from a filename given in `sys.argv[1]`) and printing the results. First, the portion of code that checks whether the user has given a file name (and produces some help text if not) and reads the data in. Again, we are storing a collection of `Chromosome` objects in a `chrnames_to_chrs` dictionary. For each VCF line, we determine whether a `Chromosome` with that name already exists: if so, we ask that object to `.add_snp()`. If not, we create a new `Chromosome` object, ask it to `.add_snp()`, and add it to the dictionary.

```
# ...

## Check usage syntax, read filename
if len(sys.argv) != 2:
    print("This program parses a VCF 4.0 file and counts")
    print("transitions and transversions on a per-chromosome basis.")
    print("")
    print("Usage: ./snps_ex.py <input_vcf_file>")
    quit()

filename = sys.argv[1]


## Create chrnames_to_chrs dictionary, parse the input file
chrnames_to_chrs = dict()
fhandle = io.open(filename, "rU")

for line in fhandle:
    # don't attempt to parse header lines (^# matches # at start of string)
    if not(re.search(r"^#", line)):
        line_stripped = line.strip()
        line_list = re.split(r"\s+", line_stripped)

        chrname = line_list[0]      # e.g. "1"
        pos = int(line_list[1])     # e.g. 25135
        snpid = line_list[2]        # e.g. "rs4970461" or "."
        refallele = line_list[3]    # e.g. "A"
        altallele = line_list[4]    # e.g. "C"

        ## Put the data in the dictionary
        if chrnames_to_chrs.has_key(chrname):
            chr_obj = chrnames_to_chrs[chrname]
            chr_obj.add_snp(chrname, pos, snpid, refallele, altallele)
        else:
            chr_obj = Chromosome(chrname)
            chr_obj.add_snp(chrname, pos, snpid, refallele, altallele)
            chrnames_to_chrs[chrname] = chr_obj
```

In the `chr_obj = chrnames_to_chrs[chrname]` line above, we are defining a variable referring to the `Chromosome` object in the dictionary, and after that we are asking that object to add the SNP with `.add_snp()`. We could have combined these two with syntax like `chrnames_to_chrs[chrname].add_snp()`.

Finally, a small block of code prints out the results by looping over the keys in the dictionary, accessing each `Chromosome` object and asking it the number of transitions and transversions:

```
# ...

## Print the results!
print("chromosome" + "\t" + "transitions" + "\t" + "transversions")
for chrname in chrnames_to_chrs.keys():
    chr_obj = chrnames_to_chrs[chrname]
    trs = chr_obj.count_transitions()
    trv = chr_obj.count_transversions()
    print(chrname + "\t" + str(trs) + "\t" + str(trv))
```

We'll have to remove or comment out the testing code (particularly the tests we expected to fail) to see the results. But once we do that, we can run the program (called **snps_ex.py**).

```
oneils@atmosphere ~/apcb/py$ chmod +x snps_ex.py
oneils@atmosphere ~/apcb/py$ ./snps_ex.py trio.sample.vcf
chromosome     transitions     transversions
20      2656     1187
21      1773     848
22      1539     639
1      9345     4262
3      8708     4261
2      10309     5130
5      7586     3874
4      9050     4372
7      6784     3274
6      7874     3697
9      5102     2653
8      6520     3419
X      3028     1527
11      5944     2908
...
```

What we've created here is no small thing, with nearly 150 lines of code! And yet each piece of code is encapsulated in some way; even the long for-loop represents the code to parse the input file and populate the `chrnames_to_chrs` dictionary. By clearly naming our variables, methods, and classes we can quickly see what each entity does. We can reason about this program without too much difficulty at the highest level of abstraction but also delve down to understand each piece individually. As a benefit, we can easily reuse or adapt this code in a powerful way by adding or modifying methods.

An Extension: Searching for SNP-Dense Regions

Counting transitions and transversions on a per-chromosome basis for this VCF file could have been accomplished without defining classes and objects. But one of the advantages of spending some time organizing the code up front is that we can more easily answer related questions about the same data.

Suppose that, having determined the number of transitions and transversions per chromosome, we're now interested in determining the most SNP-dense region of each chromosome. There are a number of ways we could define SNP density, but we'll choose an easy one: given a region from positions *l* to *m*, the density is the number of SNPs occurring within *l* and *m* divided by the size of the region, $m - l + 1$, times 1,000 (for SNPs per 1,000 base pairs).

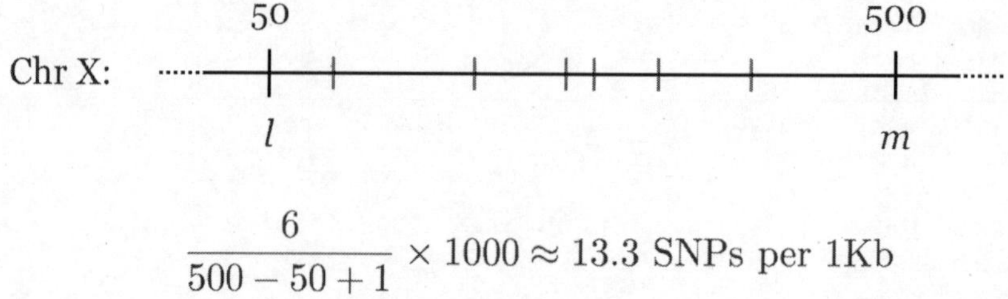

$$\frac{6}{500 - 50 + 1} \times 1000 \approx 13.3 \text{ SNPs per 1Kb}$$

For a `Chromosome` object to be able to tell us the highest-density region, it will need to be able to compute the density for any given region by counting the SNPs in that region. We can start by adding to the chromosome class a method that computes the SNP density between two positions *l* and *m*.

```
# ... (inside class Chromosome:)

## returns the number of snps between l and m, divided by
## the size of the region
def density_region(self, l, m):
    count = 0
    for location in self.locations_to_snps.keys():
        if location >= l and location <= m:
            count = count + 1

    size = m - l  + 1
    return 1000*count/float(size)
```

After debugging this method and ensuring it works, we can write a method that finds the highest-density region. But how should we define our regions? Let's say we want to consider regions of 100,000 bases. Then we might consider bases 1 to 100,000 to be a region, 100,001 to 200,000 to be a region, and so on, up until the start of the region considered is past the last SNP location. We can accomplish this with a while-loop. The strategy will be to keep information on the densest region found so far (including its density as well as start and end location), and update this answer as needed in the loop.[71]

```
# ... (inside class Chromosome:)

## given a region size, looks at non-overlapping windows
## of that size and returns a list of three elements for
## the region with the highest density:
## [density of region, start of region, end of region]
def max_density(self, region_size):
    region_start = 1
    ## default answer if no SNPs exist [density, start, end]:
    best_answer = [0.0, 1, region_size - 1]

    ## todo: implement this method
    last_snp_position = self.get_last_snp_position()
    while region_start < last_snp_position:
        region_end = region_start + region_size - 1
        region_density = self.density_region(region_start, region_end)
        # if this region has a higher density than any we've seen so far:
        if region_density > best_answer[0]:
            best_answer = [region_density, region_start, region_end]

        region_start = region_start + region_size

    return best_answer
```

In the above, we needed to access the position of the last SNP on the chromosome (so that the code could stop considering regions beyond the last SNP). Rather than write that code directly in the method, we decided that should be its own method, and marked it with a "todo" comment. So, we need to add this method as well:

71. This strategy is not the fastest we could have devised. In order to determine the highest-density region, we have to consider each region, and the computation for each region involves looping over all of the SNP positions (the vast majority of which lie outside the region). More sophisticated algorithms exist that would run much faster, but they are outside the scope of this book. Nevertheless, a slow solution is better than no solution!

```
# ... (inside class Chromosome:)

## returns the position of the last SNP known
def get_last_snp_position(self):
    locations = self.locations_to_snps.keys()
    locations.sort()
    return locations[len(locations) - 1]
```

In the code that prints the results, we can add the new call to `.max_density(100000)` for each chromosome, and print the relevant information.

```
# ...

## Print the results!
print("chromosome" + "\t" + "transitions" + "\t" + "transversions" + "\t" +
        "density" + "\t" + "region")
for chrname in chrnames_to_chrs.keys():
    chr_obj = chrnames_to_chrs[chrname]
    trs = chr_obj.count_transitions()
    trv = chr_obj.count_transversions()

    max_dens_list = chr_obj.max_density(100000)
    density = max_dens_list[0]
    region_start = max_dens_list[1]
    region_end = max_dens_list[2]
    print(chrname + "\t" + str(trs) + "\t" + str(trv) + "\t" +
            str(density) + "\t" + str(region_start) + ".." + str(region_end))
```

Let's call our new **`snps_ex_density.py`** (piping the result through `column -t` to more easily see the tab-separated column layout):

```
oneils@atmosphere ~/apcb/py$ chmod +x snps_ex_density.py
oneils@atmosphere ~/apcb/py$ ./snps_ex_density.py trio.sample.vcf | column -t
chromosome  transitions  transversions  density  region
20          2656         1187           0.22     15000001..15100000
21          1773         848            0.26     19100001..19200000
22          1539         639            0.22     47400001..47500000
1           9345         4262           0.25     105900001..106000000
3           8708         4261           0.26     166900001..167000000
2           10309        5130           0.24     225700001..225800000
5           7586         3874           0.24     8000001..8100000
4           9050         4372           0.27     162200001..162300000
```

Again, none of the individual methods or sections of code are particularly long or complex, yet together they represent a rather sophisticated analysis program.

Summary

Perhaps you find these examples using classes and objects for problem solving to be elegant, or perhaps not. Some programmers think that this sort of organization results in overly verbose and complex code. It is certainly easy to get too ambitious with the idea of classes and objects. Creating custom classes for every little thing risks confusion and needless hassle. In the end, it is up to each programmer to decide what level of encapsulation is right for the project; for most people, good separation of concepts by using classes is an art form that requires practice.

When should you consider creating a class?

- When you have many different types of data relating to the same concept, and you'd like to keep them organized into single objects as instance variables.
- When you have many different functions related to the same concept, and you'd like to keep them organized into single objects as methods.
- When you have a concept that is simple now, but you suspect might increase in complexity in the future as you add to it. Like functions, classes enable code to be reused, and it is easy to add new methods and instance variables to classes when needed.

Inheritance and Polymorphism

Despite this discussion of objects, there are some unique features of the object-oriented paradigm that we haven't covered but are sometimes considered integral to the topic. In particular, most object-oriented languages (Python included) support inheritance and polymorphism for objects and classes.

Inheritance is the idea that some types of classes may be represented as special cases of other types of classes. Consider a class defining a `Sequence`, which might have instance variables for `self.seq` and `self.id`. Sequences might be able to report their length, and so might have a `.length_bp()` method, returning `len(self.seq)`. There may also be many other operations a generic `Sequence` could support, like `.get_id()`. Now, suppose we wanted to implement an `OpenReadingFrame` class; it too should have a `self.id` and a `self.seq` and be able to report its `.length_bp()`. Because an object of this type would represent an open reading frame, it probably should also have a `.get_translation()` method returning the amino-acid translation of its `self.seq`. By using inheritance, we can define the `OpenReadingFrame` class as a type of `Sequence` class, saving us from having to re-implement `.length_bp()`—we'd only need to implement the class-specific `.get_translation()` method and any other methods would be automatically inherited from the `Sequence` class.

```
class Sequence:
    def __init__(self, gid, seq):
        self.gid = gid
        self.seq = seq

    def get_id(self):
        return self.gid

    def length_bp(self):
        return len(self.seq)

# OpenReadingFrame *inherits* from Sequence
# as well as defines an new method
class OpenReadingFrame(Sequence):
    def __init__(self, gid, seq):
        self.gid = gid
        self.seq = seq

    def get_translation(self):
        return dna_to_aa(self.seq)


geneA = Sequence("SeqA", "CATGAG")
geneB = OpenReadingFrame("SeqB", "ATGCCCTGA")
print(geneA.length_bp())    # prints 6
print(geneB.length_bp())    # prints 9
```

Polymorphism is the idea that inheriting class types don't have to accept the default methods inherited, and they are free to re-implement (or "override") specific methods even if their "parent" or "sibling" classes already define them. For example, we might consider another class called `AminoAcidSequence` that inherits from `Sequence`, so it too will have a `.get_id()` and `.length_bp()`; in this case, though, the inherited `.length_bp()` would be wrong, because `len(self.seq)` would be three times too short. So, an `AminoAcidSequence` could override the `.length_bp()` method to return `3*len(self.seq)`. The interesting feature of polymorphism is that given an object like `gene_A`, we don't even need to know what "kind" of `Sequence` object it is: running `gene_A.length_bp()` will return the right answer if it is any of these three kinds of sequence.

```
# ...

## AminoAcidSequence *inherits* from Sequence
## as well as overrides an existing method
class AminoAcidSequence(Sequence):
    def __init__(self, gid, seq):
        self.gid = gid
        self.seq = seq

    def length_bp(self):
        return(3*len(self.seq))


geneC = AminoAcidSequence("SeqC", "RQVDYW")
print(geneC.length_bp())     # prints 18
print(geneC.get_id())        # prints "SeqC"
```

These ideas are considered by many to be the defining points of "object-oriented design," and they allow programmers to structure their code in hierarchical ways (via inheritance) while allowing interesting patterns of flexibility (via polymorphism). We haven't covered them in detail here, as making good use of them requires a fair amount of practice. Besides, the simple idea of encapsulating data and functions into objects provides quite a lot of benefit in itself!

Exercises

1. Modify the `snps_ex_density.py` script to output, for each 100,000bp region of each chromosome, the percentage of SNPs that are transitions and the number of SNPs in each window. The output should be a format that looks like so:

   ```
   chromosome  region  percent_transitions num_snps
   1   1..100000    0.5646  12
   1   100001..200000  0.5214  16
   1   200001..300000  0.4513  7
   1   300001..400000  0.3126  19
   ...
   ```

   In the section on R programming (chapter 37, "Plotting Data and `ggplot2`"), we'll discover easy ways to visualize this type of output.

2. The `random` module (used with `import random`) allows us to make random choices; for example, `random.random()` returns a random float between `0.0` and `1.0`. The `random.randint(a, b)` function returns a random integer between `a` and `b` (inclusive); for example, `random.randint(1, 4)` could return `1`, `2`, `3`, or `4`. There's also a `random.choice()`

function; given a list, it returns a single element (at random) from it. So, if `bases = ["A", "C", "T", "G"]`, then `random.choice(bases)` will return a single string, either `"A"`, `"C"`, `"T"`, or `"G"`.

Create a program called `pop_sim.py`. In this program write a `Bug` class; a "bug" object will represent an individual organism with a genome, from which a fitness can be calculated. For example, if `a = Bug()`, perhaps `a` will have a `self.genome` as a list of 100 random DNA bases (e.g. `["G", "T", "A", "G", ...`; these should be created in the constructor). You should implement a `.get_fitness()` method which returns a float computed in some way from `self.genome`, for example the number of G or C bases, plus 5 if the genome contains three `"A"` characters in a row. Bug objects should also have a `.mutate_random_base()` method, which causes a random element of `self.genome` to be set to a random element from `["A", "C", "G", "T"]`. Finally, implement a `.set_base()` method, which sets a specific index in the genome to a specific base: `a.set_base(3, "T")` should set `self.genome[3]` to `"T"`.

Test your program by creating a list of 10 `Bug` objects, and in a for-loop, have each run its `.mutate_random_base()` method and print its new fitness.

3. Next, create a `Population` class. Population objects will have a list of `Bug` objects (say, 50) called `self.bug_list`.

   This `Population` class should have a `.create_offspring()` method, which will: 1) create a `new_pop` list, 2) for each element `oldbug` of `self.bug_list`: a) create a new `Bug` object `newbug`, b) and set the genome of `newbug` (one base at a time) to be the same as that of `oldbug`, c) call `newbug.mutate_random_base()`, and d) add `oldbug` and `newbug` to `new_pop`. Finally, this method should 3) set `self.bug_pop` to `new_pop`.

   The `Population` class will also have a `.cull()` method; this should reduce `self.bug_pop` to the top 50% of bug objects by fitness. (You might find the exercise above discussing `.__lt__()` and similar methods useful, as they will allow you to sort `self.bug_pop` by fitness if implemented properly.)

   Finally, implement a `.get_mean_fitness()` method, which should return the average fitness of `self.bug_pop`.

To test your code, instantiate a `p = Population()` object, and in a for-loop: 1) run `p.create_offspring()`, 2) run `p.cull()`, and 3) print `p.get_mean_fitness()`, allowing you to see the evolutionary progress of your simulation.

4. Modify the simulation program above to explore its dynamics. You could consider adding a `.get_best_individual()` method to the `Population` class, for example, or implement a "mating" scheme whereby offspring genomes are mixes of two parent genomes. You could also try tweaking the `.get_fitness()` method. This sort of simulation is known as a *genetic algorithm*, especially when the evolved individuals represent potential solutions to a computational problem.[72]

72. The idea of simulating populations "in silico" is not only quite fun, but has also produced interesting insights into population dynamics. For an example, see Hinton and Nowlan, "How Learning Can Guide Evolution," Complex Systems 1 (1987): 495–502. Simulation of complex systems using random sampling is commonly known as a Monte Carlo method. For a more fanciful treatment of simulations of natural systems, see Daniel Shiffman, The Nature of Code (author, 2012). The examples there use the graphical language available at http://processing.org, though a Python version is also available at http://py.processing.org.

Chapter 24

Application Programming Interfaces, Modules, Packages, Syntactic Sugar

We know that objects like lists have methods such as `.sort()` and `.append()`. Do they have others? They do, and one way to find them is to run Python in the interactive mode on the command line.[73]

To do this, we execute the `python` interpreter without specifying a file to interpret. The result is a Python prompt, `>>>`, where we can type individual lines of Python code and see the results.

```
[oneils@atmosphere ~/apcb/py]$ python
Python 2.7.8 (default, Aug  7 2014, 09:56:10)
[GCC 4.4.7 20120313 (Red Hat 4.4.7-4)] on linux2
Type "help", "copyright", "credits" or "license" for more information.
>>> a = 3 + 4
>>> print(a)
7
>>>
```

The interactive mode, aside from providing an interface for running quick tests of Python functionality, includes a help system! Simply run the `help()` function on either a class name (like `help(list)`) or an instance of the class.

```
>>> a = list()
>>> help(a)
```

This command opens an interactive viewer in which we can scroll to see all the methods that an object of that type provides. Here's a sample from the help page:

73. Much of this information is also available online at http://docs.python.org.

```
 |
 |  append(...)
 |      L.append(object) -- append object to end
 |
 |  count(...)
 |      L.count(value) -> integer -- return number of occurrences of value
 |
```

Browsing through this documentation reveals that Python lists have many methods, including `.append()`, `.count()`, and others. Sometimes the documentation isn't as clear as one would hope. In these cases, some experimentation may be required. Between the description above and some code tests, for example, can you determine what a list's `.count()` method does and how to use it effectively?

The set of methods and instance variables belonging to an object or class is known as its *API*, or Application Programming Interface. The API is the set of functions, methods, or variables provided by an encapsulated collection of code. APIs are an important part of programming because they represent the "user interface" for programming constructs that others have written.

Sometimes it can be useful to determine what class or "type" to which a variable refers, especially for looking up the API for objects of that type. The `type()` function comes in handy when combined with a `print()` call—it prints the type (the class) of data referenced by a variable. The result can then be investigated with `help()`.

```
oneils@atmosphere ~$ python
Python 2.7.9 (default, Feb 21 2015, 21:09:31)
[GCC 4.6.3] on linux2
Type "help", "copyright", "credits" or "license" for more information.
>>> seq = "ACTAGA"
>>> print(type(seq))
<type 'str'>
>>> help(str)
```

Modules

Consider this chunk of code from the example in chapter 23, "Objects and Classes," which makes use of the `Chromosome` and `SNP` class definitions:

```
## Create chrnames_to_chrs dictionary, parse the input file
chrnames_to_chrs = dict()
fhandle = io.open(filename, "rU")

for line in fhandle:
    # don't attempt to parse header lines (^# matches # at start of string)
    if not(re.search(r"^#", line)):
        line_stripped = line.strip()
        line_list = re.split(r"\s+", line_stripped)

        chrname = line_list[0]
        pos = int(line_list[1])
        snpid = line_list[2]
        refallele = line_list[3]
        altallele = line_list[4]

        ## Put the data in the dictionary
        if chrnames_to_chrs.has_key(chrname):
            chr_obj = chrnames_to_chrs[chrname]
            chr_obj.add_snp(chrname, pos, snpid, refallele, altallele)
        else:
            chr_obj = Chromosome(chrname)
            chr_obj.add_snp(chrname, pos, snpid, refallele, altallele)
            chrnames_to_chrs[chrname] = chr_obj
```

This segment of code takes a file name and produces a dictionary with nicely organized contents of the file. Because this functionality is succinctly defined, it makes sense to turn this into a function that takes the file name as a parameter and returns the dictionary. The code is almost exactly the same, except for being wrapped up in a function definition.

```
## Create chrnames_to_chrs dictionary, given an input VCF file name
## returns the dictionary
def vcf_to_chrnames_dict(filename):
    chrnames_to_chrs = dict()
    fhandle = io.open(filename, "rU")

    for line in fhandle:
        # don't attempt to parse header lines (^# matches # at start of string)
        if not(re.search(r"^#", line)):
            line_stripped = line.strip()
            line_list = re.split(r"\s+", line_stripped)

            chrname = line_list[0]
            pos = int(line_list[1])
            snpid = line_list[2]
            refallele = line_list[3]
            altallele = line_list[4]

            ## Put the data in the dictionary
            if chrnames_to_chrs.has_key(chrname):
                chr_obj = chrnames_to_chrs[chrname]
                chr_obj.add_snp(chrname, pos, snpid, refallele, altallele)
            else:
                chr_obj = Chromosome(chrname)
                chr_obj.add_snp(chrname, pos, snpid, refallele, altallele)
                chrnames_to_chrs[chrname] = chr_obj

    return chrnames_to_chrs
```

Later, we can call this function to do all the work of parsing a given file name:

```
filename = sys.argv[1]
chrnames_to_chrs = vcf_to_chrnames_dict(filename)

## Print the results!
print("chromosome" + "\t" + "transitions" + "\t" + "transversions")
for chrname in chrnames_to_chrs.keys():
    chr_obj = chrnames_to_chrs[chrname]
    trs = chr_obj.count_transitions()
    trv = chr_obj.count_transversions()
    print(chrname + "\t" + str(trs) + "\t" + str(trv))
```

Now, what if this function and the two class definitions were things that we wanted to use in other projects? We may wish to make a *module* out of them—a file containing Python code (usually related to a single topic, such as a set of functions or class definitions related to a particular kind of data or processing).

We've seen a number of modules already, including `io`, `re`, and `sys`. To use a module, we just need to run `import modulename`. As it turns out, modules are simply files of Python code ending in `.py`! Thus, when we use `import modulename`, the interpreter searches for a `modulename.py` containing the various function and class definitions. Module files may either exist in a system-wide location, or be present in the working directory of the program importing them.[74]

Importing the module in this way creates a namespace for the module. This namespace categorizes all of the functions and classes contained within that module so that different modules may have functions, classes, and variables with the same name without conflicting.

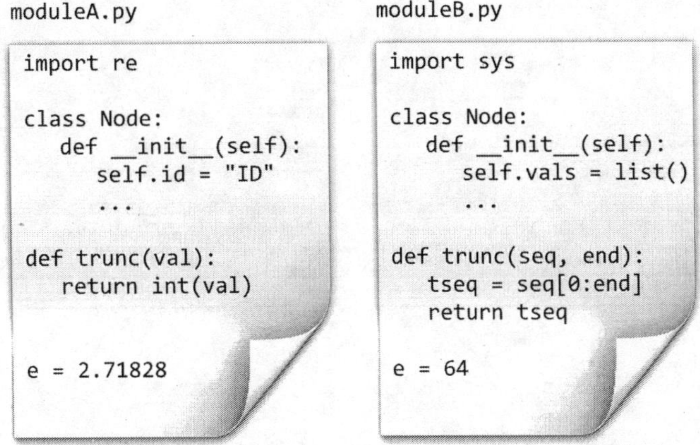

moduleA.py
import re
class Node:
def __init__(self):
self.id = "ID"
...
def trunc(val):
return int(val)
e = 2.71828
moduleB.py
import sys
class Node:
def __init__(self):
self.vals = list()
...
def trunc(seq, end):
tseq = seq[0:end]
return tseq
e = 64

If we have the above two files in the same directory as our program, source code like the following would print `4`, `64`, `"TAC"`, and `2.71828`.

```
import moduleA
import moduleB

print(moduleA.trunc(4.56))          # 4
print(moduleB.e)                    # 64
print(moduleA.e)                    # 2.71828
print(moduleB.trunc("TACTAA", 3))  # "TAC"
a_node = moduleA.Node()             # Node object defined bn moduleA
b_node = moduleB.Node()             # Node object defined by moduleB
```

74. Python modules and packages need not be installed system-wide or be present in the same directory as the program that uses them. The environment variable `$PYTHONPATH` lists the directories that are searched for modules and packages, to which each user is free to add their own search paths.

Unfortunately, Python uses the `.` operator in multiple similar ways, such as indicating a method belonging to an object (as in `a_list.append("ID3")`) and to indicate a function, variable, or class definition belonging to a namespace (as above).

Thus, as a "container" for names, the *namespace* allows different modules to name functions, classes, and variables the same way without conflicting. Just as a class may be documented by its API, so may modules: a module's API describes the functions, classes, and variables it defines. These may also be accessed through the interactive help menu (and online). Try running `import re` followed by `help(re)` to see all the functions provided by the `re` module.

```
[oneils@atmosphere ~/apcb/py]$ python
Python 2.7.8 (default, Aug  7 2014, 09:56:10)
[GCC 4.4.7 20120313 (Red Hat 4.4.7-4)] on linux2
Type "help", "copyright", "credits" or "license" for more information.
>>> import re
>>> help(re)
```

There are a huge number of modules (and packages) available for download online, and Python also comes with a wide variety of useful modules by default. Aside from `re`, `sys`, and `io`, discussed previously, some of the potentially more interesting modules include:

- `string`: common string operations (changing case, formatting, etc.)
- `time`, `datetime` and `calendar`: operations on dates and times
- `random`: generating and working with random numbers
- `argparse`: user-friendly parsing of complex arguments on the command line
- `Tkinter`: creating graphical user interfaces (with buttons, scroll bars, etc.)
- `unittest`: automating the creation and running of unit tests
- `turtle`: a simple interface for displaying line-based graphics[75]

Tutorials for these specific modules may be found online, as may lists of many other packages and modules that come installed with Python or that are available for download.

75. The `turtle` graphics module is a fun way to understand and visualize computational processes (and similar packages are available for many programming languages). While this book isn't the place for such explorations, the reader can learn more about them in Jeffrey Elkner, Allen Downey, and Chris Meyers's excellent book, *How to Think Like a Computer Scientist* (Green Tea Press, 2002). As of this writing, a version of the book is available online at http://interactivepython.org/runestone/static/thinkcspy/index.html.

Creating a Module

Let's create a module called `MyVCFModule.py` to hold the two class definitions and parsing function from the last example. While we're doing so, we'll also convert the comments that we had created for our classes, methods, and functions into "docstrings." Docstrings are triple-quoted strings that provide similar functionality to comments, but can be parsed later to produce the module's API. They may span multiple lines, but should occur immediately inside the corresponding module, class, or function definition. Here we leave out the contents of the methods and functions (which are the same as in the previous chapter) to save space:

```
""" Personal module for parsing VCF files. """

import io
import sys
import re

class SNP:
    """ A class representing simple SNPs"""   # Docstring

    def __init__(self, chrname, pos, snpid, refallele, altallele):
        """ Constructor method """
        # ...

    def is_transition(self):
        """ Returns True if refallele/altallele is A/G, G/A, C/T, or T/C """
        # ...

    def is_transversion(self):
        """ Returns True if the snp is a transversion (ie, not a transition) """
        # ...

class Chromosome:
    """ A class representing a chromosome, which has a collection of SNPs """

    def __init__(self, chrname):
        # ...

    def get_name(self):
        # ...

    def add_snp(self, chrname, pos, snpid, refallele, altallele):
        # ...

    def count_transitions(self):
        # ...

    def count_transversions(self):
        # ...

def vcf_to_chrnames_dict(filename):
    # ...
```

The program that makes use of this module, which exists in a separate file in the same directory (perhaps called **snps_ex2.py**), is short and sweet.[76]

```
#!/usr/bin/env python
import MyVCFModule
import sys

## Check usage syntax, read filename
if len(sys.argv) != 2:
    print("This program parses a VCF 4.0 file and counts")
    print("transitions and transversions on a per-chromosome basis.")
    print("")
    print("Usage: ./snps_ex.py <input_vcf_file>")
    quit()

filename = sys.argv[1]

chrnames_to_chrs = MyVCFModule.vcf_to_chrnames_dict(filename)

## Print the results!
print("chromosome" + "\t" + "transitions" + "\t" + "transversions")
for chrname in chrnames_to_chrs.keys():
    chr_obj = chrnames_to_chrs[chrname]
    trs = chr_obj.count_transitions()
    trv = chr_obj.count_transversions()
    print(chrname + "\t" + str(trs) + "\t" + str(trv))
```

Should other projects need to parse VCF files, this module may again be of use. By the way, because we documented our module with docstrings, its API is readily accessible through the interactive help interface.

```
oneils@atmosphere ~/apcb/py$ python
Python 2.7.9 (default, Feb 21 2015, 21:09:31)
[GCC 4.6.3] on linux2
Type "help", "copyright", "credits" or "license" for more information.
>>> import MyVCFModule
>>> help(MyVCFModule)
```

Part of the API help page:

76. In some Python scripts, you may find a line `if __name__ == '__main__':`. The code within this if-statement will run, but *only* if the file is run as a script. If it is imported as a module, the block will not run. This lets developers write files that can serve both purposes, and many developers include it by default so that their script's functions and class definitions can be utilized in the future by importing the file as a module.

```
NAME
    MyVCFModule - Personal module for parsing .VCF files.

FILE
    /home/oneils/apcb/py/MyVCFModule.py

CLASSES
    Chromosome
    SNP

    class Chromosome
     |  A class representing a chromosome, which has a collection of SNPs
     |
     |  Methods defined here:
     |
     |  __init__(self, chrname)
     |      Constructor
     |
     |  add_snp(self, chrname, pos, snpid, refallele, altallele)
     |      Given all necessary information to add a new SNP, create
     |      a new SNP object and add it to the SNPs dictionary. If
     |      a SNP already exists at that location, or
     |      the given chrname doesn't match self.chrname, an error is reported.
     |
```

Packages

As if Python didn't already have enough "boxes" for encapsulation—functions, objects, modules, and so on—there are also packages. In Python, a "package" is a directory containing some module files.[77]

A package directory must also contain a special file called `__init__.py`, which lets Python know that the directory should be treated as a package from which modules may be imported. (One could put code in this file that would be executed when the `import` statement is run, but we won't explore this feature here.)

77. Packages may also contain other files, like files containing specific data sets or even code written in other languages like C. Thus some modules may need to be installed by more complicated processes than simply decompressing them in the right location.

As an example, suppose that along with our `MyVCFModule.py`, we also had created a module for parsing gene ontology files called `GOParseModule.py`. We could put these together into a package (directory) called `MyBioParsers`.

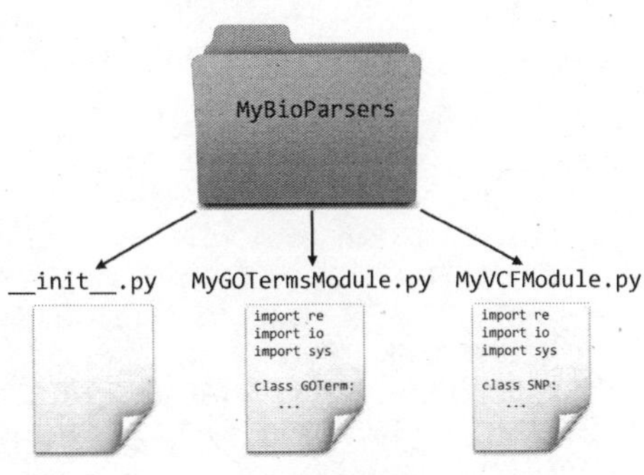

To use a module contained in a package, the syntax is `from packagename import modulename`.[78] Our Python program could live in the same directory in which the `MyBioParsers` directory was found, and might begin like so:

```
import sys
from MyBioParsers import MyVCFModule
```

Later, the module itself can be used just as before.

```
crnames_to_chrs = MyVCFModule.vcf_to_chrnames_dict(filename)
```

Parsing a FASTA File

Up until this point, we've skipped something that is widely considered a "basic" in computational biology: reading a FASTA file. For the most part, the previous examples needing sequence data read that data from simple row/column formatted files.

```
PZ7180000024555 ATAAACTGATCTTAAACTAATTGTCATGTTGAGTTCATAACGAGGTGCATTTTCGATAAATAGT
PZ7180000000678_B       CATAGTAATGTATAATAATCATATATTTATATGTTAAACCTTCCAAAAATATCTAT
PZ7180000000003_KK      AACAAGTGCACATTAATAGCAGTGTATCAACATGGGTGTGTGGCTAGAGAACTGAA
PZ7180000000005_NW      AAATGTACCCGAGTGTTTCGGTTGTGCACACGGGTGTCTAGTTTACCGCAGTATCG
...
```

78. You might occasionally see a line like `from modulename import *`. This allows one to use the functions and other definitions inside of `modulename` without prefixing them with `modulename.`. For example, we can `import math` and then use `print(math.sqrt(3.0))`, or we can `from math import *` and then `print(sqrt(3.0))`. The latter is usually avoided, because if multiple modules contain the same function or other names, then it is not possible to disambiguate them.

Most sequence data, however, appear in so-called FASTA format, where each sequence has a header line starting with > and an ID, and the sequence is broken up over any number of following lines before the next header line appears.

```
>PZ7180000024555s
ATAAACTGATCTTAAACTAATTGTCATGTTGAGTTCATAACGAGGTGCATTTTCGATAAATAGTGAAAAT
TGCAGTATTTTCTATTTAGGCAGTAATAAATATAAGGCTTGCTTTGTGCACATGTTAATATCTACTCTGA
TAAAATCCTTAACTTAAAAGCAACTACAGCGACACACCTTGAGT
>PZ7180000000678s_B
CATAGTAATGTATAATAATCATATATTTATATGTTAAACCTTCCAAAAATATCTATGTACGGATGTAGTG
TGTATACATCATGGCTGCTCCGTCCCGGACTCTGTCCCACTGAGCTGCATTATCAATAAAACAGGTTATA
TATA
>PZ7180000000003s_KK
TTGCTAGGCATTGGAGATGAGGGAGAAGATGATGGTTACCACACCCTATCTGTACAACAGGTAAAGATAG
TTGACACAGAAGGTAATTTAAAGAGTGTGTACCCATCGAGGACGGACTGTGTGTACGACGGCACTAATAT
TAAAGTGTACCGTCTGCCAAAGAAATAAACTTCATTATTAAATAACGACATTACAGTTCTAATTAACGAT
TTCGGAATTTGAAAAGTTCCAACAAATATGCTTTTGTCAATCTCATTTTGTTACAATCATATATATCTTT
...
```

Parsing data in FASTA format isn't too difficult, but it requires an extra preprocessing step (usually involving a loop and a variable that keeps track of the "current" ID; so long as subsequent lines don't match the > character, they must be sequence lines and can be appended to a list of strings to be later joined for that ID). While reinventing the FASTA-parsing wheel is good practice for novice coders, the BioPython package provides functions to parse these and many other formats.[79]

BioPython is a large package with many modules and APIs, which you can read more about at http://biopython.org. For the specific application of parsing a FASTA file, a good place to start is a simple web search for "biopython fasta," which leads us to a wiki page describing the SeqIO module, complete with an example.

79. Depending on the system you are working on, you may already have the BioPython package. If not, you can use the `pip` install utility to install it: `pip install BioPython`, or if you don't have administrator privileges, `pip install --user BioPython`. Some older versions of Python don't come with the `pip` utility, but this can be installed with an older tool: `easy_install pip` or `easy_install --user pip`.

Sequence Input

The main function is **Bio.SeqIO.parse()** which takes a file handle and format name, and returns a SeqRecord iterator. This lets you do things like:

```
from Bio import SeqIO
handle = open("example.fasta", "rU")
for record in SeqIO.parse(handle, "fasta") :
    print record.id
handle.close()
```

In the above example, we opened the file using the built-in python function **open**. The argument 'rU' means open for **r**eading using **u**niversal readline mode - this means you don't have to worry if the file uses Unix, Mac or DOS/Windows style newline characters.

This screenshot shows a few differences from the way we've been writing code, for example, the use of `open()` as opposed to `io.open()`, and the call to `print record.id` rather than our usual `print(record.id)` (the former of which will work only in older versions of Python, whereas the latter will work in all reasonably recent versions of Python). Nevertheless, given our knowledge of packages, modules, and objects, we can deduce the following from this simple code example:

1. The `SeqIO` module can be imported from the `Bio` package.
2. A file handle is opened.
3. The `SeqIO.parse()` function is called, and it takes two parameters: the file handle from which to read data, and a specification string describing the file format.
4. This function returns something iterable (similar to a list, or a regular file handle), so it can be looped over with a for-loop.
5. The loop accesses a series of objects of some kind, and each gets associated with the variable name `record`.
6. The `record.id` is printed, which is (apparently) an instance variable belonging to the `record` object.
7. The file handle is closed (always good practice!).

If we do some more reading on the `SeqIO` wiki page, we'd find that the `record` objects actually have the class `SeqRecord`, and they also have a `seq` instance variable providing access to the full sequence.

Putting this all together, here's a short program that converts a FASTA file to the row/column format we've been dealing with.

```
#!/usr/bin/env python
import sys
import io
from Bio import SeqIO

if len(sys.argv) < 2:
    print("Converts FASTA format to row/column format,")
    print("printing the results on standard out.")
    print("Usage: ./fasta2cols.py <fastafile>")
    quit()

filename = sys.argv[1]
fhandle = io.open(filename, "rU")

for record in SeqIO.parse(fhandle, "fasta"):
    print(record.id + "\t" + record.seq)

fhandle.close()
```

Syntactic Sugar

In Python, nearly everything is an object, even simple integers! By looking at an integer's API, for example, we discover that they provide a method called `.bit_length()`.

```
>>> a = 7
>>> help(a)
```

Here's a portion of the API view:

```
 |
 |  bit_length(...)
 |      int.bit_length() -> int
 |
 |      Number of bits necessary to represent self in binary.
 |      >>> bin(37)
 |      '0b100101'
 |      >>> (37).bit_length()
 |      6
 |
```

We can try it like so, to discover that the integer `7` can be represented with three binary bits (as **`111`**):

```
>>> print(a.bit_length())
3
```

If you were to try and view the API for an integer as we've done, you'd see a number of odd-looking methods, such as `.__add__()` and `.__abs__()`:

```
 |
 |  __abs__(...)
 |      x.__abs__() <==> abs(x)
 |
 |  __add__(...)
 |      x.__add__(y) <==> x+y
 |
```

This seems to indicate that we can get the absolute value of an integer by using a method call or by using the standard function call syntax. Similarly, we can apparently add an integer to another by using method call or the standard + operator. These are indeed true, and we can use the standard functions and operators or their method versions:

```
>>> b = -5
>>> print(b.__abs__())
5
>>> print(abs(b))
5
>>> print(b.__add__(3))
-2
>>> print(b + 3)
-2
```

Operations like addition look like basic, fundamental operations, and Python accomplishes such operations through method calls on objects.[80] A statement like `a = b + c` is converted to `a = b.__add__(c)` behind the scenes. Although we can run such method-oriented operations ourselves, the awkward double-underscore syntax is how the designers let us know that those methods are for internal use, and we should stick with the standard syntax. This automatic syntax conversion is known as *syntactic sugar*, and it is present in many programming languages so that they can be internally consistent but more pleasant ("sweeter") for humans to work with.

Exercises

1. Create a module called `mymodule.py` that includes at least a class definition and function definition, as well as documents the module, class, methods, and functions with docstrings. Create a `myprogram.py` that imports this module and makes use of the functionality in it. Also try viewing the API for your module in the interactive Python interpreter with `import mymodule` and `help(mymodule)`.

2. Browse through the online API at http://docs.python.org/2/library for the "Python Standard Library," and especially anything related to strings. Also, review the API documentation for the `io`, `sys`, `re`, `math`, and `random` modules.

3. For this exercise, we're going to install a module from the web and make use of it to parse a VCF file `trio.sample.vcf`. The VCF-reading module we're going to install is located on GitHub (an increasingly popular place to host software): https://github.com/jamescasbon/PyVCF.

   The first thing to do is to clone this repository from the `.git` link (which requires we have the `git` tool installed). This cloning should download a `PyVCF` folder; `cd` to there and use `ls` to view the contents:

80. It is safe to say that Python isn't "merely" object oriented, it is *very* object oriented.

```
oneils@atmosphere ~/apcb/py$ git clone https://github.com/jamescasbon/PyVCF.git
Cloning into 'PyVCF'...
remote: Counting objects: 2178, done.
remote: Total 2178 (delta 0), reused 0 (delta 0), pack-reused 2178
Receiving objects: 100% (2178/2178), 1.62 MiB | 264.00 KiB/s, done.
Resolving deltas: 100% (1243/1243), done.
Checking connectivity... done.
oneils@atmosphere ~/apcb/py$ cd PyVCF
oneils@atmosphere ~/apcb/py/PyVCF$ ls
LICENSE       README.rst  requirements  setup.py  vcf
MANIFEST.in   docs        scripts       tox.ini
oneils@atmosphere ~/apcb/py/PyVCF$
```

Next, we need to install it—this will set up the module and its needed files, and put them into a hidden folder called `.local` inside of your home directory (Python also looks there for modules and packages by default). To get it into `.local` in your own home directory rather than using a system-wide install (which you may not have permission to do), we have to add `--user` to the command that runs the `setup.py` script:

```
oneils@atmosphere ~/apcb/py/PyVCF$ python setup.py install --user
```

Once this is done, go back to your original directory, and then you can fire up the Python interactive interpreter, load the module, and view its API:

```
oneils@atmosphere ~/apcb/py/PyVCF$ cd ..
oneils@atmosphere ~/apcb/py$ python
Python 2.7.10 (default, Jul 14 2015, 19:46:27)
[GCC 4.2.1 Compatible Apple LLVM 6.0 (clang-600.0.39)] on darwin
Type "help", "copyright", "credits" or "license" for more information.
>>> import vcf
>>> help(vcf)
```

(Sadly, the API documentation in the package itself is not that great—the online version at http://pyvcf.rtfd.org is better, specifically in the introduction.)

4. Write a program called `vcf_module_use.py` that imports this `vcf` module and uses it to count and print how many lines in **`trio.sample.vcf`** have a reference allele of `"A"`. (Note that you should import and use the module, rather than parsing each line yourself.)

Chapter 25

Algorithms and Data Structures

Having learned so many programming concepts—variables and the data to which they refer, functions, loops, and so on—it would be a shame not to talk about some of the surprising and elegant methods they enable. Computer science students spend years learning about these topics, and the themes in this chapter underlie much of bioinformatics.

We'll start with algorithms, which according to a classic book on the topic—*Introduction to Algorithms* by Thomas H. Cormen, Charles E. Leiserson, Ronald L. Rivest, and Clifford Stein—are:

> any well-defined computational procedure that takes some value, or set of values, as input and produces some value, or set of values, as output.

With such a broad definition, all of the Python coding we've done could accurately be categorized as practice in algorithms. The word "algorithm," by the way, derives from the name of the eighth-century Persian mathematician al-Khwārizmī, who developed systematic approaches for solving equations (among other accomplishments). Although most functioning code may be characterized as an algorithm, algorithms usually involve more than just collating data and are paired with well-defined *problems*, providing a specification for what valid inputs are and what the corresponding outputs should be. Some examples of problems include:

1. Given a list of n numbers in arbitrary order, return it or a copy of it in sorted order. (The "sorting" problem.)

2. Given a list of n numbers in sorted order and a query number `q`, return `True` if `q` is present in the list and `False` if it is not. (The "searching" problem.)
3. Given two strings of length m and n, line them up against each other (inserting dashes where necessary to make them the same length) to maximize a score based on the character matches. (The "string alignment" problem.)

```
ACTAGCCAG          ACTAGCCAG
                   || || ||
ACAGTCA     →      AC-AGTCA-
                   +2 +2 -3 +2 +2 -1 +2 +2 -3

Input              Output
                   (Best Score: +5)
```

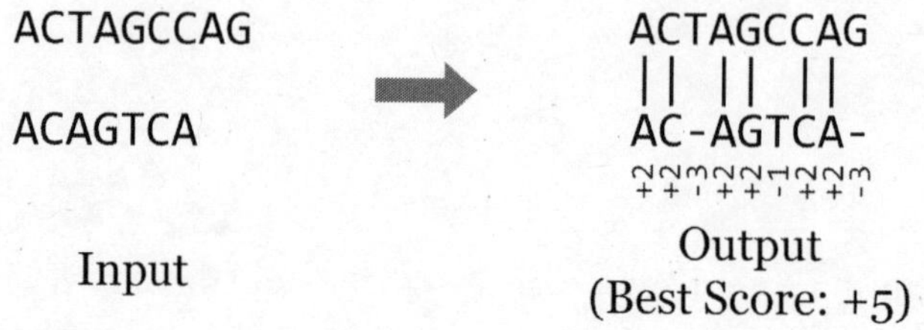

Clearly, some problems, like the string alignment problem, are of special interest for life scientists. Others, like sorting and searching, are ubiquitous. No matter who cares about the problem, a good algorithm for solving it should do so efficiently. Usually, "efficiently" means "in a short amount of time," though with the large data sets produced by modern DNA-sequencing technologies, it often also means "using a small amount of memory."

Most bioinformatics problems like string alignment require rather sophisticated methods built from simpler concepts, so we'll start with the usual introductory topic for algorithms design and analysis: sorting. After all, *someone* had to write Python's `sorted()` function for lists, and it does form a good basis for study.

Consider a small list of five numbers, in unsorted order.

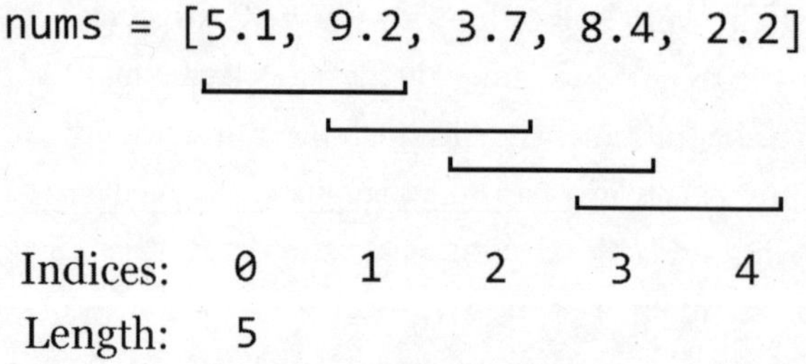

```
nums = [5.1, 9.2, 3.7, 8.4, 2.2]

Indices:  0    1    2    3    4
Length:   5
```

What code might we write to sort these numbers? Whatever it is, it might make sense to put it into a function that takes an unsorted list, and returns a copy of it in sorted order (to stay consistent with Python's use of the principle of command-query separation). To start, we'll make a copy of the list, and then we'll have the function merely "fix up" some of the out-of-order numbers in this copy by looking at a sliding window of size 2—pairs of adjacent numbers—and switching their placement if they are out of order.

```
## Returns a sorted copy of a list of numbers (almost)
def get_sorted(nums):
    ## make a copy that we can sort and return
    c_nums = list()
    for num in nums:
        c_nums.append(num)

    ## Look neighboring pairs, swapping where needed
    for index in range(0, len(c_nums) - 1):
        leftnum = c_nums[index]
        rightnum = c_nums[index + 1]
        if leftnum > rightnum:
            c_nums[index] = rightnum
            c_nums[index + 1] = leftnum

    return c_nums
```

Just fixing up a handful of pairs in this way won't result in the list being sorted, but it should be obvious that because the pairs of fix-ups overlap, the largest number will have found its way to the last position. In a sense, it "bubbled" to the top. If we were to repeat this process, the second-to-largest number must also find its way to the correct spot. In fact, if we repeat this process as many times as there are numbers, the list will be fully sorted.

```
## Returns a sorted copy of a list of numbers (almost)
def get_sorted(nums):
    ## make a copy that we can sort and return
    c_nums = list()
    for num in nums:
        c_nums.append(num)

    for repetition in range(0, len(c_nums)):
        ## Look neighboring pairs, swapping where needed
        for index in range(0, len(c_nums) - 1):
            leftnum = c_nums[index]
            rightnum = c_nums[index + 1]
            if leftnum > rightnum:
                c_nums[index] = rightnum
                c_nums[index + 1] = leftnum

    return c_nums
```

```
nums = [3, 4, 5, 1, 6, 3, 7, 9]
print(get_sorted(nums))                  # prints [1, 3, 3, 4, 5, 6, 7, 9]
```

This algorithm for sorting is affectionately known as "bubblesort." In a rough sense, how many operations will this algorithm perform, in the worst case, for a list of size n? Suppose the if-statement always finds `True`, and numbers need to be swapped. In this case, each operation of the inner loop requires five or so "time steps" (four assignments and an `if` check). The inner loop runs $n - 1$ times (if there are n numbers), which is itself repeated n times in the outer loop. The copy step to produce the new `c_nums` list also uses n steps, so the total number of basic operations is, more or less,

$$n(n-1)5 + n = 5n^2 - 4n \ .$$

When analyzing the running time of an algorithm, however, we only care about the "big stuff."

Order Notation

In computing an algorithm's run time, we only care about the highest-order terms, and we tend to ignore constants that aren't related to the "size" of the input. We say that the run time is *order* of the highest-order term (without constants), which we denote with an uppercase O: $5n^2 - 4n$ is $O(n^2)$.

We read this as "five n squared minus four n is order n squared" or "five n squared minus four n is big-oh n squared." Because this run time talks about a particular algorithm, we might also say "bubblesort is order n squared." Big-O notation implies that, roughly, an algorithm will run in time *less than or equal to* the equation interpreted as a function of the input size, in the *worst case* (ignoring constants and small terms).

Worst-case analysis assumes that we get unlucky with every input, but in many cases the algorithm might run faster in practice. So, why do we do worst-case analysis? First, it gives the user of an algorithm a guarantee—nobody wants to hear that software *might* use a certain amount of time. Second, while it is sometimes possible to do average-case analysis, it requires knowledge of the distribution of input data in practice (which is rare) or more sophisticated analysis techniques.

Why do we use order notation to characterize algorithm run time, dropping small terms and constants? First, although we stated previously that each basic operation is one "computer step," that's not really true: the line `c_nums[index + 1] = leftnum` requires both an addition and an assignment. But no one wants to count each CPU cycle individually, especially if the result won't change how we compare two algorithms for large data sets, which is the primary purpose of order notation. Dropping off constant terms and lower-order terms makes good mathematical sense in these cases. For large enough input (i.e., all n greater than some size c), even terms that seem like they would make a big difference turn out not to matter, as the following hypothetical comparison demonstrates.

$$
\begin{aligned}
0.01n^2 - 100n > 200n^{1.5} + 20n, \text{for } n > \text{some } c, \\
O(n^2) > O(n^{1.5}), \\
\text{Time Alg A} > \text{Time Alg B (for large inputs)}.
\end{aligned}
$$

(Technically in the above $O(n^2) > O(n^{1.5})$ is an abuse of notation; we should say $O(n^{1.5})$ is $O(n^2)$ based on the definition of the notation.) In this case, $0.01n^2 - 100n$ is larger than $200n^{1.5} + 20n$ when n is larger than 400,024,000 (which is this example's c).

So, although order notation seems like a fuzzy way of looking at the efficiency of an algorithm, it's a rigorously defined and reasonable way to do so.[81] Because some programming languages are faster than others—but only by a constant factor (perhaps compiled C code is 100 times faster than Python)—a good algorithm implemented in a slow language often beats a mediocre algorithm implemented in a fast language!

Now, a run time of $O(n^2)$ isn't very good: the time taken to sort a list of numbers will grow quadratically with the size of the list.

81. When analyzing an algorithm in Python, however, not all lines are a single computational step. For example, Python has a built-in `sorted()` function for sorting, and although it is not as slow as bubblesort, it takes much longer than one step to sort millions of numbers. If an algorithm makes use of a function like `sorted()`, the run time of that (based on the size of the input given) also needs to be considered. A function that calls our bubblesort function n times, for example, would run in time $O(n^2)$.

$O(n^2)$ isn't great, but for the sorting problem we can do better, and we'll return to the study of algorithms and sorting a bit later. First, let's detour and talk about interesting ways we can organize data, using variables and the objects to which they refer.

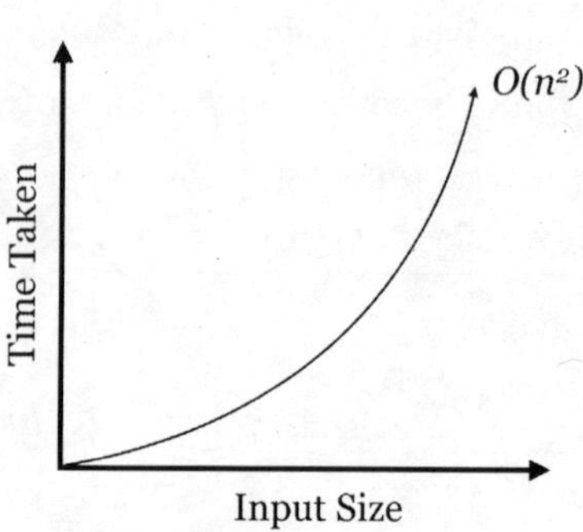

Data Structures

A data structure is an organization for a collection of data that ideally allows for fast access and certain operations on the data. We've seen a couple of data structures already in Python: lists and dictionaries. Given such a structure, we might ask questions like:

- Does the data structure keep its elements in a specific order (such as sorted order)? If we use `.append()` to add elements to lists, lists are kept in the order items are added. If we wanted the list to stay sorted, we'd have to ensure new elements are inserted in right spot.
- How long does it take to add a new element? For Python lists, the `.append()` method operates in time $O(1)$, which is to say a single element can be added to the end, and the time taken does not depend on how big the list is. Unfortunately, this won't be true if we need to keep the list sorted, because inserting into the middle requires rearranging the way the data are stored.
- How long does it take to find the smallest element? Because using `.append()` adds elements to the end of the list, unless we keep the list sorted, we need to search the entire list for the smallest element. This operation takes time $O(n)$, where n is the length of the list. If the list is sorted, though, it takes only $O(1)$ time to look at the front of it.

There are trade-offs that we can make when thinking about how to use data structures. Suppose, for example, that we wanted to quickly find the smallest element of a list, even as items are added to it. One possibility would be to sort the list after each addition; if we did that, then the smallest element would always be at index `0` for quick retrieval. But this is a lot of work, and the total time spent sorting would start to outweigh the benefits of keeping the list sorted in the first place! Even better, we could write a function that (1) inserts the new item to the end and then (2) bubbles it backward to its correct location; only one such bubbling would be needed for each insertion, but each is an $O(n)$ operation (unless we could somehow guarantee that only large elements are added).

```
sorted_nums = [1, 4, 5, 6, 9]
sorted_nums.append(3)
sorted_nums = bubble_back_last(sorted_nums)
```

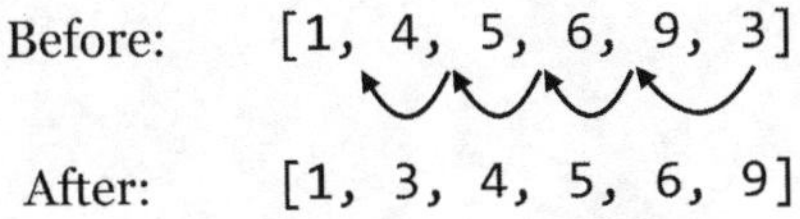

Before: `[1, 4, 5, 6, 9, 3]`

After: `[1, 3, 4, 5, 6, 9]`

Finding the smallest item is easy, as we know it is always present at index `0`.

Structure	Insert an Item	Get Smallest
Sorted Simple List	$O(n)$	$O(1)$

Instead, let's do something much more creative, and build our own data structure from scratch. While this data structure is a lot of trouble to create for only a little benefit, it does have its uses and serves as a building block for many other sophisticated structures.

Sorted Linked Lists

Our data structure, called a "sorted linked list," will keep a collection of items in sorted order, and will do the work of inserting new items in the correct location in that order. These items could be integers, floats, or strings, anything that can be compared with the < operator (which, recall, compares strings in lexicographic order; as discussed in previous chapters, we can even define our own object types that are comparable with > by implementing `.__lt__()`, `.__eq__()` and similar methods).

To accomplish this task, we're going to make heavy use of classes and objects, and the fact that a variable (even an instance variable) is a name that refers to an object.

The strategy will be to have two types of objects: the first, which we'll call a `LinkedList`, will be the main type of object with which our programs will interact (much like we interacted with `Chromosome` objects in previous chapters, while `SNP` objects were dealt with by the `Chromosome` objects). The `LinkedList` object will have methods like `.insert_item()` and `.get_smallest()`. The other objects will be of type `Node`, and each of these will have an instance variable `self.item` that will refer to the item stored by an individual `Node`. So, drawing just the objects in RAM, our sorted linked list of three items (`4`, `7`, and `9`) might look something like this:

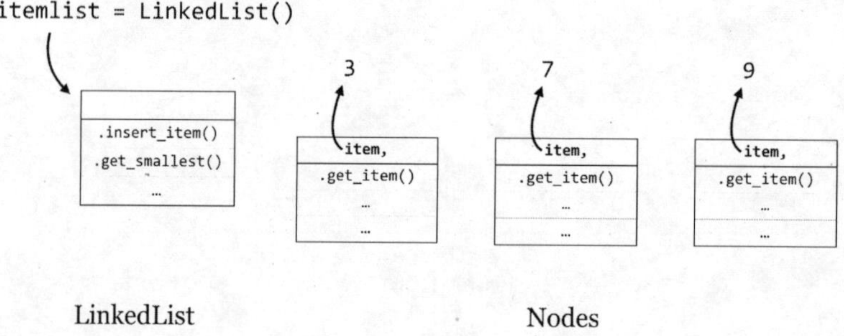

(The objects are unlikely to be neatly organized this way in RAM—we're showing them in a line for illustration only.) The arrows in this figure indicate that after we create a new `LinkedList` object called `itemlist`, this variable is a name that refers to an object, and each `Node` object has a `self.item` instance variable that refers to a "data" object of some type, like an integer or string.[82]

Now, if this collection of objects were to exist as shown, it wouldn't be that useful. Our program would only be able to interact with the `itemlist` object, and in fact there are no variables that refer to the individual `Node` objects, so they would be deleted via garbage collection.

Here's the trick: if an instance variable is just a variable (and hence a reference to an object), we can give each `Node` object a `self.next_n` instance variable that will refer to the next node in line. Similarly, the `LinkedList` object will have a `self.first_n` instance variable that will refer to the first one.

82. There is no reason a `Node` object couldn't also hold other information. For example, nodes could have a `self.snp` to hold an `SNP` object as well, with `self.item` being the location of the SNP, so SNPs are stored in order of their locations.

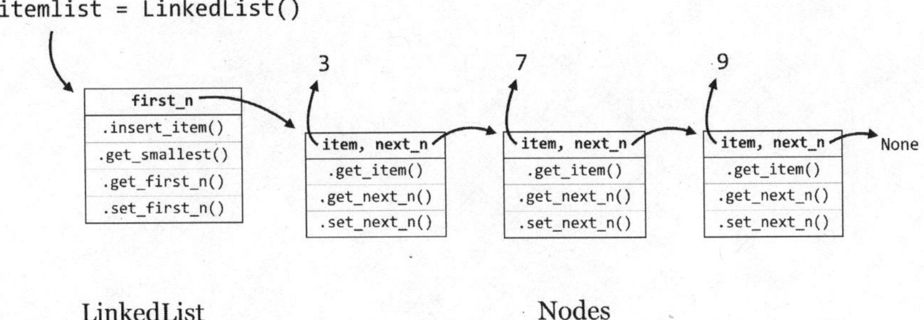

The last `Node` object's `self.next_n` refers to `None`, a placeholder that we can use to allow a variable to reference "nothing here." Actually, `None` will be the initial value for `self.next_n` when a new node is created, and we'll have to add methods for `get_next_n()` and `set_next_n()` that allow us to get or change a `Node`'s `next_n` variable at will. The LinkedLists's `first_n` variable will similarly be initialized as `None` in the constructor.

Suppose that we have this data structure in place, and we want to add the number `2`; this would be the new "smallest" item. To do this we'd need to run `itemlist.insert_item(2)`, and this method would consider the following questions to handle all of the possible cases for inserting an item (by using if-statements):

1. Is `self.first_n` equal to `None`? If so, then the new item is the *only* item, so create a new `Node` holding the new item and set `self.first_n` to that node.
2. If `self.first_n` is *not* equal to `None`:
    a. Is the new item smaller than `self.first_n`'s item? If so, then (1) create a new `Node` holding the new item, (2) set its `next_n` to `self.first_n`, and then (3) set `self.first_n` to the new node. Here's an illustration for this case:

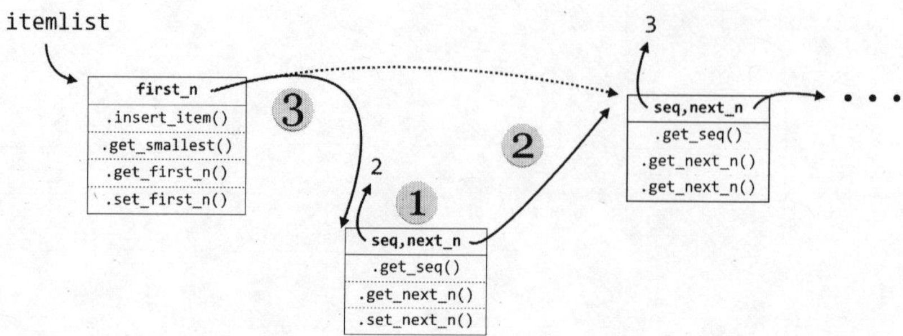

b. Otherwise, the new node does not go between the `LinkedList` object and the first `Node` object. In this case, we could treat the `self.first_n` object as though it were itself a `LinkedList`, if only *it* had an `.insert_item()` method.

This case (b) is really the heart of the linked list strategy: each `Node` object will also have an `.insert_item()` method. Further, each node's `insert_item()` will follow the same logic as above: if `self.next_n` is `None`, the new node goes after that node. If not, the node needs to determine if the new node should go between itself and the next node, or if it should "pass the buck" to the node next in line.

Now we can turn to the code. First, here's the code for the `LinkedList` class.

```
#!/usr/bin/env python

class LinkedList:

    def __init__(self):
        self.first_n = None

    def get_smallest(self):
        if self.first_n != None:
            return self.first_n.get_item()
        else:
            return None

    def insert_item(self, item):
        if self.first_n == None:                            # 1.
            newnode = Node(item)
            self.first_n = newnode
        else:
            if item < self.first_n.get_item():              # 2a.
                newnode = Node(item)                           # 1)
                newnode.set_next_n(self.first_n)               # 2)
                self.first_n = newnode                         # 3)
            else:                                           # 2b.
                self.first_n.insert_item(item)
```

The highlighted lines above are those illustrated in step 2a and are crucial; in particular, the order in which the various variables are set makes all the difference. What would happen if `self.first_n = newnode` was called *before* `newnode.set_next(self.first_n)`? We would lose all references to the rest of the list: `itemlist.first_n` would refer to `newnode`, the new node's `.next_n` would refer to `None`, and no variable would refer to the node holding 3—it would be lost in RAM and eventually garbage collected.

As mentioned above, the class for a `Node` is quite similar. In fact, it is possible to build linked lists with only one object type, but this would necessitate a "dummy" node to represent an empty list anyway (perhaps storing `None` in its `self.item`).

```
class Node:

    def __init__(self, item):
        self.item = item
        self.next_n = None

    def get_item(self):
        return self.item

    def get_next_n(self):
        return self.next_n

    def set_next_n(self, newnext):
        self.next_n = newnext

    def insert_item(self, item):
        if self.next_n == None:
            newnode = Node(item)
            self.next_n = newnode
        else:
            if item < self.next_n.get_item():
                newnode = Node(item)
                newnode.set_next_n(self.next_n)
                self.next_n = newnode
            else:
                self.next_n.insert_item(item)
```

Our new data structure is relatively easy to use, and it keeps itself sorted nicely:

```
numlist = LinkedList()
numlist.insert_item(9)
numlist.insert_item(3)
numlist.insert_item(7)
print(numlist.get_smallest())        # prints 3
numlist.insert_item(2)
print(numlist.get_smallest())        # prints 2
```

Linked List Methods

This idea of "passing the buck" between nodes is pretty clever, and it allows us to write sophisticated queries on our data structure with ease. Suppose we wanted to ask whether a given item is already present in the list.

To solve a problem like this, we can think of each node as implementing a decision procedure (using a method, like `.is_item_present(query)`). The `LinkedList` interface object would return

`False` if its `self.first_n` is `None` (to indicate the list is empty, so the query item can't be present). If its `self.first_n` is not `None`, it calls `self.first_n.is_item_present(query)`, expecting *that* node to either return `True` or `False`.

```
# ... (inside class LinkedList:)

def is_item_present(self, query):
    if self.first_n == None:
        return False
    else:
        answer = self.first_n.is_item_present(query)
        return answer
```

For a node, the decision procedure is only slightly more complex:

1. Check whether `self.item` is equal to the `query`. If so, a `True` can safely be returned.
2. Otherwise:
   a. If `self.next_n` is `None`, then `False` can be returned, because if the buck got passed to the end of the list, no node has matched the `query`.
   b. If `self.next_n` does exist, on the other hand, just pass the buck down the line, and rely on the answer to come back, which can be returned.

```
# ... (inside class Node:)

def is_item_present(self, query):
    if self.item == query:                          # 1.
        return True
    else:
        if self.next_n == None:                     # 2a.
            return False
        else:                                       # 2b.
            answer = self.next_n.is_item_present(query)
            return answer
```

Here is a quick demonstration of the usage (the whole script can be found in the file **`linkedlist.py`**):

```
numlist = LinkedList()
numlist.insert_item(9)
numlist.insert_item(3)
numlist.insert_item(7)
print(numlist.get_smallest())       # prints 3
numlist.insert_item(2)
print(numlist.get_smallest())       # prints 2
print(numlist.is_item_present(7))   # prints True
print(numlist.is_item_present(6))   # prints False
```

Notice the similarity in all of these methods: each node first determines whether it can answer the problem—if so, it computes and returns the answer. If not, it checks for a node to pass the problem on to, and if one exists, the buck is passed. Notice that at each buck pass the method being called is the same—it's just being called for a different object. And each time the overall "problem" gets smaller as the number of nodes left to pass the buck to decreases.

How much time does it take to insert an item into a list that is already of length n? Because the new item might have to go at the end of the list, the buck might need to be passed n times, meaning an insertion is $O(n)$. What about getting the smallest element? In the `LinkedList`'s `.get_smallest()` method, it only needs to determine whether `self.first_n` is `None`, and if not, it returns the element stored in that node. Because there is no buck passing, the time is $O(1)$.

Structure	Insert an Item	Get Smallest
Sorted Simple List	$O(n)$	$O(1)$
Sorted Linked List	$O(n)$	$O(1)$

The creation of the sorted linked list structure didn't get us much over a more straightforward list kept in sorted order via bubbling, but the ideas implemented here pave the way for much more sophisticated solutions.

Exercises

1. How much time would it take to insert n sequences into a Python list, and then at the end sort it with bubblesort in the worst-case scenario (using order notation)?
2. How much time would it take to insert n elements into a sorted linked list that starts empty, in the worst-case scenario (using order notation)? (Note that the first insertion is quick, but

the second item might take two buck-passes, the third may take three, and so on.)

3. Add "pass the buck" methods to the `LinkedList` and `Node` classes that result in each item being printed in order.
4. Write "pass the buck" methods that cause the list of items to be printed, but in *reverse* order.
5. Add methods to the `LinkedList` and `Node` classes so that the linked list can be converted into a normal list (in any order, though reverse order is most natural). For example, `print(itemlist.collect_to_list())` should print something like `['9', '3', '7']`.

Divide and Conquer

So far, both the algorithm (bubblesort) and the data structure (sorted linked list) we've studied have been linear in nature. Here, we'll see how these ideas can be extended in "bifurcating" ways.

Let's consider the sorted linked list from the last section, which was defined by a "controlling" class (the `LinkedList`) and a number of nearly identical `Node`s, each with a reference to a "next" `Node` object in the line. So long as certain rules were followed (e.g., that the list was kept in sorted order), this allowed each node to make *local* decisions that resulted in *global* answers to questions.

What if we gave each node a bit more power? Rather than a single `self.next_n` instance variable, what if there were two: a `self.left_n` and a `self.right_n`? We will need a corresponding rule to keep things organized: smaller items go toward the left, and larger (or equal-sized) items go toward the right. This data structure is the well-known *binary tree*.

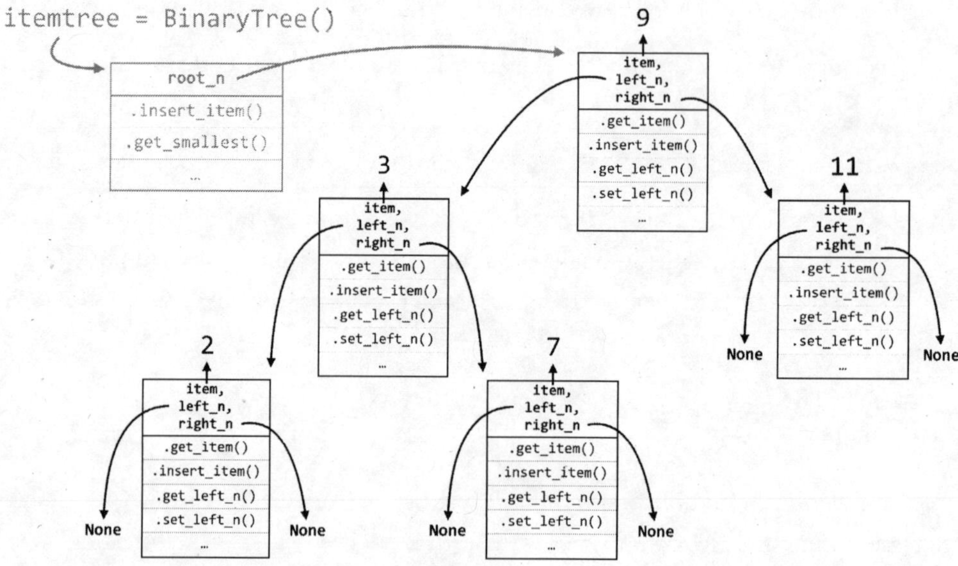

The illustration above looks quite a bit more complicated. But if we inspect this structure closely, it's quite similar to the linked list:[83] there is a controlling class of `BinaryTree`, and instead of a `self.first_n` instance variable, it has an instance variable called `self.root_n`, because the node it references is the "root" of the tree. Before any items have been inserted, `self.root_n` will be `None`. Upon an item insertion, if `self.root_n` is `None`, the new item goes there; otherwise, the buck is necessarily passed to `self.root_n`. We'll see why in a moment.

83. If we ignore all the nodes' `self.left_n` references (i.e., the entire left side of the tree), then the `self.right_n` path from top to the lower right is a sorted linked list! Similarly, the path from the top to the lower left is a reverse-sorted linked list.

```
#!/usr/bin/env python

class BinaryTree:

    def __init__(self):
        self.root_n = None

    def insert_item(self, item):
        if self.root_n == None:
            newnode = Node(item)
            self.root_n = newnode
        else:
            self.root_n.insert_item(item)
```

Now, for the `Node` class, we'll need a constructor, as well as "get" and "set" methods for both `left_n` and `right_n`, which initially are set to `None`.

```
class Node:

    def __init__(self, item):
        self.item = item
        self.left_n = None
        self.right_n = None

    def get_item(self):
        return self.item

    def get_left_n(self):
        return self.left_n

    def set_left_n(self, newleft):
        self.left_n = newleft

    def get_right_n(self):
        return self.right_n

    def set_right_n(self, newright):
        self.right_n = newright
```

What about a node's `.insert_item()` method? What sort of decision-making process needs to happen? The process is even simpler than for a sorted linked list. If each node always follows the rule that smaller items can be found to the left and larger or equal items can always be found to the right, then new items can always be inserted at the bottom of the tree. In the tree above, for example, a node holding `8` would be placed to the right of (and "below") the node holding `7`. The decision process for a node is thus as follows:

1. Is the new item to insert less than our `self.item`? If so, the new item goes to the left:
    a. Is `self.left_n` equal to `None`? If so, then we need to create a new node holding the new item, and set `self.left_n` to that node.
    b. If not, we can pass the buck to `self.left_n`.
2. Otherwise, the item must be larger than or equal to `self.item`, so it needs to go to the right:
    a. Is `self.right_n` equal to `None`? If so, then we need to create a new node holding the new item, and set `self.right_n` to that node.
    b. If not, we can pass the buck to `self.right_n`.

In the previous figure of a tree, `8` would go to the right of `7`, `6` would go to the left of `7`, `18` would go the right of `11`, and so on. The logic for inserting a new item is thus quite simple, from a node's point of view:

```
# ... (inside class Node:)

def insert_item(self, item):
    if item < self.item:
        if self.left_n == None:                         # 1a.
            newnode = Node(item)
            self.left_n = newnode
        else:                                           # 1b.
            self.left_n.insert_item(item)
    else:
        if self.right_n == None:                        # 2a.
            newnode = Node(item)
            self.right_n = newnode
        else:                                           # 2b.
            self.right_n.insert_item(newnode)
```

The remaining method to consider is the tree's `.get_smallest()`. In the case of a linked list, the smallest item (if present) was the first node in the list, and so could be accessed with no buck passing. But in the case of a binary tree, this isn't true: the smallest item can be found all the way to the left. The code for `.get_smallest()` in the `Tree` class and the corresponding `Node` class reflects this.

```
# ... (inside class Tree:)

def get_smallest(self):
    if self.root_n == None:
        return None
    else:
        answer = self.root_n.get_smallest()
        return answer
```

```
# ... (inside class Node:)

def get_smallest(self):
    if self.left_n == None:
        return self.item
    else:
        answer = self.left_n.get_smallest()
        return answer
```

In the case of a node, if `self.left_n` is `None`, then *that* node's item must be the smallest one, because it can assume the message was only ever passed toward it to the left. Similar usage code as for the linked list demonstrates that this wonderful structure (**`binarytree.py`**) really does work:

```
numtree = BinaryTree()
numtree.insert_item(9)
numtree.insert_item(3)
numtree.insert_item(7)
print(numtree.get_smallest())          # prints 3
numtree.insert_item(2)
print(numtree.get_smallest())          # prints 2
```

The most interesting question is: how much time does it take to insert a new item into a tree that already contains n items? The answer depends on the shape of the tree. Suppose that the tree is nice and "bushy," meaning that all nodes except those at the bottom have nodes to their left and right.

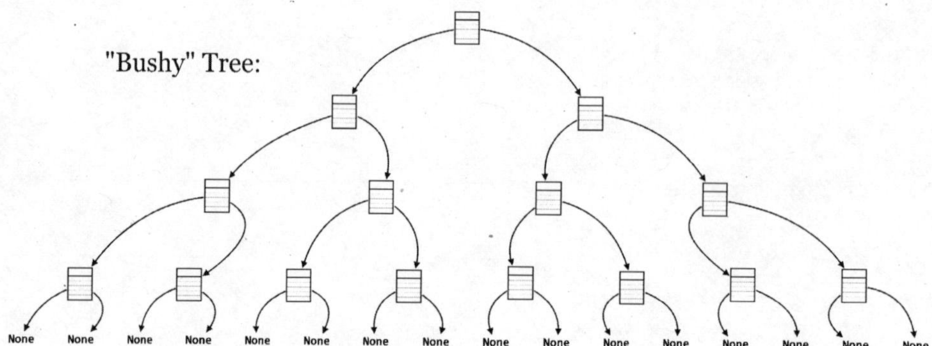

The time taken to insert an item is the number of times the buck needs to be passed to reach the bottom of the tree. In this case, at each node, the total number of nodes in consideration is reduced by half; first n, then $n/2$, then $n/4$, and so on, until there is only a single place the new item could go. How many times can a number n be divided in half until reaching a value of 1 (or smaller)? The formula is $log_2(n)$. It takes the same amount of time to find the smallest item for a bushy tree, because the length down the left-hand side is the same as any other path to a "leaf" in the tree.

Structure	Insert an Item	Get Smallest
Sorted Simple List	$O(n)$	$O(1)$
Sorted Linked List	$O(n)$	$O(1)$
"Bushy" Binary Tree	$O(\log_2(n))$	$O(\log_2(n))$

In general, the logarithm of n is much smaller than n itself, so a binary tree trades off some speed in finding the smallest element for speed in insertion.

Note, however, that the shape of a tree depends on the order in which the items are inserted; for example if `10` is inserted into an empty tree, followed by `9`, the `9` will go to the left. Further inserting `8` will put it all the way to the left of `9`. Thus, it is possible that a tree isn't in fact bushy, but rather very unbalanced. For an extreme example, if the numbers from $n, n-1, n-2, \ldots, 3, 2$ were inserted in that order, the tree would look like so:

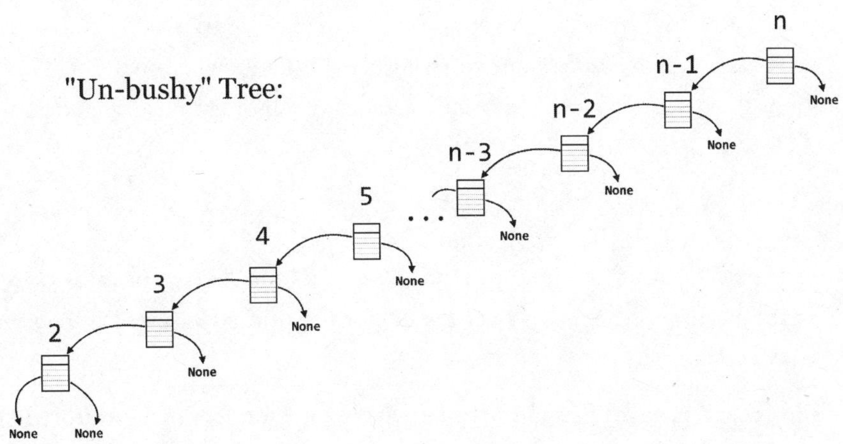

In this case, the tree has degenerated into a reverse-sorted linked list. So, the insertion time (for `1`, for example) is $O(n)$, and finding the smallest item is also, because the tree is heavily unbalanced in a leftward direction. Unfortunately, in practice, we can't guarantee the order in which data will be inserted, and such runs of consecutive insertions aren't uncommon in real-world data.

More sophisticated structures called "balanced" binary trees have modifications to their insertion methods that ensure the tree stays bushy, no matter what order items are inserted. This is tricky, as it requires manipulating the structure of the tree after insertions, but the structure manipulation itself can't take too long or the benefit of insertion is lost. Examples of balanced binary trees include so-called red-black trees, and AVL trees (named after Georgy Adelson-Velsky and Evgenii Landis, who first described them).

Structure	Insert an Item	Get Smallest
Sorted Simple List	$O(n)$	$O(1)$
Sorted Linked List	$O(n)$	$O(1)$
"Bushy" Binary Tree	$O(\log_2(n))$	$O(\log_2(n))$
"Un-bushy" Binary Tree	$O(n)$	$O(n)$
Balanced Binary Tree	$O(\log_2(n))$	$O(\log_2(n))$

In practice, we don't make a distinction between "bushy" and "un-bushy" binary trees: simple binary-search trees are $O(n)$ for both `.insert_item()` and `.get_smallest()`, because we cannot guarantee bushiness (recall that we assume worst-case performance when analyzing an algorithm). Real applications thus use AVL trees, red-black trees, or other more sophisticated data structures.

Exercises

1. Add "pass the buck" methods to `BinaryTree` and its `Node` class for `.print_in_order()` and `.print_reverse_order()`, causing the items to be printed in sorted and reverse-sorted order, respectively.
2. Add `.count_nodes()` methods that return the total number of items stored in the tree. How long does it take, in order notation? Does this depend on whether the tree is bushy or not? If so, what would be the run time for a bushy versus un-bushy tree?
3. Add `.count_leaves()` methods that return the total number of "leaves" (nodes with `None` in `left_n` and `right_n`).
4. Binary search trees are so called because they can easily and efficiently determine whether a query item is present. Add `.is_item_present()` methods that return `True` if a query item exists in the tree and `False` otherwise (similar to the `LinkedList` `.is_item_present()`). How long does it take, in order notation? Does this depend on whether the tree is bushy or not? If so, what would be the run time for a bushy versus un-bushy tree?
5. Modify the binary tree code so that duplicate items can't be stored in separate nodes.

Back to Sorting

We previously left the topic of sorting having developed `bubblesort`, an $O(n^2)$ method for sorting a simple list of items. We can certainly do better.

One of the interesting features of the `insert_item()` method used by nodes in both the tree and linked list is that this method, for any given node, calls itself, but in another node. In reality, there aren't multiple copies of the method stored in RAM; rather, a single method is shared between them, and only the `self` parameter is really changing. So, this method (which is just a function associated with a class) is actually calling *itself.*

In a related way, it turns out that any function (not associated with an object) can call itself. Suppose we wanted to compute the factorial function, defined as

$$factorial(n) = n \times (n-1) \times (n-2) \times \cdots \times 2 \times 1\ .$$

One of the most interesting features of the factorial function is that it can be defined in terms of itself:

$$factorial(n) = \begin{cases} 1, & \text{if } n = 1, \\ n \times factorial(n-1), & \text{otherwise.} \end{cases}$$

If we wanted to compute `factorial(7)`, a logical way to think would be: "first, I'll compute the factorial of 6, then multiply it by 7." This reduces the problem to computing `factorial(6)`, which we can logically solve in the same way. Eventually we'll want to compute `factorial(1)`, and realize that is just 1. The code follows this logic impeccably:

```
#!/usr/bin/env python

def factorial(n):
    if n == 1:
        return 1
    else:
        subanswer = factorial(n-1)
        answer = subanswer * n
        return answer

print(factorial(7))                     # prints 5040
print(factorial(8))                     # prints 40320
print(factorial(9))                     # prints 362880
```

As surprising as it might be, this bit of code really works.[84] The reason is that the parameter `n` is a local variable, and so in each call of the function it is independent of any other `n` variable that might exist.[85] The call to `factorial(7)` has an `n` equal to 7, which calls `factorial(6)`, which in turn gets its own `n` equal to `6`, and so on. Each call waits at the `subanswer = factorial(n-1)` line, and only

84. Factorials can be easily (and slightly more efficiently) computed with a loop, but we're more interested in illustrating the concept of a self-calling function.

85. Additionally, the operations of defining a function and executing it are disjoint, so there's nothing to stop a function being defined in terms of itself.

when `factorial(1)` is reached do the returns start percolating back up the chain of calls. Because calling a function is a quick operation ($O(1)$), the time taken to compute `factorial(n)` is $O(n)$, one for each call and addition computed at each level.

```
fac(5)
fac(5) -> fac(4)
fac(5) -> fac(4) -> fac(3)
fac(5) -> fac(4) -> fac(3) -> fac(2)
fac(5) -> fac(4) -> fac(3) -> fac(2) -> fac(1)
fac(5) -> fac(4) -> fac(3) -> fac(2) <- 1
fac(5) -> fac(4) -> fac(3) <- 2
fac(5) -> fac(4) <- 6
fac(5) <- 24
<- 120
```

This strategy—a function that calls itself—is called *recursion*. There are usually at least two cases considered by a recursive function: (1) the *base case*, which returns an immediate answer if the data are simple enough, and (2) the *recursive case*, which computes one or more subanswers and modifies them to return the final answer. In the recursive case, the data on which to operate must get "closer" to the base case. If they do not, then the chain of calls will never finish.

Applying recursion to sorting items is relatively straightforward. The more difficult part will be determining how fast it runs for a list of size n. We'll illustrate the general strategy with an algorithm called `quicksort`, first described in 1960 by Tony Hoare.

For an overall strategy, we'll implement the recursive sort in a function called `quicksort()`. The first thing to check is whether the list is of length 1 or 0: if so, we can simply return it since it is already sorted. (This is the base case of the recursive method.) If not, we'll pick a "pivot" element from the input list; usually, this will be a random element, but we'll use the first element as the pivot to start with. Next, we'll break the input list into three lists: `lt`, containing the elements less than the pivot; `eq`, containing the elements equal to the pivot; and `gt`, containing elements greater than the pivot. Next, we'll sort `lt` and `gt` to produce `lt_sorted` and `gt_sorted`. The answer, then, is a new list containing first the elements from `lt_sorted`, then the elements from `eq`, and finally the elements from `gt_sorted`.

The interesting parts of the code below are the highlighted lines: how do we sort `lt` and `gt` to produce `lt_sorted` and `gt_sorted`?

```
def quicksort(nums):
    if len(nums) <= 1:                 # base case
        return nums
    else:                              # recursive case
        pivot = nums[0]                # O(1)
        lt = list()
        eq = list()
        gt = list()

        for num in nums:               # O(n)
            if num < pivot:
                lt.append(num)
            elif num == pivot:
                eq.append(num)
            else:
                gt.append(num)

        ## producing lt_sorted and gt_sorted
        lt_sorted = quicksort(lt)  # O(?)
        gt_sorted = quicksort(gt)  # O(?)

        answer = list()
        for num in lt_sorted:         # O(n)
            answer.append(num)
        for num in eq:                # O(n)
            answer.append(num)
        for num in gt_sorted:         # O(n)
            answer.append(num)

    return answer
```

We could use Python's built-in `sorted()` function, but that's clearly cheating. We could use bubblesort, but doing so would result in our function suffering the same time bound as for bubblesort (we say "time bound" because the order notation essentially provides an upper bound on time usage). The answer is to use recursion: because `lt` and `gt` must both be smaller than the input list, as subproblems they are getting closer to the base case, and we can call `quicksort()` on them!

```
nums = [4, 6, 8, 1, 2, 8, 9, 2]
print(quicksort(nums))                # prints [1, 2, 2, 4, 6, 8, 8, 9]
```

Let's attempt to analyze the running time of `quicksort()`—will it be better than, equal to, or worse than bubblesort? As with the binary trees covered previously, it turns out the answer is "it depends."

First, we'll consider how long the function takes to run—not counting the subcalls to `quicksort()`—on a list of size n. The first step is to pick a pivot, which is quick. Next, we split the input list into three sublists. Because appending to a list is quick, but all n elements need to be appended to one of the three, the time spent in this section is $O(n)$. After the subcalls, the sorted versions of the lists need to be appended to the `answer` list, and because there are n things to append, the time there is also $O(n)$. Thus, not counting the recursive subcalls, the run time is $O(n)$ plus $O(n)$ which is $O(n)$.

But that's not the end of the story—we can't just ignore the time taken to sort the sublists, even though they are smaller. For the sake of argument, let's suppose we always get "lucky," and the pivot happens to be the median element of the input list, so `len(lt)` and `len(gt)` are both approximately $n/2$. We can visualize the execution of the function calls as a sort of "call tree," where the size of the nodes represents how big the input list is.

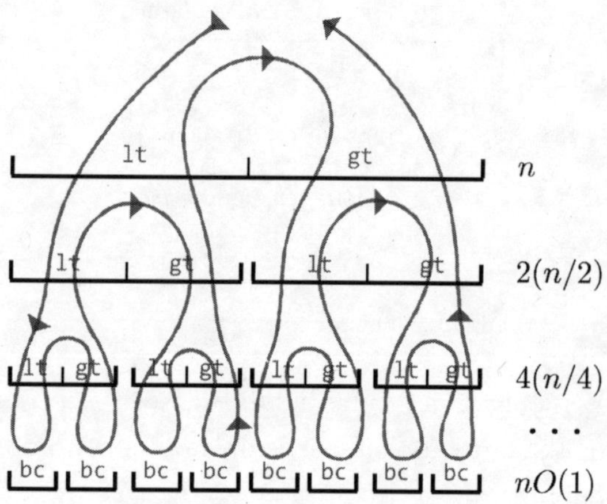

Here, the top "node" represents the work done by the first call; before it can finish, it must call to sort the `lt` list on the second layer, which must again call to sort its own `lt` list, and so on, until the bottom layer, where a base case is reached. The path of execution can be traced in the figure along the line. To analyze the run time of the algorithm, we can note that the amount of work done at each layer is $O(n)$, so the total amount of work is this value times the number of layers. In the case where the pivots split the lists roughly in half, the result is the same as the number of layers in the bushy binary tree: $O(\log_2(n))$. Thus, in this idealized scenario, the total run time is $O(n \log_2(n))$, which is much better than the $O(n^2)$ of bubblesort.[86]

86. This proof is a visually oriented one, and not terribly rigorous. A more formalized proof would develop what is known as a recurrence relation for the behavior of the algorithm, in the form of $T(n) = 2T(n/2) + c, \;\; T(1) = d$, to be solved for $T(n)$, representing the time taken to sort n items.

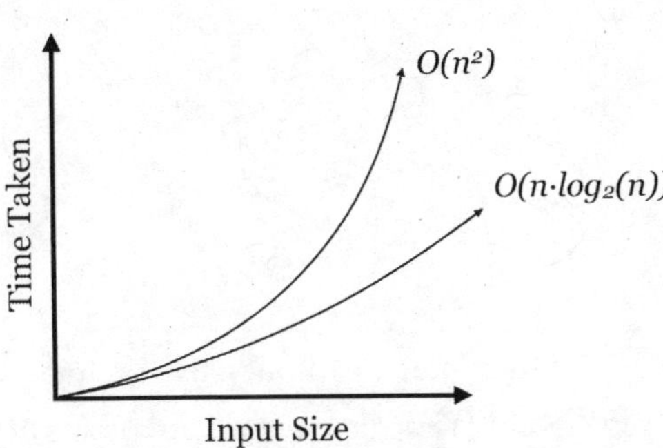

This equation assumes that the pivots split the input lists more or less in half, which is likely to be the case (on average), if we choose the pivots at random. But we didn't choose at random; we used `nums[0]`. What would happen if the input to the list happened to be already sorted? In this case, the pivot would always be the first element, `lt` would always be empty, and `gt` would have one fewer element. The "work tree" would also be different.

In this "unlucky" case, the work on each level is approximately $n, n-1, n-2$, and so on, so the total work is $n + (n-1) + (n-2) + \cdots + 1 = n(n+1)/2$ which is $O(n^2)$. Does this mean that in the worst case `quicksort()` is as slow as bubblesort? Unfortunately, yes. But some very smart people have analyzed `quicksort()` (assuming the pivot element is chosen at random, rather than using the first element) and proved that in the average case the run time is $O(n \log_2(n))$, and the chances of significantly worse performance are astronomically small. Further, a similar method known as "mergesort" operates by guaranteeing an even 50/50 split (of a different kind).

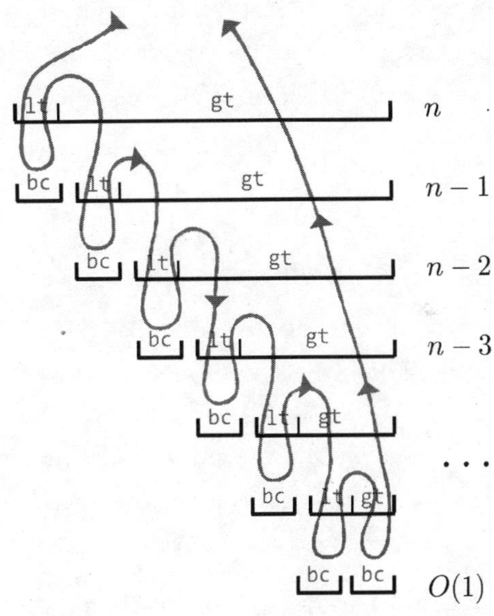

Algorithm	Average Case	Worst Case
Bubblesort	$O(n^2)$	$O(n^2)$
Quicksort	$O(n \log_2(n))$	$O(n^2)$
Mergesort	$O(n \log_2(n))$	$O(n \log_2(n))$

Using random selection for pivot, `quicksort()` is fast in practice, though mergesort and similar algorithms are frequently used as well. (Python's `.sort()` and `sorted()` use a variant of mergesort called "Timsort.") Although as mentioned above worst-case analysis is most prevalent in algorithms analysis, `quicksort()` is one of the few exceptions to this rule.

These discussions of algorithms and data structures might seem esoteric, but they should illustrate the beauty and creativeness possible in programming. Further, recursively defined methods and sophisticated data structures underlie many methods in bioinformatics, including pairwise sequence alignment, reference-guided alignment, and Hidden Markov Models.

Exercises

1. The first step for writing `mergesort()` is to write a function called `merge()`; it should take two sorted lists (together comprising n elements) and return a merged sorted version. For example, `merge([1, 2, 4, 5, 8, 9, 10, 15], [2, 3, 6, 11, 12, 13])` should return the list `[1, 2, 2, 3, 4, 5, 6, 8, 9, 10, 11, 12, 13, 15]`, and it should do so in time $O(n)$, where n is the total number of elements from both lists. (Note that `.append()` on a Python list is time $O(1)$, as are mathematical expressions like `c = a + b`, but other list operations like `.insert()` are not.)

   The `mergesort()` function should first split the input list into two almost-equal-sized pieces (e.g., `first_half = input_list[0:len(input_list)/2]`); these can then be sorted recursively with `mergesort()`, and finally the `merge()` function can be used to combine the sorted pieces into a single answer. If all steps in the function (not counting the recursive calls) are $O(n)$, then the total time will be $O(n \log_2(n))$.

   Implement `merge()` and `mergesort()`.

2. The Fibonacci numbers (1, 1, 2, 3, 5, 8, 13, etc.) are, like the factorials, recursively defined:

$$\text{Fib}(n) = \begin{cases} 1 & \text{if } n = 1, \\ 1 & \text{if } n = 2, \\ \text{Fib}(n-1) + \text{Fib}(n-2) & \text{otherwise.} \end{cases}$$

Write a recursive function `fib()` that returns the nth Fibonacci number (`fib(1)` should return `1`, `fib(3)` should return `2`, `fib(10)` should return `55`).

3. Next, write a function called `fib_loop()` that returns the nth Fibonacci by using a simple loop. What is the run time, in order notation, in terms of n? Compare how long it takes to compute `fib(35)` versus `fib_loop(35)`, and then try `fib(40)` versus `fib_loop(40)`. Why do you think `fib()` takes so much longer? Try drawing the "call trees" for `fib(1)`, `fib(2)`, `fib(3)`, `fib(4)`, `fib(5)`, and `fib(6)`. Could you make a guess as to what the run time of this function is in order notation? Can you imagine any ways it could be sped up?

Part III: Programming in R

Chapter 26

An Introduction

The R programming language has a rich history, tracing its roots to the S language originally developed for statistical computing in the mid-1970s at (where else?) Bell Laboratories. Later, the open-source R project extended the capabilities of S while incorporating features of languages like LISP and Scheme.

Many features of R are shared with Python: both are high-level, interpreted languages. (For a discussion of interpreted vs. compiled languages, see chapter 13, "Hello, World".) Both languages provide a wide array of features and functions for common tasks, and both languages are buttressed by a staggering variety of additional packages for more specialized analyses. Superficially, much of their syntax is similar, though below the surface lie significant (and fascinating) differences.

Practically, the major difference between the two languages lies in what built-in features and functions are available, and what packages are available for download. Where Python is considered a "general purpose" language, R specializes in statistical analyses. Need to build a mixed nonlinear model for a large table of numerical values from a multifactorial experiment? R is probably the tool of choice. Need to count potential promoter motifs in a large sequence set? Python is likely a better candidate. R does support functionality for the types of string analyses covered in the section on Python (such as DNA sequence analysis and regular expressions), but these are currently easier to work with in Python. Python provides excellent data plotting through the `matplotlib` library, but R's `ggplot2` library quickly became one of the dominant tools for data visualization since its initial release in 2005.

Where the analysis of biological data is concerned, both languages have grown rapidly. The `bioconductor` packages in R provide many statistical bioinformatics tools, while `BioPython` focuses on some statistical methods and many sequence-oriented methods such as multiple alignment. As of this writing, both languages appear be heading toward a common feature set: relatively recent Python packages such as `pandas`, `numpy`, `scipy`, and `statsmodels` add functionality that has been present in R for decades, while R has grown in general functionality and popularity. For now, though, both languages still make great additions to a computational biologist's repertoire, and both are supported by large and engaged communities.

So which of these two languages (and of course Python and R are far from the only two choices) should an aspiring computational biologist learn first? Well, the placement of Python in this book is no accident. For most users, Python is a better "introductory programming experience," even if the experience is brief, for a couple of reasons. First, much of Python was designed with education and ease of use in mind, easing the transition to computational thinking and explaining its current popularity in Computer Science departments around the world. Second, Python shares more similarity with other "mainstream" languages like Java, C, and C++ than does R, easing transference of concepts should one wish to continue on the programming journey. Further, R contains a much larger menagerie of data types and specialized syntax for working with them, as well as multiple frameworks for things like variable assignment and object orientation. Effective R programmers arguably have more to keep in mind as they work.

R is a remarkably flexible language. With so much flexibility comes both power and interesting ways of thinking about programming. While Python emphasizes the use of for-loops and if-statements to control program flow, R provides an alternative syntax for manipulation of data through sophisticated logical statements. (For-loops and if-statements are discussed late in this section.) Functions are quite important in Python, but in R they take on such significance that we are required to think about them at a higher level (as types of data that can be operated on by other functions). For many of the statistical tasks in which R excels, the underlying interpreter code is highly optimized or parallelized so that analyses of millions or billions of data points can be completed quickly. Finally, many excellent packages are available only for R.

Ultimately, though, the answer to "which language should I learn?" is as dynamic as "which language should I use?" There are good arguments to be made for (and against) all tools, and the types of skills you wish to acquire and situational needs will play a large role at any given time. Some advice: *eventually, learn to program in multiple languages.* The benefits of learning more than one language are easily on par with learning to program in the first place!

Hello, World

R is an interpreted language, meaning that an R program is a text file (or multiple text files, in some cases) with commands that are interpreted by another program interacting with the CPU and RAM through the operating system. On the command line, the R interpreter is simply `R`, which we can run and send commands to one at a time.

```
oneils@atmosphere ~/apcb/r$ R

R version 3.1.2 (2014-10-31) -- "Pumpkin Helmet"
Copyright (C) 2014 The R Foundation for Statistical Computing
Platform: x86_64-pc-linux-gnu (64-bit)

R is free software and comes with ABSOLUTELY NO WARRANTY.
You are welcome to redistribute it under certain conditions.
Type 'license()' or 'licence()' for distribution details.

  Natural language support but running in an English locale

R is a collaborative project with many contributors.
Type 'contributors()' for more information and
'citation()' on how to cite R or R packages in publications.

Type 'demo()' for some demos, 'help()' for on-line help, or
'help.start()' for an HTML browser interface to help.
Type 'q()' to quit R.

> print("Hello world!")
[1] "Hello world!"
>
```

When we are done working with the R interpreter this way, we can run `quit(save = "no")` to exit, instructing that any temporary data that we haven't already explicitly saved should not be saved.

```
> quit(save = "no")
oneils@atmosphere ~/apcb/r$
```

We will occasionally need to work with the R interpreter in this way, particularly when we need to install packages. For the most part, however, we will run R programs as executable scripts, much like we did for Python. In this case, we use the `Rscript` interpreter via the familiar `#!/usr/bin/env Rscript` line, which as always must be the first line of the file. (See chapter 5, "Permissions and Executables," for more information on creating executable script files on the command line.)

```
  GNU nano 2.2.6              File: helloworld.R

#!/usr/bin/env Rscript

print("Hello World")



^G Get Help  ^O WriteOut  ^R Read File ^Y Prev Page ^K Cut Text  ^C Cur Pos
^X Exit      ^J Justify   ^W Where Is  ^V Next Page ^U UnCut Text^T To Spell
```

As with other script types, we can make this script executable with `chmod` and execute it.

```
oneils@atmosphere ~/apcb/r$ chmod +x helloworld.R
oneils@atmosphere ~/apcb/r$ ./helloworld.R
[1] "Hello World"
oneils@atmosphere ~/apcb/r$
```

RStudio

Programming in R on the command line works as well as any other language, but the most common way to program in R today is using an integrated development environment (IDE) known as RStudio. Graphical IDEs like RStudio help the programmer to write code, manage the various files associated with a project, and provide hints and documentation during the programming process. Many IDEs (like Eclipse and Xcode) are complicated pieces of software in their own right. The RStudio IDE is moderately complex, but the developers have worked hard to focus on ease of use specifically for the R programming language. It is available for Windows, OS X, and Linux, at http://rstudio.com. Installing RStudio requires first installing the R interpreter from http://www.r-project.org.

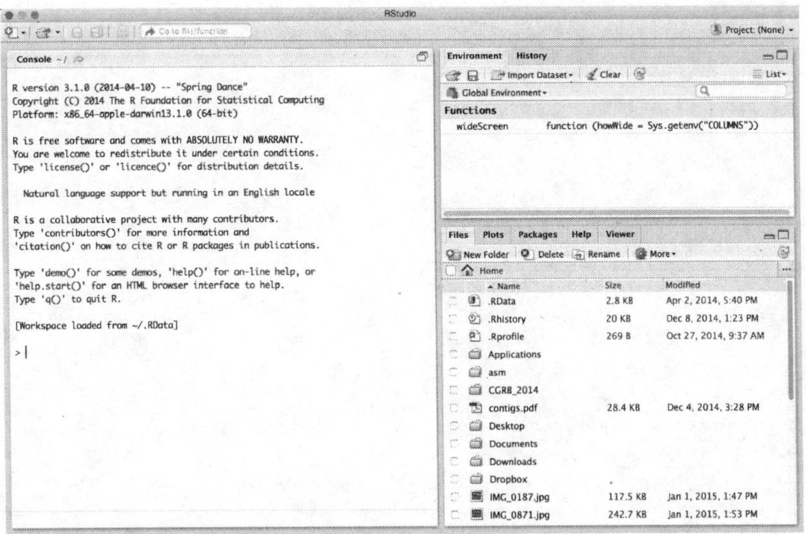

When first opened, RStudio presents three panes. On the left is the same interface with the interpreter that we saw by running `R` on the command line. On the lower right is a pane presenting tabs for a file browser, a help browser, and a panel where plots can be viewed. The upper right pane shows the history of commands that the interpreter has run since it opened and the "global environment," illustrating some information about which variables and data the interpreter currently has stored in memory.

None of these three panes, however, is the one we are primarily interested in! To open up the most important pane, we need to create a new "R script" file—a text file of R commands, just like the executable script on the command line. To do this, we use the button with a green plus sign.

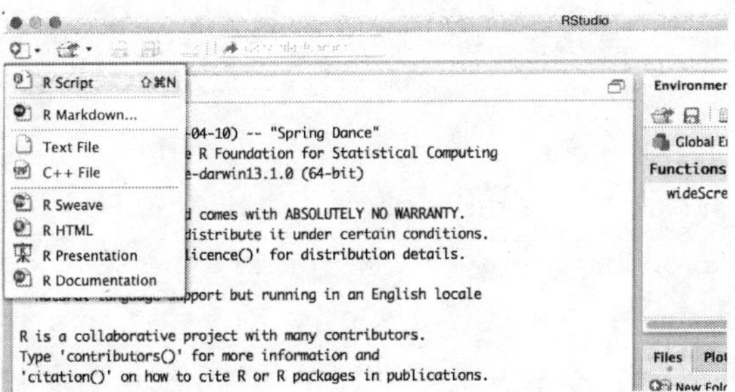

The new pane is an editor for our code file. Here we've entered three lines of code (a line like `#!/usr/bin/env Rstudio` is only necessary for running R scripts on the command line).

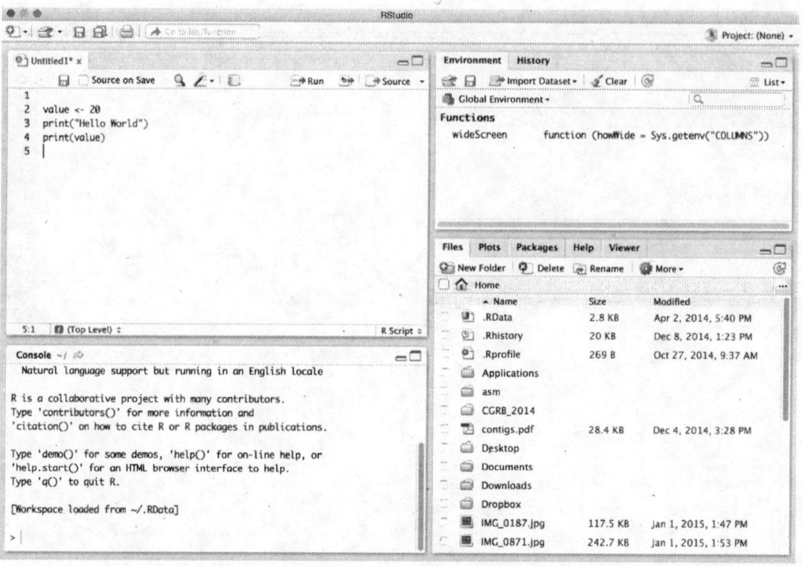

The file editor pane contains a number of buttons, four of which are worth discussing immediately. First, the save button (the small blue diskette) saves the file—R script files traditionally get the file extension `.R`. Farther to the right, the `Run` button sends the highlighted section of code to the interpreter window, even if that section is multiple lines (which are executed by the interpreter in sequence, as with most languages). The next button (with the loopy blue arrow) reruns the most recently run section of code, even if it is not highlighted. Finally, the `Source` button runs all the code in the file, just as would the `Rscript` version on the command line. The outputs of `Run` and `Source` are shown in the interpreter pane below, in black text.

Note that the `Run` button allows the programmer to execute lines of code out of their natural order—we could just as easily run lines 2, 3, and 4 in that order (by highlighting them with the mouse and clicking `Run`) as we could 4 followed by 3 followed by 2 (by highlighting and running each in turn). As it turns out, programs are usually sensitive to the order in which lines of code are executed! So, as much as possible, avoid the `Run` button and instead use the `Source` button. This means that sections of code will be rerun as you develop your programs. The benefit is that you can be sure that your code will run in a non-RStudio environment as well, and you will be less likely to create confusing code.[87] For the most part, we won't illustrate code directly in RStudio, but rather as simple text files and snippets of them.

87. On the other hand, some sections of code might run for many minutes or even hours, so you could consider carefully avoiding rerunning those sections of code when needed.

Libraries/Packages

While a default installation of the R interpreter includes a huge number of functions and packages, many additional libraries have been made available on CRAN (the Comprehensive R Archive Network), providing important new functions. Fortunately, installing these libraries from CRAN is easy and requires only the interactive R interpreter and an internet connection.[88]

As an example, we'll install the `stringr` package, which provides additional functions for working with character-like data types (this will be the first additional package we'll need in later chapters). To install it at the interactive R console, we need to run `install.packages("stringr")`. You may be asked whether the package should be installed in a "personal library" and whether to create such a personal library, to which you can answer `y`. You may also be prompted to select a nearby geographical location from which to download the package, also known as a "mirror."

Once the package has been installed, using it in an R script is as easy as first calling `library("stringr")` or `library(stringr)`, after which the functions provided by the library are available. In this example, we're using the `str_split()` function provided by the `stringr` package; the printed output would be `"Hello" "world"` rather than `"Hello world"`.

```
library(stringr)

print(str_split("Hello world", " "))
```

Note that `install.packages()` needs to be run only once per package, and should usually be done in the interactive R interpreter. The `library()` function will need to be used (once) for each library in each R script that uses it. These calls are usually collected near the top of the script.

Exercises

1. If you are working on the command line, create an executable file interpreted by `Rscript` and have it print some information. If you prefer to try RStudio, install it and create a new R script, having it print several lines of text and using the `Source` button to run the entire script. Experiment with the "Run" and "Re-Run" buttons as well.
2. If you are working in RStudio, use the interface to create a new "Project" (represented by a

88. In R, the terms "library" and "package" are frequently used synonymously, but technically they are distinct. The library is the directory where packages (collections of code and data providing the functionality) are stored.

directory in the file system housing data, R scripts, and other files related to an analysis project) and create several R scripts. For one of the scripts you create, try clicking the "Compile Notebook" icon (it looks like a small notepad) to create an HTML report, and save it in the project directory.

3. If you are using RStudio, try creating a new "R Markdown" file rather than an R script file. R Markdown files allow you to mix "chunks" of R code along with text documentation, which can then be "knitted" into a nicely formatted HTML report. Save the report in the project directory.
4. Install the `stringr` library via the interactive console and write a script that uses the `str_split()` function in it. If you are using RStudio, libraries can also be installed in the "Packages" tab of the RStudio interface.

Chapter 27

Variables and Data

Like most languages, R lets us assign data to variables. In fact, we can do so using either the = assignment operator or the <- operator, though the latter is most commonly found and generally preferred.

```
alpha <- -4.4
print(alpha)                              # prints [1] -4.4
```

Here, `print()` is a function, which prints the contents of its parameter (to the interpreter window in RStudio, or standard output on the command line). This function has the "side effect" of printing the output but doesn't return anything.[89] By contrast, the `abs()` function returns the absolute value of its input without any other effects.

```
alpha_abs <- abs(alpha)
print(alpha_abs)                          # prints [1] 4.4
```

The interpreter ignores # characters and anything after them on a single line, so we can use them to insert comments in our code for explanation or to improve readability. Blank lines are ignored, so we can add them to improve readability as well.

You might be curious why the extra `[1]` is included in the printed output; we'll return to that point soon, but for now, let it suffice to say that the number `4.4` is the first (and only) of a collection of values being printed.

89. The R interpreter will also print the contents of any variable or value returned without being assigned to a variable. For example, the lines `alpha` and `3 + 4` are equivalent to `print(alpha)` and `print(3 + 4)`. Such "printless" prints are common in R code, but we prefer the more explicit and readable call to the `print()` function.

The right-hand side of an assignment is usually evaluated first, so we can do tricky things like reuse variable names in expressions.

```
gamma <- -6.6
gamma <- abs(gamma)
print(gamma)                                        # prints [1] 6.6

count <- 22
count <- count + 1
print(count)                                        # prints [1] 23
```

Variable and function names in R deserve some special discussion. There are a variety of conventions, but a common one that we'll use is the same convention we used for Python: variable names should (1) consist of only letters and numbers and underscores, (2) start with a lowercase letter, (3) use underscores to separate words, and (4) be meaningful and descriptive to make code more readable.

In R, variable and function names are also allowed to include the `.` character, which contains no special meaning (unlike in many other languages). So, `alpha.abs <- abs(alpha)` is not an uncommon thing to see, though we'll be sticking with the convention `alpha_abs <- abs(alpha)`. R variables may be almost anything, so long as we are willing to surround the name with back-tick characters. So, `` `alpha abs` <- abs(alpha) `` would be a valid line of code, as would a following line like `` print(`alpha abs`) ``, though this is *not* recommended.

Numerics, Integers, Characters, and Logicals

One of the most basic types of data in R is the "numeric," also known as a float, or floating-pointing number in other languages.[90] R even supports scientific notation for these types.

```
gc_content <- 0.34                  # numeric 0.34
evalue <- 1e-7                      # numeric 0.0000001
```

R also provides a separate type for integers, numbers that don't have a fractional value. They are important, but less commonly seen in R primarily because numbers are created as numerics, even if they look like integers.

90. This reflects the most common use of the term "numeric" in R, though perhaps not the most accurate. R has a `double` type which implements floating-point numbers, and technically both these and integers are subtypes of `numeric`.

```
seq_len <- 215                          # actually numeric 215.0
```

It is possible to convert numeric types to actual integer types with the `as.integer()` function, and vice versa with the `as.numeric()` function.

```
seq_len_int <- as.integer(seq_len)  # integer 215
seq_len2 <- as.numeric(seq_len_int) # numeric 215.0
```

When converting to an integer type, decimal parts are removed, and thus the values are rounded toward `0` (`4.8` becomes `4`, and `-4.8` would become `-4`.)

The "character" data type holds a string of characters (though of course the string may contain only a single character, or no characters as in `''`). These can be specified using either single or double quotes.

```
name <- 'Shawn'
last_name <- "O'Neil"
```

Concatenating character strings is trickier in R than in some other languages, so we'll cover that in chapter 32, "Character and Categorical Data." (The `cat()` function works similarly, and allows us to include special characters like tabs and newlines by using `\t` and `\n`, respectively; `cat("Shawn\tO'Neil")` would output something like `Shawn  O'Neil`.)

Character types are different from integers and numerics, and they can't be treated like them even if they look like them. However, the `as.character()` and `as.numeric()` functions will convert character strings to the respective type if it is possible to do so.

```
value_chr <- "6.2"
asum <- 5.4 + value_chr                     # Error!
asum2 <- 5.4 + as.numeric(value_chr)        # numeric 11.6
```

By default, the R interpreter will produce a warning (`NAs induced by conversion`) if such a conversion doesn't make sense, as in `as.numeric("Shawn")`. It is also possible to convert a numeric or integer type to a character type, using `as.character()`.

```
asum2_char <- as.character(asum2)           # character, "11.6"
```

The "logical" data type, known as a Boolean type in other languages, is one of the more important types for R. These simple types store either the special value `TRUE` or the special value `FALSE` (by default, these can also be represented by the shorthand `T` and `F`, though this shorthand is less preferred because some coders occasionally use `T` and `F` for variable names as well). Comparisons between other types return logical values (unless they result in a warning or error of some kind). It is possible to compare character types with comparators like < and >; the comparison is done in lexicographic (dictionary) order.

```
sun_is_yellow <- TRUE                              # logical TRUE
result <- 3 < 5                                    # logical TRUE

char_test <- "AACT" < "CGTAC"                      # logical TRUE
```

But beware: in R (and Python), such comparisons also work when they should perhaps instead result in an error: character types can be validly compared to numeric types, and character values are always considered larger. This particular property has resulted in a number of programming mistakes.

```
mix_test <- "AACT" > -20.4                         # logical TRUE
```

R supports <, >, <=, >=, ==, and != comparisons, and these have the same meaning as for the comparisons in Python (see chapter 17, "Conditional Control Flow," for details). For numeric types, R suffers from the same caveat about equality comparison as Python and other languages: rounding errors for numbers with decimal expansions can compound in dangerous ways, and so comparing numerics for equality should be done with care. (You can see this by trying to run `print(0.2 * 0.2 / 0.2 == 0.2)`, which will result in `FALSE`; again, see chapter 17 for details.[91]) The "official" way to compare two numerics for approximate equality in R is rather clunky: `isTRUE(all.equal(a, b))` returns `TRUE` if `a` and `b` are approximately equal (or, if they contain multiple values, all elements are). We'll explore some alternatives in later chapters.

```
a <- 0.2 * 0.2 / 0.2
b <- 0.2
result <- isTRUE(all.equal(a, b))                  # logical TRUE
```

91. Because whole numbers are by default stored as numerics (rather than integers), this may cause some discomfort when attempting to compare them. But because whole numbers can be stored exactly as numerics (without rounding), statements like `4 + 1 == 5`, equivalent to `4.0 + 1.0 == 5.0`, would result in `TRUE`. Still, some cases of division might cause a problem, as in `(1/5) * (1/5) / (1/5) == (1/5)`.

Speaking of programming mistakes, because `<-` is the preferred assignment operator but `=` is also an assignment operator, one must be careful when coding with these and the `==` or `<` comparison operators. Consider the following similar statements, all of which have different meanings.

```
val <- 5                                          # assigns 5 to val
val < -5                                          # compares val to -5 (less than)
val = -5                                          # assigns -5 to val
val == -5                                         # compares val to -5 (equality)
```

R also supports logical connectives, though these take on a slightly different syntax than most other languages.

Connective	Meaning	Example (with `a <- 7`, `b <- 3`)
`&`	and: `True` if both sides are `True`	`a < 8 & b == 3      # True`
`|`	or: `True` if one or both sides are `True`	`a < 8 | b == 9      # True`
`!`	not: `True` if following is `False`	`! a < 3      # True`

These can be grouped with parentheses, and usually should be to avoid confusion.

```
base1 <- "A"
base2 <- "T"
val1 <- 3.5
val2 <- 4.7

result <- val1 < val2 & (base2 == "T" | !(base1 < base2))    # TRUE
```

When combining logical expressions this way, each side of an ampersand or `|` must result in a logical—the code `a == 9 | 7` is *not* the same as `a == 9 | a == 7` (and, in fact, the former will always result in `TRUE` with no warning).

Because R is such a dynamic language, it can often be useful to check what type of data a particular variable is referring to. This can be accomplished with the `class()` function, which returns a character string of the appropriate type.

```
id <- "REG3A"
len <- as.integer(525)
gc_content <- 0.67
id_class <- class(id)
len_class <- class(len)

print(id_class)                          # prints [1] "character"
print(len_class)                         # prints [1] "integer"
print(class(gc_content))                 # prints [1] "numeric"
```

We'll do this frequently as we continue to learn about various R data types.

Exercises

1. Given a set of variables, `a`, `b`, `c`, and `d`, find assignments of them to either `TRUE` or `FALSE` such that the `result` variable holds `TRUE`.

```
a <- # TRUE or FALSE?
b <- # TRUE or FALSE?
c <- # TRUE or FALSE?
d <- # TRUE or FALSE?
## We want the below code to print TRUE
result <- ( c | b ) & ( !b | d ) & ( !c | a ) & ( !c | !a )
print(result)
```

2. Without running the code, try to reason out what `print(class(class(4.5)))` would result in.

3. Try converting a character type like `"1e-50"` to a numeric type with `as.numeric()`, and one like `"1x10^5"`. What are the numeric values after conversion? Try converting the numeric value `0.00000001` to a character type—what is the string produced? What are the smallest and largest numerics you can create?

4. The `is.numeric()` function returns the logical `TRUE` if its input is a numeric type, and `FALSE` otherwise. The functions `is.character()`, `is.integer()`, and `is.logical()` do the same for their respective types. Try using these to test whether specific variables are specific types.

5. What happens when you run a line like `print("ABC"* 4)`? What about `print("ABC" + 4)`? Why do you think the results are what they are? How about `print("ABC" + "DEF")`? Finally, try the following: `print(TRUE + 5)`, `print(TRUE + 7)`, `print(FALSE + 5)`, `print(FALSE + 7)`, `print(TRUE * 4)`, and `print(FALSE * 4)`. What do you think is happening here?

Chapter 28

Vectors

Vectors (similar to single-type arrays in other languages) are ordered collections of simple types, usually numerics, integers, characters, or logicals. We can create vectors using the `c()` function (for concatenate), which takes as parameters the elements to put into the vector:

```
samples <- c(3.2, 4.7, -3.5)              # 3-element numeric vector
```

The `c()` function can take other vectors as parameters, too—it will "deconstruct" all subvectors and return one large vector, rather than a vector of vectors.

```
samples2 <- c(20.4, samples, 37.6)       # 5-element numeric vector
print(samples2)                          # prints [1] 20.4, 3.2, 4.7, -3.5, 37.6
```

We can extract individual elements from a vector using `[]` syntax; though note that, unlike many other languages, the first element is at index 1.

```
second_sample <- samples2[2]             # numeric 3.2
```

The `length()` function returns the number of elements of a vector (or similar types, like lists, which we'll cover later) as an integer:

```
num_samples <- length(samples2)          # integer 5
```

We can use this to extract the last element of a vector, for example.

```
last_sample <- samples2[num_samples]                # numeric 37.6
# OR
last_sample <- samples2[length(samples2)]         # numeric 37.6
```

No "Naked Data": Vectors Have (a) Class

So far in our discussion of R's data types, we've been making a simplification, or at least we've been leaving something out. Even individual values like the numeric `4.6` are actually vectors of length one. Which is to say, `gc_content <- 0.34` is equivalent to `gc_content <- c(0.34)`, and in both cases, `length(gc_content)` will return `1`, which itself is a vector of length one. This applies to numerics, integers, logicals, and character types. Thus, at least compared to other languages, R has no "naked data"; the vector is the most basic unit of data that R has. This is slightly more confusing for character types than others, as each individual element is a string of characters of any length (including potentially the "empty" string `""`).

```
char_vec <- c("one", "two", "three", "four", "five")
```

char_vec[1] char_vec[2] char_vec[3] char_vec[4] char_vec[5]

This explains quite a lot about R, including some curiosities such as why `print(gc_content)` prints `[1] 0.34`. This output is indicating that `gc_content` is a vector, the first element of which is `0.34`. Consider the `seq()` function, which returns a vector of numerics; it takes three parameters:[92] (1) the number at which to start, (2) the number at which to end, and (3) the step size.

```
range <- seq(1, 20, 0.5)
print(range)
```

When we print the result, we'll get output like the following, where the list of numbers is formatted such that it spans the width of the output window.

```
 [1]  1.0  1.5  2.0  2.5  3.0  3.5  4.0  4.5  5.0  5.5  6.0  6.5  7.0  7.5  8.0
[16]  8.5  9.0  9.5 10.0 10.5 11.0 11.5 12.0 12.5 13.0 13.5 14.0 14.5 15.0 15.5
[31] 16.0 16.5 17.0 17.5 18.0 18.5 19.0 19.5 20.0
```

92. Most R functions take a large number of parameters, but many of them are optional. In the next chapter, we'll see what such optional parameters look like, and how to get an extensive list of all the parameters that built-in R functions can take.

The numbers in brackets indicate that the first element of the printed vector is `1.0`, the sixteenth element is `8.5`, and the thirty-first element is `16.0`.

By the way, to produce a sequence of integers (rather than numerics), the step-size argument can be left off, as in `seq(1,20)`. This is equivalent to a commonly seen shorthand, `1:20`.

If all of our integers, logicals, and so on are actually vectors, and we can tell their type by running the `class()` function on them, then vectors must be the things that we are examining the class of. So, what if we attempt to mix types within a vector, for example, by including an integer with some logicals?

```
mix <- c(TRUE, FALSE, as.integer(20))
```

Running `print(class(mix))` will result in `"integer"`. In fact, if we attempt to print out `mix` with `print(mix)`, we'd find that the logicals have been converted into integers!

```
[1]  1  0 20
```

R has chosen to convert `TRUE` into `1` and `FALSE` into `0`; these are standard binary values for true and false, whereas there is no standard logical value for a given integer. Similarly, if a numeric is added, everything is converted to numeric.

```
mix <- c(TRUE, FALSE, as.integer(20), 3.5)
print(class(mix))                              # [1] "numeric"
print(mix)                                     # [1] 1.0 0.0 20.0 3.5
```

And if a character string is added, everything is converted into a character string (with `3.5` becoming `"3.5"`, `TRUE` becoming `"TRUE"`, and so on).

```
mix <- c(TRUE, FALSE, as.integer(20), 3.5, "A")
print(class(mix))                              # [1] "character"
print(mix)                                     # [1] "TRUE" "FALSE" "20" "3.5" "A"
```

In summary, vectors are the most basic unit of data in R, and they cannot mix types—R will autoconvert any mixed types in a single vector to a "lowest common denominator," in the order of logical (most specific), integer, numeric, character (most general). This can sometimes result in difficult-to-find bugs, particularly when reading data from a file. If a file has a column of what appears to be numbers, but a single element cannot be interpreted as a number, the entire vector may be converted to a character type with no warning as the file is read in. We'll discuss reading data in from text files after examining vectors and their properties.

Subsetting Vectors, Selective Replacement

Consider the fact that we can use `[]` syntax to extract single elements from vectors:

```
numbers <- c(10, 20, 30, 40, 50)
second_el <- numbers[2]                    # 20
```

Based on the above, we know that the `20` extracted is a vector of length one. The `2` used in the brackets is also a vector of length one; thus the line above is equivalent to `second_el <- nums[c(2)]`. Does this mean that we can use longer vectors for extracting elements? Yes!

```
subvector <- numbers[c(3,2)]
print(subvector)                           # [1] 30 20
```

In fact, the extracted elements were even placed in the resulting two-element vector in the order in which they were extracted (the third element followed by the second element). We can use a similar syntax to selectively replace elements by specific indices in vectors.

```
numbers[c(3,2)] <- c(35, 25)
print(numbers)                             # [1] 10 25 35 40 50
```

Selective replacement is the process of replacing selected elements of a vector (or similar structure) by specifying which elements to replace with `[]` indexing syntax combined with assignment `<-`.[93]

R vectors (and many other data container types) can be named, that is, associated with a character vector of the same length. We can set and subsequently get this names vector using the `names()` function, but the syntax is a little odd.

93. The term "selective replacement" is not widely used outside of this book. In some situations, the term "conditional replacement" is used, but we wanted to define some concrete terminology to capture the entirety of the idea.

```
# create vector
scores <- c(89, 94, 73)
# set names for the elements
names(scores) <- c("Student A", "Student B", "Student C")

print("Printing the vector:")
print(scores)

print("Printing the names:")
names_scores <- names(scores)
print(names_scores)
```

Named vectors, when printed, display their names as well. The result from above:

```
[1] "Printing the vector:"
Student A Student B Student C
       89        94        73
[1] "Printing the names:"
[1] "Student A" "Student B" "Student C"
```

Named vectors may not seem that helpful now, but the concept will be quite useful later. Named vectors give us another way to subset and selectively replace in vectors: by name.

```
ca_scores <- scores[c("Student C", "Student A")]  # 2 element vector: 73 98
# OR
ca_names <- c("Student C", "Student A")
ca_scores <- scores[ca_names]

scores[c("Student A", "Student C")] <- c(93, 84)
print(scores)
```

Although R doesn't enforce it, the names should be unique to avoid confusion when selecting or selectively replacing this way. Having updated Student A's and Student B's score, the change is reflected in the output:

```
Student A Student B Student C
       93        94        84
```

There's one final and extremely powerful way of subsetting and selectively replacing in a vector: by logical vector. By indexing with a vector of logicals of the same length as the vector to be indexed, we can extract only those elements where the logical vector has a `TRUE` value.

```
select_vec <- c(TRUE, FALSE, TRUE)
ac_scores <- scores[select_vec]                    # 2 element vector: 93 84
# OR
ac_scores <- scores[c(TRUE, FALSE, TRUE)]
```

While indexing by index number and by name allows us to extract elements in any given order, indexing by logical doesn't afford us this possibility.

We can perform selective replacement this way as well; let's suppose Students A and C retake their quizzes and moderately improve their scores.

```
scores[c(TRUE, FALSE, TRUE)] <- c(94, 86)
print(scores)
```

And the printed output:

```
Student A Student B Student C
       94        94        86
```

In this case, the length of the replacement vector (`c(159, 169)`) is equal to the number of `TRUE` values in the indexing vector (`c(TRUE, FALSE, TRUE)`); we'll explore whether this is a requirement below.

In summary, we have three important ways of indexing into/selecting from/selectively replacing in vectors:

1. by index number vector,
2. by character vector (if the vector is named), and
3. by logical vector.

Vectorized Operations, NA Values

If vectors are the most basic unit of data in R, all of the functions and operators we've been working with—`as.numeric()`, `*`, and even comparisons like `>`—implicitly work over entire vectors.

```
numeric_chars <- c("6", "3.7", "9b3x")
numerics <- as.numeric(numeric_chars)
print(numerics)                                   # [1] 6.0 3.7 NA
```

In this example, each element of the character vector has been converted, so that `class(numerics)` would return `"numeric"`. The final character string, `"9b3x"`, cannot be reasonably converted to a numeric type, and so it has been replaced by `NA`. When this happens, the interpreter produces a warning message: `NAs introduced by coercion`.

`NA` is a special value in R that indicates either missing data or a failed computation of some type (as in attempting to convert `"9b3x"` to a numeric). Most operations involving `NA` values return `NA` values; for example, `NA + 3` returns `NA`, and many functions that operate on entire vectors return an `NA` if any element is `NA`. A canonical example is the `mean()` function.

```
ave <- mean(numerics)
print(ave)                                # [1] NA
```

Such functions often include an optional parameter that we can give, `na.rm = TRUE`, specifying that `NA` values should be removed before the function is run.

```
ave <- mean(numerics, na.rm = TRUE)
print(ave)                                # [1] 4.85
```

While this is convenient, there is a way for us to remove `NA` values from any vector (see below).

Other special values in R include `NaN`, for "Not a Number," returned by calculations such as the square root of -1, `sqrt(-1)`, and `Inf` for "Infinity," returned by calculations such as `1/0`. (`Inf/Inf`, by the way, returns `NaN`.)

Returning to the concept of vectorized operations, simple arithmetic operations such as `+`, `*`, `/`, `-`, `^` (exponent), and `%%` (modulus) are vectorized as well, meaning that an expression like `3 * 7` is equivalent to `c(3) * c(7)`. When the vectors are longer than a single element, the operation is done on an element-by-element basis.

```
values <- c(10, 20, 30, 40)
mult <- c(1, 2, 3, 4)

result <- values * mult                  # 4 element vector: 10 40 90 160
```

If we consider the * operator, it takes two inputs (numeric or integer) and returns an output (numeric or integer) for each pair from the vectors. This is quite similar to the comparison >, which takes two inputs (numeric or integer or character) and returns a logical.

```
10  20  30  40
 *   *   *   *
 1   2   3   4
       =
10  40  90  160
```

```
values <- c(10, 20, 30, 40)
comparison_values <- c(25, 10, 25, 35)

result <- values > comparison_values   # 4 element vector: FALSE TRUE TRUE TRUE
```

Vector Recycling

What happens if we try to multiply two vectors that aren't the same length? It turns out that the shorter of the two will be reused as needed, in a process known as *vector recycling*, or the reuse of the shorter vector in a vectorized operation.

```
values <- c(10, 20, 30, 40)
mult <- c(10, -10)

result <- values * mult                # 4 element vector: 100 -200 300 -400
```

This works well when working with vectors of length one against longer vectors, because the length-one vector will be recycled as needed.

```
result <- values * 2                   # same as values * c(2)
print(result)                          # [1] 20 40 60 80
```

If the length of the longer vector is not a multiple of the length of the shorter, however, the last recycle will go only partway through.

```
values <- c(3, 5, 7)
mult <- c(10, -10)

result <- values * mult                # 3 element vector: 30 -50 70
```

When this happens, the interpreter prints a warning: `longer object length is not a multiple of shorter object length`. There are few situations where this type of partial recycling is not an accident, and it should be avoided.

Vector recycling also applies to selective replacement; for example, we can selectively replace four elements of a vector with elements from a two-element vector:

```
values <- c(10, 20, 30, 40, 50, 60)
values[c(TRUE, TRUE, FALSE, TRUE, FALSE, TRUE)] <- c(5, -5)

print(values)                          # [1] 5 -5 30 5 50 -5
```

More often we'll selectively replace elements of a vector with a length-one vector.

```
values <- c(10, 20, 30, 40, 50, 60)
values[c(TRUE, TRUE, FALSE, TRUE, FALSE, TRUE)] <- 0 # same as ... <- c(0)

print(values)                          # [1] 0 0 30 0 50 0
```

These concepts, when combined with vector indexing of various kinds, are quite powerful. Consider that an expression like `values > 35` is itself vectorized, with the shorter vector (holding just `35`) being recycled such that what is returned is a logical vector with `TRUE` values where the elements of `values` are greater than `35`. We could use this vector as an indexing vector for selective replacement if we wish.

```
values <- c(10, 20, 30, 40, 50, 60)
select_vec <- values > 35              # TRUE TRUE TRUE FALSE FALSE FALSE
values[select_vec] <- 0

print(values)                          # [1] 10 20 30 0 0 0
```

More succinctly, rather than create a temporary variable for `select_vec`, we can place the expression `values > 35` directly within the brackets.

```
values <- c(10, 20, 30, 40, 50, 60)
values[values > 35] <- 0

print(values)                          # [1] 10 20 30 0 0 0
```

Similarly, we could use the result of something like `mean(values)` to replace all elements of a vector greater than the mean with `0` easily, no matter the order of the elements!

```
values <- c(30, 10, 60, 20, 40, 50)
values[values > mean(values)] <- 0

print(values)                          # [1] 30 10 0 20 0 0
```

More often, we'll want to extract such values using logical selection.

```
values <- c(30, 10, 60, 20, 40, 50)
gt_mean <- values[values > mean(values)]

print(gt_mean)                         # [1] 60 40 50
```

These sorts of vectorized selections, especially when combined with logical vectors, are a powerful and important part of R, so study them until you are confident with the technique.

Exercises

1. Suppose we have `r` as a range of numbers from 1 to 30 in steps of 0.3; `r<- seq(1, 30, 0.3)`. Using just the `as.integer()` function, logical indexing, and comparisons like `>`, generate a sequence `r_decimals` that contains all values of `r` that are not round integers. (That is, it should contain all values of `r` except 1.0, 2.0, 3.0, and so on. There should be 297 of them.)
2. We briefly mentioned the `%%`, or "modulus," operator, which returns the remainder of a number after integer division (e.g., `4 %% 3 == 1` and `4 %% 4 == 0`; it is also vectorized). Given any vector `r`, for example `r <- seq(1, 30, 0.3)`, produce a vector `r_every_other` that contains every other element of `r`. You will likely want to use `%%`, the `==` equality comparison, and you might also want to use `seq()` to generate a vector of indices of the same length as `r`.

   Do the same again, but modify the code to extract every third element of `r` into a vector called `r_every_third`.
3. From chapter 27, "Variables and Data," we know that comparisons like `==`, `!=`, `>=` are available as well. Further, we know that `!` negates the values of a logical vector, while `&` combines two logical vectors with "and," and `|` combines two logical vectors with "or." Use these, along with the `%%` operator discussed above, to produce a vector `div_3_4` of all integers between 1 and 1,000 (inclusive) that are evenly divisible by 3 and evenly divisible by 4. (There are 83 of them.) Create another, `not_div_5_6`, of numbers that are not evenly divisible by 5 or 6. (There are 667 of them. For example, 1,000 should not be included because it is divisible by 5, and 18

should not be included because it is divisible by 6, but 34 should be because it is divisible by neither.)

Common Vector Functions

As vectors (specifically numeric vectors) are so ubiquitous, R has dozens (hundreds, actually) of functions that do useful things with them. While we can't cover all of them, we can quickly cover a few that will be important in future chapters.

First, we've already seen the `seq()` and `length()` functions; the former generates a numeric vector comprising a sequence of numbers, and the latter returns the length of a vector as a single-element integer vector.

```
range <- seq(0, 7, 0.2)                        # 0.0 0.2 0.4 ... 7.0
len_range <- length(range)                     # 36
```

Presented without an example, `mean()`, `sd()`, and `median()` return the mean, standard deviation, and median of a numeric vector, respectively. (Provided that none of the input elements are `NA`, though all three accept the `na.rm = TRUE` parameter.) Generalizing `median()`, the `quantile()` function returns the *Y*th percentile of a function, or multiple percentiles if the second argument has more than one element.

```
quantiles_range <- quantile(range, c(0.25, 0.5, 0.75))
print(quantiles_range)
```

The output is a named numeric vector:

```
 25%  50%  75% 
1.75 3.50 5.25
```

The `unique()` function removes duplicates in a vector, leaving the remaining elements in order of their first occurrence, and the `rev()` function reverses a vector.

```
values <- c(20, 40, 30, 20, 10, 50, 10)
values_uniq <- unique(values)          # 20 40 30 10 50

rev_uniq <- rev(values_uniq)           # 50 10 30 40 20
```

There is the `sort()` function, which sorts a vector (in natural order for numerics and integers, and lexicographic (dictionary) order for character vectors). Perhaps more interesting is the `order()` function, which returns an integer vector of indices describing where the original elements of the vector would need to be placed to produce a sorted order.

```
order_rev_uniq <- order(rev_uniq)       # 2 5 3 4 1
```

In this example, the order vector, `2 5 3 4 1`, indicates that the second element of `rev_uniq` would come first, followed by the fifth, and so on. Thus we could produce a sorted version of `rev_uniq` with `rev_uniq[order_rev_uniq]` (by virtue of vectors' index-based selection), or more succinctly with `rev_uniq[order(rev_uniq)]`.

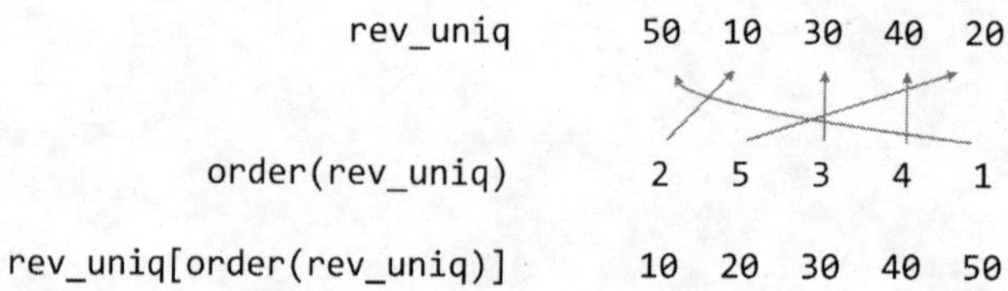

Importantly, this allows us to rearrange multiple vectors with a common order determined by a single one. For example, given two vectors, `id` and `score`, which are related element-wise, we might decide to rearrange both sets in alphabetical order for `id`.

```
id <- c("cc4", "aa6", "bb3")
score <- c(20.05, 35.62, 42.71)

id_sorted <- id[order(id)]
score_sorted <- score[order(id)]

print(id_sorted)                         # [1] "aa6" "bb3" "cc4"
print(score_sorted)                      # [1] 35.62 42.71 20.05
```

The `sample()` function returns a random sampling from a vector of a given size, either with replacement or without as specified with the `replace =` parameter (`FALSE` is the default if unspecified).

```
values <- c(5, 10, 15, 20, 25, 30)

sample_1 <- sample(values, 3, replace = FALSE)        # 15 5 30
sample_2 <- sample(values, 3, replace = TRUE)         # 15 30 15
```

The `rep()` function repeats a vector to produce a longer vector. We can repeat in an element-by-element fashion, or over the whole vector, depending on whether the `each =` parameter is used or not.

```
count <- c(1, 2)

count_rep1 <- rep(count, 3)                     # 1 2 1 2 1 2
count_rep2 <- rep(count, each = 3)              # 1 1 1 2 2 2
```

Last (but not least) for this discussion is the `is.na()` function: given a vector with elements that are possibly `NA` values, it returns a logical vector whole elements are `TRUE` in indices where the original was `NA`, allowing us to easily indicate which elements of vectors are `NA` and remove them.

```
values_char <- c("5.7", "4.3", "a9b3", "2.4")
values <- as.numeric(values_char)               # 5.7 4.3 NA 2.4

values_na <- is.na(values)                      # FALSE FALSE TRUE FALSE

values_no_nas <- values[!values_na]             # 5.7 4.3 2.4
# OR
values_no_nas <- values[!is.na(values_na)]      # 5.7 4.3 2.4
```

Notice the use of the exclamation point in the above to negate the logical vector returned by `is.na()`.

Generating Random Data

R excels at working with probability distributions, including generating random samples from them. Many distributions are supported, including the Normal (Gaussian), Log-Normal, Exponential, Gamma, Student's *t*, and so on. Here we'll just look at generating samples from a few for use in future examples.

First, the `rnorm()` function generates a numeric vector of a given length sampled from the Normal distribution with specified mean (with `mean =`) and standard deviation (with `sd =`).

```
sample_norm <- rnorm(5, mean = 6, sd = 2)     # e.g. 7.07 2.4 4.5 6.2 5.1
```

Similarly, the `runif()` function samples from a uniform distribution limited by a minimum and maximum value.

```
sample_unif <- runif(5, min = 2, max = 6)       # e.g. 2.1 4.06 2.48 4.67 5.80
```

The `rexp()` generates data from an Exponential distribution with a given "rate" parameter, controlling the rate of decay of the density function (the mean of large samples will approach `1.0/rate`).

```
sample_exp <- rexp(5, rate = 1.5)               # e.g. 0.24 0.50 0.01 0.55 0.30
```

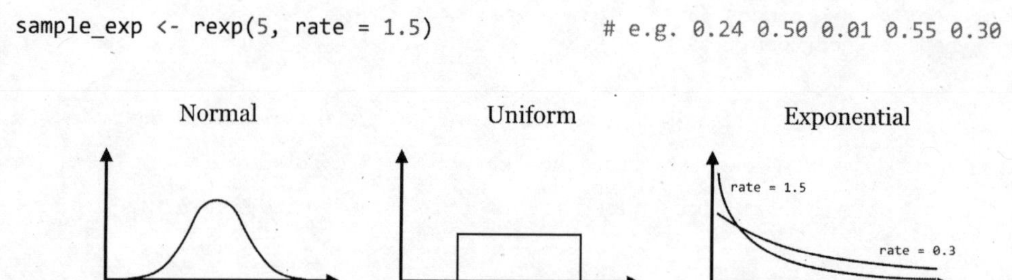

R includes a large number of statistical tests, though we won't be covering much in the way of statistics other than a few driving examples. The `t.test()` function runs a two-sided student's *t*-test comparing the means of two vectors. What is returned is a more complex data type with class `"htest"`.

```
sample_1 <- rnorm(100, mean = 10, sd = 4)
sample_2 <- rnorm(100, mean = 12, sd = 4)

ttest_result <- t.test(sample_1, sample_2)
print(class(ttest_result))                      # [1] "htest"
print(ttest_result)
```

When printed, this complex data type formats itself into nice, human-readable output:

```
Welch Two Sample t-test

data:  sample_1 and sample_2
t = -2.6847, df = 193.503, p-value = 0.007889
alternative hypothesis: true difference in means is not equal to 0
95 percent confidence interval:
 -2.690577 -0.411598
sample estimates:
mean of x mean of y
 10.03711  11.58819
```

Reading and Writing Tabular Data, Wrapping Long Lines

Before we go much further, we're going to want to be able to import data into our R programs from external files (which we'll assume to be rows and columns of data in text files). We'll do this with `read.table()`, and the result will be a type of data known as a "data frame" (or `data.frame` in code). We'll cover the nuances of data frames later, but note for now that they can be thought of as a collection of vectors (of equal length), one for each column in the table.

As an example, let's suppose we have a tab-separated text file in our present working directory called **`states.txt`**.[94] Each row represents one of the US states along with information on population, per capita income, illiteracy rate, murder rate (per 100,000), percentage of high school graduates, and region (all measured in the 1970s). The first row contains a "header" line with column names.

```
name    population      income  murder  hs_grad region
Alabama 3615    3624    15.1    41.3    South
Alaska  365     6315    11.3    66.7    West
Arizona 2212    4530    7.8     58.1    West
Arkansas        2110    3378    10.1    39.9    South
California      21198   5114    10.3    62.6    West
Colorado        2541    4884    6.8     63.9    West
...
```

Later in the file, someone has decided to annotate Michigan's line, indicating it as the "mitten" state:

```
...
Massachusetts   5814    4755    3.3     58.5    Northeast
Michigan        9111    4751    11.1    52.8    North Central   # mitten
Minnesota       3921    4675    2.3     57.6    North Central
...
```

Like most functions, `read.table()` takes many potential parameters (23, in fact), but most of them have reasonable defaults. Still, there are five or so that we will commonly need to set. Because of the need to set so many parameters, using `read.table()` often results in a long line of code. Fortunately, the R interpreter allows us to break long lines over multiple lines, so long as each line

94. When running on the command line, the present working directory is inherited from the shell. In RStudio, the present working directory is set to the "project" directory if the file is part of a project folder. In either case, it is possible to change the working directory from within R using the `setwd()` directory, as in `setwd("/home/username/rproject")` in Unix/Linux and `setwd("C:/Documents and Settings/username/My Documents/rproject")` in Windows. It is also possible to specify file names by absolute path, as in `/home/username/rproject/states.txt`, no matter the present working directory.

ends on a character that doesn't complete the expression (so the interpreter knows it needs to keep reading following lines before executing them). Common character choices are the comma and plus sign. When we do wrap a long line in this way, it's customary to indent the following lines to indicate their continuance in a visual way.

```
states <- read.table(file = "states.txt",
                     header = TRUE,
                     sep = "\t",
                     stringsAsFactors = FALSE,
                     comment.char = "#")
```

When reading `states.txt`, the `file =` parameter specifies the file name to be read, while `header = TRUE` indicates to the interpreter that the first line in the file gives the column names (without it, the column names will be `"V1"`, `"V2"`, `"V3"` and so on). The `sep = "\t"` parameter indicates that tab characters are used to separate the columns in the file (the default is any whitespace), and `comment.char = "#"` indicates that `#` characters and anything after them should be ignored while reading the file (which is appropriate, as evident by the `# mitten` annotation in the file). The `stringsAsFactors = FALSE` parameter is more cryptic: it tells the interpreter to leave the character-vector columns (like `region` in this example) as character vectors, rather than convert them to the more sophisticated `factor` data type (to be covered in later chapters).

At this point, the `states` variable contains the data frame holding the columns (vectors) of data. We can print it with `print(states)`, but the result is quite a lot of output:

```
         name population income murder hs_grad        region
1     Alabama       3615   3624   15.1    41.3         South
2      Alaska        365   6315   11.3    66.7          West
3     Arizona       2212   4530    7.8    58.1          West
4    Arkansas       2110   3378   10.1    39.9         South
5  California      21198   5114   10.3    62.6          West
6    Colorado       2541   4884    6.8    63.9          West
...
```

It might make better sense to extract just the first 10 rows of data and print them, which we can do with the `head()` function (`head()` can also extract just the first few elements of a long vector).

```
first_10 <- head(states, n = 10)

print(first_10)
# OR
print(head(states, n = 10))
```

The functions `nrow()` and `ncol()` return the number of rows and columns of a data frame, respectively (which is preferred over `length()`, which returns the number of columns); the `dim()` function returns a two-element vector with number of rows (at index 1) and number of columns (at index 2).

As mentioned previously, individual columns of a data frame are (almost always) vectors. To access one of these individual vectors, we can use a special `$` syntax, with the column name following the `$`.

```
incomes <- states$"income"
print(incomes)                    # [1] 3624 6315 4530 3378 5114 4884 ...
```

So long as the column name is sufficiently simple (in particular, so long as it doesn't have any spaces), then the quote marks around the column name can be (and often are) omitted.

```
incomes <- states$income
print(incomes)                    # [1] 3624 6315 4530 3378 5114 4884 ...
```

Although this syntax can be used to extract a column from a data frame as a vector, note that it refers to the vector within the data frame as well. In a sense, `states$income` *is* the vector stored in the `states` data frame. Thus we can use techniques like selective replacement to work with them just like any other vectors. Here, we'll replace all instances of "North Central" in the `states$region` vector with just the term "Central," effectively renaming the region.[95]

```
nrth_cntrl_logical <- states$region == "North Central"  # Logical vector
states$region[nrth_cntrl_logical] <- "Central"          # Selective replacement
# OR
states$region[states$region == "North Central"] <- "Central"
```

95. If you have any familiarity with R, you might have run across the `attach()` function, which takes a data frame and results in the creation of a separate vector for each column. Generally, "disassembling" a data frame this way is a bad idea—after all, the columns of a data frame are usually associated with each other for a reason! Further, this function results in the creation of many variables with names based on the column names of the data frame. Because these names aren't clearly delimited in the code, it's easy to create hard-to-find bugs and mix up columns from multiple data frames this way.

Writing a data frame to a tab-separated file is accomplished with the `write.table()` function.[96] As with `read.table()`, `write.table()` can take quite a few parameters, most of which have reasonable defaults. But there are six or so we'll want to set more often than others. Let's write the modified `states` data frame to a file called `states_modified.txt` as a tab-separated file.

```
write.table(states,
            file = "states_modified.txt",
            quote = FALSE,
            sep = "\t",
            row.names = FALSE,
            col.names = TRUE)
```

The first two parameters here are the data frame to write and the file name to write to. The `quote = FALSE` parameter specifies that quotation marks shouldn't be written around character types in the output (so the `name` column will have entries like `Alabama` and `Alaska` rather than `"Alabama"` and `"Alaska"`). The `sep = "\t"` indicates that tabs should separate the columns, while `row.names = FALSE` indicates that row names should not be written (because they don't contain any meaningful information for this data frame), and `col.names = TRUE` indicates that we do want the column names output to the first line of the file as a "header" line.

R and the Unix/Linux Command Line

In chapter 26, "An Introduction," we mentioned that R scripts can be run from the command line by using the `#!/usr/bin/env Rscript` executable environment. (Older versions of R required the user to run a command like `R CMD BATCH scriptname.R`, but today using `Rscript` is preferred.) We devoted more discussion to interfacing Python with the command line environment than we will R, partially because R isn't as frequently used that way, but also because it's quite easy.

When using `read.table()`, for example, data can be read from standard input by using the file name `"stdin"`. Anything that is printed from an R script goes to standard output by default. Because R does a fair amount of formatting when printing, however, it is often more convenient to print data frames using `write.table()` specifying `file = ""`.

96. There are also more specialized functions for both reading and writing tabular data, such as `read.csv()` and `write.csv()`. We've focused on `read.table()` and `write.table()` because they are flexible enough to read and write tables in a variety of formats, including comma separated, tab separated, and so on.

Finally, to get command line parameters into an R script as a character vector, the line `args <- commandArgs(trailingOnly = TRUE)` will do the trick. Here's a simple script that will read a table on standard input, write it to standard output, and also read and print out any command line arguments:

```
#!/usr/bin/env Rscript

# read args from command-line params
args <- commandArgs(trailingOnly = TRUE)
print(args)

# read data frame from stdin
input_df <- read.table("stdin",
                       header = FALSE,
                       stringsAsFactors = FALSE)
# write data frame to stdout
write.table(input_df,
            file = "",
            row.names = FALSE,
            col.names = FALSE,
            sep = "\t")
```

Try making this script executable on the command line, and running it on **`p450s_blastp_yeast_top1.txt`** with something like `cat p450s_blastp_yeast_top1.txt | ./stdin_stdout_ex.R arg1 'arg 2'`.

Exercises

1. Suppose we have any odd-length numeric vector (e.g., `sample<- c(3.2, 5.1, 2.5, 1.6, 7.9)` or `sample <- runif(25, min = 0, max = 1)`). Write some lines of code that result in printing the median of the vector, *without* using the `median()` or `quantile()` functions. You might find the `length()` and `as.integer()` functions to be helpful.
2. If `sample` is a sample from an exponential distribution, for example, `sample <- rexp(1000, rate = 1.5)`, then the median of the sample is generally smaller than the mean. Generate a vector, `between_median_mean`, that contains all values of `sample` that are larger than (or equal to) the median of the sample, and less than (or equal to) the mean of the sample.
3. Read in the **`states.txt`** file into a data frame as described. Extract a numeric vector called `murder_lowincome` containing murder rates for just those states with per capita incomes less

than the median per capita income (you can use the `median()` function this time). Similarly, extract a vector called `murder_highincome` containing murder rates for just those states with greater than (or equal to) the median per capita income. Run a two-sample `t.test()` to determine whether the mean murder rates are different between these two groups.

4. Let `states` be the state information data frame described above. Describe what the various operations below do in terms of indexing, selective replacement, vector recycling, and the types of data involved (e.g., numeric vectors and logical vectors). To get you started, the first line adds a new column to the `states` data frame called `"newpop"` that contains the same information as the `"population"` column.

```
states$newpop <- states$population
highmurder <- states$murder >= median(states$murder)
states$newpop[highmurder] <- states$population[highmurder] * 0.9
states$newpop[!highmurder] <- states$population[!highmurder] * 1.1
```

5. Determine the number of unique regions that are listed in the `states` data frame. Determine the number of unique regions represented by states with greater than the median income.

6. What does the `sum()` function report for a numeric vector `c(2, 3, 0, 1, 0, 2)`? How about for `c(1, 0, 0, 1, 1, 0)`? And, finally, how about for the logical vector `c(TRUE, FALSE, FALSE, TRUE, TRUE, FALSE)`? How could the `sum()` function thus be useful in a logical context?

Chapter 29

R Functions

While we could continue to cover R's unique and powerful vector syntax, it's time to have some fun and learn about functions. Functions in R are similar to their Python counterparts (see chapter 18, "Python Functions"): they *encapsulate* a block of code, making it reusable as well as allowing us to consider the block in isolation of the rest of the program. As with functions in most languages, R functions consist of three main parts:

1. The input (parameters given to the function).
2. The code block that is to be executed using those parameters. In R, blocks are defined by a matching pair of curly brackets, { and }.
3. The output of the function, called the return value. This may be optional if the function "does something" (like `print()`) rather than "returns something."

Let's consider the problem of determining which elements of two numeric vectors, say `vec1` and `vec2`, are close enough to equal to call them equal. As mentioned in chapter 27, "Variables and Data," the standard way to check if all elements in two equal-length vectors are approximately pairwise-equal is to use `isTRUE(all.equal(vec1, vec2))`, which returns a single `TRUE` if this is the case and a single `FALSE` if not.

```
vec1 <- c(1/5, 0.2, 0.2 * 0.2 / 0.2)
vec2 <- c(1/5, 1/5, 1/5)

eq <- isTRUE(all.equal(vec1, vec2))
print(eq)                                  # [1] TRUE
```

But perhaps we'd rather like a logical vector indicating *which* elements are approximately equal. The most straightforward way to do this is by comparing the absolute difference between the elements with some small epsilon value.

```
vec1 <- c(4.00000001, 6)
vec2 <- c(4, 2)
epsilon <- 0.00001
eq <- abs(vec1 - vec2) < epsilon
print(eq)                                  # [1] TRUE FALSE
```

As a review of the last chapter, what is happening here is that the - operation is vectorized over the left- and right-hand sides, producing a vector (using vector recycling if one of the two were shorter, which not the case here; see chapter 28), as is the `abs()` function, which takes a vector and returns a vector of absolute values. Similarly, the < operator is vectorized, and because `epsilon` is a vector of length one, so it is compared to all elements of the result of `abs(vec1 - vec2)` using vector recycling, for the final result of a logical vector.

Because this sort of operation is something we might want to perform many times, we could write a function for it. In this case, we'll call our function `equalish()`; here's the R code for defining and running such a function.

```
# Compares elements of a and b;
# returns TRUE for each within epsilon
equalish <- function(a, b, epsilon = 0.00001) {
  result <- abs(a - b) < epsilon
  return(result)
}

vec1 <- c(4.00000001, 6)
vec2 <- c(4, 2)
eq <- equalish(vec1, vec2)
print(eq)                              # [1] TRUE FALSE
```

There are many things to note here. First, when defining a function, we define the parameters it can take. Parameters in R functions have a *position* (`a` is at position 1, `b` is at position 2, and `epsilon` is at position 3) and a *name* (`a`, `b`, and `epsilon`). Some parameters may have a default value: the value they should have if unspecified otherwise, while other parameters may be required: the user of the function must specify them. Default values are assigned within the parameter list with = (not <- as in standard variable assignment).

The block that defines the operations performed by the function is enclosed in curly brackets, usually with the opening bracket on the same line as the function/parameter list definition, and the closing bracket on its own line. We've indented the lines that belong to the function block by two spaces (an R convention). Although not required, this is a good idea, as it makes code much more readable. The value that is returned by the function is specified with a call to a special `return()` function—functions can only return one value, though it might be something sophisticated like a vector or data frame.[97]

After a function has been defined, it can be called, as in `eq <- equalish(vec1, vec2)`. The variable names associated with the data *outside* the function (in this case `vec1` and `vec2`) needn't match the parameter names *inside* the function (`a` and `b`). This is an important point to which we will return.

In the call above, we let the `epsilon` parameter take its default value of `0.00001`. We could alternatively use a stricter comparison.

```
eq <- equalish(vec1, vec2, 0.000000000001)
print(eq)                                   # [1] FALSE FALSE
```

In R, arguments to functions may be specified by position (as in the example above), by name, or by a combination.

```
# by name:
eq <- equalish(a = vec1, b = vec2, epsilon = 0.000000000001)
# mix of position and name:
eq <- equalish(vec1, vec2, epsilon = 0.000000000001)
```

Many R functions take a few required parameters and many nonrequired parameters with reasonable defaults; this calling scheme allows us to specify the required parameters as well as only those nonrequired ones that we wish to change.

97. Any variable that is simply stated (without assignment) in a function will be returned. So, this definition is equivalent:

```
equalish <- function(a, b, epsilon = 0.00001) {
  result <- abs(a - b) < epsilon
  result
}
```

Some R programmers prefer this syntax; for this text, however, we'll stick to using the more explicit `return()`. This also helps differentiate between such "returnless returns" and "printless prints" (see the footnote in Chapter 27, Variables and Data).

In general, you should specify parameters by position first (if you want to specify any by position), then by name. Although the following calls will work, they're quite confusing.

```
eq <- equalish(vec1, epsilon = 0.00000000001, vec2)       # confusing!
eq <- equalish(epsilon = 0.00000000001, vec2, a = vec1)   # confusing!
```

We frequently use default parameters to specify named parameters in functions called within the function we're defining. Here is an example of a function that computes the difference in means of two vectors; it takes an optional `remove_NAs` parameter that defaults to `FALSE`. If this is specified as `TRUE`, the `na.rm` parameter in the calls to `mean()` is set to `TRUE` as well in the computation.

```
# returns the difference between mean(vec1)
# and mean(vec2)
diff_mean <- function(vec1, vec2, remove_NAs = FALSE) {
  m1 <- mean(vec1, na.rm = remove_NAs)
  m2 <- mean(vec2, na.rm = remove_NAs)
  return(m1 - m2)
}
```

For continuity with other R functions, it might have made better sense to call the parameter `na.rm`; in this case, we would modify the computation lines to read like `m1 <- mean(vec1, na.rm = na.rm)`. Although it may seem that the R interpreter would be confused by the duplicate variable names, the fact that the `mean()` parameter `na.rm` happens to have the same name as the variable being passed will cause no trouble.

Variables and Scope

Let's run a quick experiment. Inside our function, the variable `result` has been assigned with the line `result <- abs(a - b) < epsilon`. After we run the function, is it possible to access that variable by printing it?

```
vec1 <- c(4.00000001, 6)
vec2 <- c(4, 2)
eq <- equalish(vec1, vec2)                   # TRUE FALSE
print(result)
```

Printing doesn't work!

```
Error in print(result) : object 'result' not found
Execution halted
```

This variable doesn't print because, as in most languages, variables assigned within functions have a scope *local* to that function block. (A variable's scope is the context in which it can be accessed.) The same goes for the parameter variables—we would have no more success with `print(a)`, `print(b)`, or `print(epsilon)` outside of the function.

One of the best features of these local variables is that they are independent of any variables that might already exist. For example, the function creates a variable called `result` (which we now know is a local variable scoped to the function block). What if, outside of our function, we also had a `result` variable being used for an entirely different purpose? Would the function overwrite its contents?

```
result <- "Success"
vec1 <- c(4.00000001, 6)
vec2 <- c(4, 2)
eq <- equalish(vec1, vec2)                    # TRUE FALSE
print(result)
```

True to the independence of the local `result` variable inside the function, the contents of the external `result` are not overwritten.

```
[1] "Success"
```

This feature of how variables work within functions might seem somewhat strange, but the upshot is important: functions can be *fully encapsulated*. If they are designed correctly, their usage cannot affect the code context in which they are used (the only way standard R functions can affect the "outside world" is to return some value). Functions that have this property and always return the same value given the same inputs (e.g., have no random component) are called "pure." They can be treated as abstract black boxes and designed in isolation of the code that will use them, and the code that uses them can be designed without consideration of the internals of functions it calls. This type of design dramatically reduces the cognitive load for the programmer.

Now, let's try the reverse experiment: if a variable is defined *outside* of the function (before it is called), can it be accessed from within the function block definition?

```
equalish <- function(a, b, epsilon = 0.00001) {
  result <- abs(a - b) < epsilon
  print(testvar)
  return(result)
}


testvar <- "Contents of testvar"
vec1 <- c(4.00000001, 6)
vec2 <- c(4, 2)
eq <- equalish(vec1, vec2)  # TRUE FALSE
```

The lack of error in the output indicates that yes, the function block can access such external variables:

```
[1] "Contents of testvar"
```

This means that it is possible to write functions that take no parameters and simply access the external variables they will need for computation.

```
equalish <- function() {
  result <- abs(vec1 - vec2) < epsilon
  return(result)
}

vec1 <- c(4.00000001, 6)
vec2 <- c(4, 2)
epsilon <- 0.00001
eq <- equalish()                # TRUE FALSE
```

But writing such functions is fairly bad practice. Why? Because although the function still cannot affect the external environment, it is now quite dependent on the state of the external environment in which it is called. The function will only work if external variables called `vec1`, `vec2`, and `epsilon` happen to exist and have the right types of data when the function is called. Consider this: the former version of the function could be copied and pasted into an entirely different program and still be guaranteed to work (because the `a` and `b` parameters are required local variables), but that's not the case here.

The same four "rules" for designing functions in Python apply to R:

1. Functions should only access local variables that have been assigned within the function block, or have been passed as parameters (either required or with defaults).

2. Document the use of each function with comments. What parameters are taken, and what types should they be? Do the parameters need to conform to any specification, or are there any caveats to using the function? Also, what is returned?
3. Functions shouldn't be "too long." This is subjective and context dependent, but most programmers are uncomfortable with functions that are more than one page long in their editor window. The idea is that a function encapsulates a single, small, reusable idea. If you do find yourself writing a function that is hard to read and understand, consider breaking it into two functions that need to be called in sequence, or a short function that calls another short function.
4. Write lots of functions! Even if a block of code is only going to be called once, it's ok to make a function out of it (if it encapsulates some idea or well-separable block). After all, you never know if you might need to use it again, and just the act of encapsulating the code helps you ensure its correctness and forget about it when working on the rest of your program.

Argument Passing and Variable Semantics

So far, the differences we've seen between Python and R have mostly been in R's emphasis on vectorized operations. In later chapters, we'll also see that R emphasizes the creative use of functions more strongly than does Python (which should at the very least be a good reason to study them well).

There is another dramatic difference between these two languages, having to do with variables and their relationship to data. This is probably easiest to see with a couple of similar code examples. First, here's some Python code that declares a list of numbers `nums`, creates a new variable based on the original called `numsb`, modifies the first element of `numsb`, and then prints both.

```
nums = [1, 2, 3, 4, 5]
numsb = nums
numsb[0] = 1000
print(nums)
print(numsb)
```

The output indicates that `nums` and `numsb` are both variables (or "names," in Python parlance) for the same underlying data.

```
[1000, 2, 3, 4, 5]
[1000, 2, 3, 4, 5]
```

Corresponding R code and output reveals that R handles variables very differently:

```
nums <- c(1, 2, 3, 4, 5)
numsb <- nums
numsb[1] <- 1000
print(nums)
print(numsb)
```

```
[1] 1 2 3 4 5
[1] 1000    2    3    4    5
```

While in Python it's common for the same underlying data to be referenced by multiple variables, in R, unique variables are almost always associated with unique data. Often these semantics are emphasized in the context of local variables for functions. Here's the same thing, but the operation is mediated by a function call. First, the Python version and output:

```
def testfunc(param):
    param[0] = 1000
    return param

nums = [1, 2, 3, 4, 5]
numsb = testfunc(nums)
print(nums)
print(numsb)
```

```
[1000, 2, 3, 4, 5]
[1000, 2, 3, 4, 5]
```

And now the R version and output:

```
testfunc <- function(param) {
  param[1] <- 1000
  return(param)
}

nums <- c(1, 2, 3, 4, 5)
numsb <- testfunc(nums)
print(nums)
print(numsb)
```

```
[1] 1 2 3 4 5
[1] 1000    2    3    4    5
```

In the Python code, the `param` local variable is a new variable for the *same* underlying data, whereas in the R code the local `param` variable is a new variable for *new* underlying data. These two paradigms are found in a wide variety of languages; the latter is known as "pass-by-value," though one could think of it as "pass-by-copy." This doesn't mean that R always creates a copy–it uses a "copy-on-write" strategy behind the scenes to avoid excess work. As for the former, the Python documentation refers to it as "pass-by-assignment," and the effect is similar to "pass-by-reference." (The term "pass-by-reference" has a very narrow technical definition, but is often used as a catch-all for this type of behavior.)

There are advantages and disadvantages to both strategies. The somewhat more difficult scheme used by Python is both speedier and allows for more easy implementations of some sophisticated algorithms (like the structures covered in chapter 25, "Algorithms and Data Structures"). The pass-by-value scheme, on the other hand, can be easier to code with, because functions that follow rule 1 above can't surreptitiously modify data: they are "side effect free."

Getting Help

The R interpreter comes with extensive documentation for all of the functions that are built-in. Now that we know how to write functions, reading this documentation will be easy.

The help page for a particular function can be accessed by running `help("function_name")` in the interactive console, as in `help("t.test")` for help with the `t.test()` function.

```
> help("t.test")
```

Alternatively, if the function name contains no special characters, we can use the shorthand `?function_name`, as in `?t.test`. The help is provided in an interactive window in which you can use the arrow keys to scroll up and down.

```
t.test                  package:stats                  R Documentation

Student's t-Test

Description:

     Performs one and two sample t-tests on vectors of data.

Usage:

     t.test(x, ...)

     ## Default S3 method:
     t.test(x, y = NULL,
            alternative = c("two.sided", "less", "greater"),
            mu = 0, paired = FALSE, var.equal = FALSE,
            conf.level = 0.95, ...)

     ## S3 method for class 'formula'
     t.test(formula, data, subset, na.action, ...)

Arguments:

       x: a (non-empty) numeric vector of data values.

       y: an optional (non-empty) numeric vector of data values.
...
```

The help pages generally have the following sections, and there may be others:

- *Description:* Short description of the function.
- *Usage:* An example of how the function should be called, usually listing the most important parameters; parameters with defaults are shown with an equal sign.
- *Arguments:* What each parameter accepted by the function does.
- *Details:* Notes about the function and caveats to be aware of when using the function.
- *Value:* What is returned by the function.
- *References:* Any pertinent journal article or book citations. These are particularly useful for complex statistical functions.
- *Examples:* Example code using the function. Unfortunately, many examples are written for those who are familiar with the basics of the function, and illustrate more complex usage.
- *See Also:* Other related functions that a user might find interesting.

If a function belongs to a package (such as `str_split()` in the `stringr` package), one can either load the package first (with `library(stringr)`) and access the help for the function as usual (`help("str_split")`), or specify the package directly, as in `help("str_split", package = "stringr")`. An overview help page for all functions in the package can be accessed with `help(package = "stringr")`.

Finally, in the interactive window, using `help.search("average")` will search the documentation for all functions mentioning the term "average"—the shortcut for this is `??average`.

Exercises

1. We often wish to "normalize" a vector of numbers by first subtracting the mean from each number and then dividing each by the standard deviation of the numbers. Write a function called `normalize_mean_sd()` that takes such a vector and returns the normalized version. The function should work even if any values are `NA` (the normalized version of `NA` should simply be `NA`).
2. The `t.test()` function tests whether the means of two numeric vectors are unequal. There are multiple versions of *t*-tests: some assume that the variances of the input vectors are equal, and others do not make this assumption. By default, does `t.test()` assume equal variances? How can this behavior be changed?
3. Using the help documentation, generate a vector of 100 samples from a Poisson distribution with the lambda parameter (which controls the shape of the distribution) set to `2.0`.
4. The following function computes the difference in mean of two vectors, but breaks at least one of the "rules" for writing functions. Fix it so that it conforms. (Note that it is also lacking proper documentation.)

```
mean_diff <- function() {
  m1 <- mean(v1)
  m2 <- mean(v2)
  answer <- m1 - m2
  return(answer)
}

sample1 <- rnorm(100, mean = 4, sd = 3)
sample2 <- rnorm(100, mean = 6, sd = 3)
v1 <- sample1
v2 <- sample2

diff <- mean_diff()
print(diff)
```

5. The following code generates two random samples, and then computes and prints the difference in coefficient of variation for the samples (defined as the standard deviation divided by the mean). Explain how this code works, step by step, in terms of local variables, parameters, and returned values. What if immediately before `sample1 <- rnorm(100, mean = 4, sd = 2)`, we had `result <- "Test message."`, and after `print(answer)`, we had `print(result)`? What would be printed, and why?

```
## given two numeric vectors a and b,
## returns the difference in coefficient of
## variation (sd over mean) NAs are ignored.
diff_cov <- function(a, b) {
  cova <- coeff_of_var(a)
  covb <- coeff_of_var(b)
  result <- cova - covb
  return(result)
}

## given an numeric vector, returns
## the coefficient of variation (sd over mean)
## NAs are ignored
coeff_of_var <- function(a) {
  result <- sd(a, na.rm = TRUE) / mean(a, na.rm = TRUE)
  return(result)
}

sample1 <- rnorm(100, mean = 4, sd = 2)
sample2 <- rnorm(100, mean = 8, sd = 2.5)
answer <- diff_cov(sample1, sample2)
print(answer)
```

Chapter 30

Lists and Attributes

The next important data type in our journey through R is the list. Lists are quite similar to vectors—they are ordered collections of data, indexable by index number, logical vector, and name (if the list is named). Lists, however, can hold multiple different types of data (including other lists). Suppose we had three different vectors representing some information about the plant *Arabidopsis thaliana*.

```
organism <- "A. thaliana"
ecotypes <- c("C24", "Col0", "WS2")
num_chromosomes <- 5
```

We can then use the `list()` function to gather these vectors together into a single unit with class `"list"`.

```
athal <- list(organism, ecotypes, num_chromosomes)
print(class(athal))                                    # [1] "list"
```

Graphically, we might represent this list like so:

```
athal:
  Element 1: [1] "A. thaliana"
  Element 2: [1] "C24" "Col0" "WS2"
  Element 3: [1] 5
```

Here, the `[1]` syntax is indicating that the elements of the list are vectors (as in when vectors are printed). Like vectors, lists can be indexed by index vector and logical vector.

```
sublist <- athal[c(1,3)]
sublist <- athal[c(TRUE, FALSE, TRUE)]
```

Both of the above assign to the variable `sublist` a list looking like:

```
sublist:
   Element 1: [1] "A. thaliana"
   Element 2: [1] 5
```

This seems straightforward enough: subsetting a list with an indexing vector returns a smaller list with the requested elements. But this rule can be deceiving if we forget that a vector is the most basic element of data. Because `2` is the length-one vector `c(2)`, `athal[2]` returns not the second element of the `athal` list, but rather a length-one list with a single element (the vector of ecotypes).

```
eco_list <- athal[2]
print(class(eco_list))                          # [1] "list"
print(length(eco_list))                         # [1] 1
```

A graphical representation of this list:

```
eco_list:
   Element 1: [1] "C24" "Col0" "WS2"
```

We will thus need a different syntax if we wish to extract an individual element from a list. This alternate syntax is `athal[[2]]`.

```
ecotypes <- athal[[2]]
print(class(ecotypes))                          # [1] "character"
print(ecotypes)                                 # [1] "C24" "Col0" "Ws2"
```

If we wanted to extract the second ecotype directly, we would need to use the relatively clunky `second_ecotype <- athal[[2]][2]`, which accesses the second element of the vector (accessed by `[2]`) inside of the of the second element of the list (accessed by `[[2]]`).

```
print(athal)
```

When we print a list, this structure and the double-bracket syntax is reflected in the output.

```
[[1]]
[1] "A. thaliana"

[[2]]
[1] "C24"  "Col0" "WS2"

[[3]]
[1] 5
```

Named Lists, Lists within Lists

Like vectors, lists can be named—associated with a character vector of equal length—using the `names()` function. We can use an index vector of names to extract a sublist, and we can use `[[]]` syntax to extract individual elements by name.

```
names(athal) <- c("Species", "Ecotypes", "# Chromosomes")
sublist <- athal[c("Species", "# Chromosomes")]      # list of length two

ecotypes <- athal[["Ecotypes"]]                      # ecotypes vector
```

We can even extract elements from a list if the name of the element we want is stored in another variable, using the `[[]]` syntax.

```
extract_name <- "Ecotypes"
ecotypes <- athal[[extract_name]]                    # ecotypes vector
```

As fun as this double-bracket syntax is, because extracting elements from a list by name is a common operation, there is a shortcut using `$` syntax.

```
ecotypes <- athal[["Ecotypes"]]
# same as
ecotypes <- athal$"Ecotypes"
```

In fact, if the name doesn't contain any special characters (spaces, etc.), then the quotation marks can be left off.

```
ecotypes <- athal$Ecotypes
```

This shortcut is widely used and convenient, but, because the quotes are implied, we can't use `$` syntax to extract an element by name if that name is stored in an intermediary variable. For example, if `extract_name <- "ecotypes"`, then `athal$extract_name` will expand to `athal[["extract_name"]]`, and we won't get the ecotypes vector. This common error reflects a misunderstanding of the syntactic sugar employed by R. Similarly, the `$` syntax won't work for names like `"# Chromosomes"` because that name contains a space and a special character (for this reason, names of list elements are often simplified).

Frequently, `$` syntax is combined with vector syntax if the element of the list being referred to is a vector. For example, we can directly extract the third ecotype, or set the third ecotype.

```
third_ecotype <- athal$Ecotypes[3]
athal$Ecotypes[3] <- "WS2b"
```

Continuing with this example, let's suppose we have another list that describes information about each chromosome. We can start with an empty list, and assign elements to it by name.

```
chrs <- list()
chrs$Lengths <- c(34.9, 22.0, 25.4, 20.8, 31.2)
chrs$GeneCounts <- c(7078, 4245, 5437, 4124, 6318)
```

This list of two elements relates to *A. thaliana*, so it makes sense to include it somehow in the `athal` list. Fortunately, lists can contain other lists, so we'll assign this `chrs` list as element of the `athal` list.

```
athal$ChrInfo <- chrs
```

Lists are an excellent container for general collections of heterogeneous data in a single organized "object." (These differ from Python objects in that they don't have methods stored in them as well, but we'll see how R works with methods in later chapters.) If we ran `print(athal)` at this point, all this information would be printed, but unfortunately in a fairly unfriendly manner:

```
$Species
[1] "A. thaliana"

$Ecotypes
[1] "C24"  "Col0" "WS2b"

$`# Chromosomes`
[1] 5

$ChrInfo
$ChrInfo$Lengths
[1] 34.9 22.0 25.4 20.8 31.2

$ChrInfo$GeneCounts
[1] 7078 4245 5437 4124 6318
```

This output does illustrate something of interest, however. We can chain the `$` syntax to access elements of lists and contained lists by name. For example, `lengths <- athal$ChrInfo$Lengths` extracts the vector of lengths contained in the internal `ChrInfo` list, and we can even modify elements of these vectors with syntax like `athal$ChrInfo$GeneCounts[1] <- 7079` (perhaps a new gene was recently discovered on the first chromosome). Expanding the syntax a bit to use double-brackets rather than `$` notation, these are equivalent to `lengths <- athal[["ChrInfo"]][["Lengths"]]` and `athal[["ChrInfo"]][["GeneCounts"]][1] <- 7079`.

Attributes, Removing Elements, List Structure

Lists are an excellent way to organize heterogeneous data, especially when data are stored in a Name → Value association,[98] making it easy to access data by character name. But what if we want to look up some information associated with a piece of data but not represented in the data itself? This would be a type of "metadata," and R allows us to associate metadata to any piece of data using what are called *attributes*. Suppose we have a simple vector of normally distributed data:

```
sample <- rnorm(10, mean = 20, sd = 10)
```

98. R lists are often used like dictionaries in Python and hash tables in other languages, because of this easy and effective Name → Value lookup operation. It should be noted that (at least as of R 3.3), name lookups in lists are not as efficient as name lookups in Python dictionaries or other true hash tables. For an efficient and more idiomatic hash table/dictionary operation, there is also the package `hash` available for install with `install.packages("hash")`.

Later, we might want know what type of data this is: is it normally distributed, or something else? We can solve this problem by assigning the term `"normal"` as an attribute of the data. The attribute also needs a name, which we'll call `"disttype"`. Attributes are assigned in a fashion similar to names.

```
attr(sample, "disttype") <- "normal"
print(sample)
```

When printed, the output shows the attributes that have been assigned as well.

```
 [1] -3.177991  8.676695 10.623292 22.020329  5.497178 10.135908  7.113268
 [8] 26.784623 16.306509 21.237899
attr(,"disttype")
[1] "normal"
```

We can separately extract a given attribute from a data item, using syntax like `sample_dist <- attr(sample, "disttype")`. Attributes are used widely in R, though they are rarely modified in day-to-day usage of the language.[99]

To expand our *A. thaliana* example, let's assign a "kingdom" attribute to the species vector.

```
attr(athal$Species, "kingdom") <- "Plantae"
```

At this point, we've built a fairly sophisticated structure: a list containing vectors (one of which has an attribute) and another list, itself containing vectors, with the various list elements being named. If we were to run `print(athal)`, we'd see rather messy output. Fortunately, R includes an alternative to `print()` called `str()`, which nicely prints the structure of a list (or other data object). Here's the result of calling `str(athal)` at this point.

```
List of 4
 $ Species      : atomic [1:1] A. thaliana
  ..- attr(*, "kingdom")= chr "Plantae"
 $ Ecotypes     : chr [1:3] "C24" "Col0" "WS2b"
 $ # Chromosomes: num 5
 $ ChrInfo      :List of 2
  ..$ Lengths   : num [1:5] 54.2 36.4 19.7 46.3 29.2
  ..$ GeneCounts: num [1:5] 8621 7215 3124 7219 4140
```

99. For example, the names of a vector are stored as an attribute called "names"—the lines `names(scores) <- c("Student A", "Student B", "Student C")` and `attr(scores, "names") <- c("Student A", "Student B", "Student C")` are (almost) equivalent. Still, it is recommended to use specialized functions like `names()` rather than set them with `attr()` because the `names()` function includes additional checks on the sanity of the names vector.

Removing an element or attribute from a list is as simple as assigning it the special value `NULL`.

```
# delete ChrInfo and Species attribute
athal$ChrInfo <- NULL
attr(athal$Species, "kingdom") <- NULL
str(athal)
```

The printed structure reveals that this information has been removed.

```
List of 3
 $ Species      : chr "A. thaliana"
 $ Ecotypes     : chr [1:3] "C24" "Col0" "WS2b"
 $ # Chromosomes: num 5
```

What is the point of all this detailed list making and attribute assigning? It turns out to be quite important, because many R functions return exactly these sorts of complex attribute-laden lists. Consider the `t.test()` function, which compares the means of two vectors for statistical equality:

```
samp1 <- rnorm(100, mean = 10, sd = 5)
samp2 <- rnorm(100, mean = 8, sd = 5)
tresult <- t.test(samp1, samp2)
print(tresult)
```

When printed, the result is a nicely formatted, human-readable result.

```
Welch Two Sample t-test

data:  samp1 and samp2
t = 2.8214, df = 197.172, p-value = 0.005271
alternative hypothesis: true difference in means is not equal to 0
95 percent confidence interval:
 0.6312612 3.5626936
sample estimates:
mean of x mean of y
 9.307440  7.210463
```

If we run `str(tresult)`, however, we find the true nature of `tresult`: it's a list!

```
List of 9
 $ statistic  : Named num 5.41
  ..- attr(*, "names")= chr "t"
 $ parameter  : Named num 198
  ..- attr(*, "names")= chr "df"
 $ p.value    : num 1.84e-07
 $ conf.int   : atomic [1:2] 2.07 4.45
  ..- attr(*, "conf.level")= num 0.95
 $ estimate   : Named num [1:2] 10.41 7.15
  ..- attr(*, "names")= chr [1:2] "mean of x" "mean of y"
 $ null.value : Named num 0
  ..- attr(*, "names")= chr "difference in means"
 $ alternative: chr "two.sided"
 $ method     : chr "Welch Two Sample t-test"
 $ data.name  : chr "samp1 and samp2"
 - attr(*, "class")= chr "htest"
```

Given knowledge of this structure, we can easily extract specific elements, such as the *p* value with `pval <- tresult$p.value` or `pval <- tresult[["p.value"]]`.

One final note about lists: vectors (and other types) can be converted into a list with the `as.list()` function. This will come in handy later, because lists are one of the most general data types in R, and we can use them for intermediary data representations.

```
scores <- c(56.3, 91.7, 87.4)
scores_list <- as.list(scores)      # A list with 3 single-element vectors
```

Exercises

1. The following code first generates a random sample called `a`, and then a sample called `response`, wherein each element of `response` is an element of `a` times 1.5 plus some random noise:

```
a <- rnorm(100, mean = 2, sd = 3)
response <- a * 1.5 + rnorm(100, mean = 0, sd = 1)
```

Next, we can easily create a linear model predicting values of `response` from `a`:

```
model <- lm(response ~ a)
print(model)
```

When printed, the output nicely describes the parameters of the model.

```
Call:
lm(formula = response ~ a)

Coefficients:
(Intercept)            a
   -0.06706      1.53367
```

We can also easily test the significance of the parameters with the `anova()` function (to run an analysis of variance test on the model).

```
vartest <- anova(model)
print(vartest)
```

The output again shows nicely formatted text:

```
Analysis of Variance Table

Response: response
          Df  Sum Sq Mean Sq F value    Pr(>F)
a          1 2848.43 2848.43  3827.2 < 2.2e-16 ***
Residuals 98   72.94    0.74
---
Signif. codes:  0 '***' 0.001 '**' 0.01 '*' 0.05 '.' 0.1 ' ' 1
```

From the `model`, extract the coefficient of a into a variable called `a_coeff` (which would contain just the number `1.533367` for this random sample).
Next, from `vartest` extract the *p* value associated with the a coefficient into a vector called `a_pval` (for this random sample, the *p* value is `2.2e-16`).

2. Write a function called `simple_lm_pval()` that automates the process above; it should take two parameters (two potentially linearly dependent numeric vectors) and return the *p* value associated with the first (nonintercept) coefficient.

3. Create a list containing three random samples from different distributions (e.g., from `rnorm()`, `runif()`, and `rexp()`), and add an attribute for `"disttype"` to each. Use `print()` and `str()` on the list to examine the attributes you added.

4. Some names can be used with `$` notation without quotation marks; if `l <- list(values = c(20, 30))`, then `print(l$values)` will print the internal vector. On the other hand, if `l <- list("val-entries" = c(20, 30))`, then quotations are required as in `print(l$"val-entries")`. By experimentation, determine at least five different characters that require the use of quotation marks when using `$` notation.

5. Experiment with the `is.list()` and `as.list()` functions, trying each of them on both vectors

and lists.

Chapter 31

Data Frames

In chapter 28, "Vectors," we briefly introduced data frames as storing tables of data. Now that we have a clear understanding of both vectors and lists, we can easily describe data frames. (If you hastily skipped chapters 28 and 30 to learn about data frames, now's the time to return to them!) Data frames are essentially named lists, where the elements are vectors representing columns. But data frames provide a few more features than simple lists of vectors. They ensure that the component column vectors are always the same length, and they allow us to work with the data by row as well as by column. Data frames are some of the most useful and ubiquitous data types in R.

While we've already covered using the `read.table()` function to produce a data frame based on the contents of a text file, it's also possible to create a data frame from a set of vectors.

```
ids <- c("AGP", "T34", "ALQ", "IXL")
lengths <- c(256, 134, 92, 421)
gcs <- c(0.21, 0.34, 0.41, 0.65)

gene_info <- data.frame(ids, lengths, gcs, stringsAsFactors = FALSE)
print(gene_info)
```

When printed, the contents of the column vectors are displayed neatly, with the column names along the top and row names along the left-hand side.

```
  ids lengths  gcs   } Column Names
1 AGP     256 0.21
2 T34     134 0.34
3 ALQ      92 0.41
4 IXL     421 0.65
Row Names
```

As with `read.table()`, the `data.frame()` function takes an optional `stringsAsFactors` argument, which specifies whether character vectors (like `ids`) should be converted to factor types (we'll cover these in detail later). For now, we'll disable this conversion.

Running `str(gene_info)` reveals the data frame's list-like nature:

```
'data.frame': 4 obs. of  3 variables:
 $ ids     : chr  "AGP" "T34" "ALQ" "IXL"
 $ lengths: num  256 134 92 421
 $ gcs     : num  0.21 0.34 0.41 0.65
```

Like elements of lists, the columns of data frames don't have to have names, but not having them is uncommon. Most data frames get column names when they are created (either by `read.table()` or `data.frame()`), and if unset, they usually default to `V1`, `V2`, and so on. The column names can be accessed and set with the `names()` function, or with the more appropriate `colnames()` function.

```
# set column names
names(gene_info) <- c("ids", "lengths", "gcs")
# better way:
colnames(gene_info) <- c("ids", "lengths", "gcs")

print(colnames(gene_info))                         # [1] "ids" "lengths" "gcs"
```

To highlight the list-like nature of data frames, we can work with data frames by column much like lists by element. The three lines in the following example all result in `sub_info` being a two-column data frame.

```
sub_info <- gene_info[c(1,3)]
sub_info <- gene_info[c("ids", "gcs")]
sub_info <- gene_info[c(TRUE, FALSE, TRUE)]
```

An expression like `gene_info[2]` thus would not return a numeric vector of lengths, but rather a single-column data frame containing the numeric vector. We can use `[[]]` syntax and `$` syntax to refer to the vectors contained within data frames as well (the latter is much more common).

```
ids_vec <- gene_info[[1]]
ids_vec <- gene_info[["ids"]]
ids_vec <- gene_info$ids
```

We can even delete columns of a data frame by setting the element to `NULL`, as in `gene_info$lengths <- NULL`.

The real charm of data frames is that we can extract and otherwise work with them by row. Just as data frames have column names, they also have row names: a character vector of the same length as each column. Unfortunately, by default, the row names are `"1"`, `"2"`, `"3"`, and so on, but when the data frame is printed, the quotation marks are left off (see the result of `print(gene_info)` above). The row names are accessible through the `rownames()` function.

Data frames are indexable using an extended `[]` syntax: `[<row_selector>, <column_selector>]`, where `<row_selector>` and `<column_selector>` are vectors. Just as with vectors and lists, these indexing/selection vectors may be integers (to select by index), characters (to select by name), or logicals (to select logically). Also as with vectors, when indexing by index position or name, the requested order is respected.

```
# select the third and first rows, columns "lengths" and "ids"
subframe <- gene_info[c(3, 1), c("lengths", "ids")]
print(subframe)
```

Here's the resulting output, illustrating that `"3"` and `"1"` were the row names, which now occur at the first and second row, respectively.[100]

```
  lengths ids
3      92 ALQ
1     256 AGP
```

If you find this confusing, consider what would happen if we first assigned the row names of the original data frame to something more reasonable, before the extraction.

```
rownames(gene_info) <- c("g1", "g2", "g3", "g4")
subframe <- gene_info[c(3, 1), c("lengths", "ids")]
print(subframe)
```

Now, when printed, the character nature of the row names is revealed.

100. Unfortunately, when printed, quotation marks are left off of row names, commonly leading programmers to think that the row indexes are what is being displayed. Similarly, quotation marks are left off of columns that are character vectors (like the `ids` column in this example), which can cause confusion if the column is a character type with elements like `"1"`, `"2"`, and so on, potentially leading the programmer to think the column is numeric. In a final twist of confusion, using `[<row_selector>, <column_selector>]` syntax breaks a common pattern: `[]` indexing of a vector always returns a vector, `[]` indexing of a list always returns a list, and `[<row_selector>, <column_selector>]` indexing of a data frame always returns a data frame, *unless* it would have only one column, in which case the column/vector is returned. The remedy for this is `[<row_selector>, <column_selector>, drop = FALSE]`, which instructs R not to "drop" a dimension.

```
   lengths ids
g3      92 ALQ
g1     256 AGP
```

Finally, if one of <row_selector> or <column_selector> are not specified, then all rows or columns are included. As an example, gene_info[c(3,1),] returns a data frame with the third and first rows and all three columns, while gene_info[, c("lengths", "ids")] returns one with only the "lengths" and "ids" columns, but all rows.

Data Frame Operations

Because data frames have much in common with lists and rows—and columns can be indexed by index number, name, or logical vector—there are many powerful ways we can manipulate them. Suppose we wanted to extract only those rows where the lengths column is less than 200, or the gcs column is less than 0.3.

```
row_selector <- gene_info$lengths < 200 | gene_info$gcs < 0.3
selected <- gene_info[row_selector, ]
print(selected)
```

This syntax is concise but sophisticated. While gene_info$lengths refers to the numeric vector named "lengths" in the data frame, the < logical operator is vectorized, with the single element 200 being recycled as needed. The same process happens for gene_info$gcs < 0.3, and the logical-or operator | is vectorized, producing a logical vector later used for selecting the rows of interest. An even shorter version of these two lines would be selected <- gene_info[gene_info$lengths < 200 | gene_info$gcs < 0.3,]. The printed output:

```
   ids lengths  gcs
g1 AGP     256 0.21
g2 T34     134 0.34
g3 ALQ      92 0.41
```

If we wished to extract the gcs vector from this result, we could use something like selected_gcs <- selected$gcs. Sometimes more compact syntax is used, where the $ and column name are appended directly to the [] syntax.

```
selected_gcs <- gene_info[row_selector, ]$gcs
```

sub-dataframe column

Alternatively, and perhaps more clearly, we can first use `$` notation to extract the column of interest, and then use `[]` logical indexing on the resulting vector.

```
selected_gcs <- gene_info$gcs[row_selector]
```

column selected elements

Because subsetting data frame rows by logical condition is so common, there is a specialized function for this task: `subset()`. The first parameter is the data frame from which to select, and later parameters are logical expressions based on column names *within* that data frame (quotation marks are left off). For example, `selected <- subset(gene_info, lengths < 200 | gcs < 0.3)`. If more than one logical expression is given, they are combined with `&` (and). Thus `subset(gene_info, lengths < 200, gcs < 0.3)` is equivalent to `gene_info[gene_info$lengths < 200 & gene_info$gcs < 0.3 ,]`.

While the `subset()` function is convenient for simple extractions, knowing the ins and outs of `[]` selection for data frames as it relates to lists and vectors is a powerful tool. Consider the `order()` function, which, given a vector, returns an index vector that can be used for sorting. Just as with using `order()` to sort a vector, we can use `order()` to sort a data frame based on a particular column.

```
gene_info_sorted <- gene_info[order(gene_info$lengths), ]
print(gene_info_sorted)
```

The result is a data frame ordered by the lengths column:

```
   ids lengths  gcs
g3 ALQ      92 0.41
g2 T34     134 0.34
g1 AGP     256 0.21
g4 IXL     421 0.65
```

Because data frames force all column vectors to be the same length, we can create new columns by assigning to them by name, and relying on vector recycling to fill out the column as necessary. Let's create a new column called gc_categories, which is initially filled with NA values, and then use selective replacement to assign values "low" or "high" depending on the contents of the gcs column.

```
gene_info$gc_category <- NA
gene_info$gc_category[gene_info$gcs < 0.5] <- "low"
gene_info$gc_category[gene_info$gcs >= 0.5] <- "high"
print(gene_info)
```

While there are more automated approaches for categorizing numerical data, the above example illustrates the flexibility and power of the data frame and vector syntax covered so far.

```
   ids lengths  gcs gc_category
g1 AGP     256 0.21         low
g2 T34     134 0.34         low
g3 ALQ      92 0.41         low
g4 IXL     421 0.65        high
```

One final note: while the head() function returns the first few elements of a vector or list, when applied to a data frame, it returns a similar data frame with only the first few rows.

Matrices and Arrays

Depending on the type of analysis, you might find yourself working with matrices in R, which are essentially two-dimensional vectors. Like vectors, all elements of a matrix must be the same type, and attempts to mix types will result in autoconversion. Like data frames, they are two dimensional, and so can be indexed with [<row_selector>, <column_selector>] syntax. They also have rownames() and colnames().

```
m <- matrix(c(1, 2, 3, 4, 5, 6), nrow = 2, ncol = 3)
print(m)
```

```
     [,1] [,2] [,3]
[1,]    1    3    5
[2,]    2    4    6
```

There are a number of interesting functions for working with matrices, including `det()` (for computing the determinant of a matrix) and `t()` (for transposing a matrix).

Arrays generalize the concept of multidimensional data; where matrices have only two dimensions, arrays may have two, three, or more. The line `a <- array(c(1,1,1,1,2,2,2,2,3,3,3,3), dim = c(2,2,3))`, for example, creates a three-dimensional array, composed of three two-by-two matrices. The upper left element can be extracted as `a[1,1,1]`.

Exercises

1. Read the `states.txt` file into a data frame, as discussed in Chapter 28, Vectors. Extract a new data frame called `states_name_pop` containing only the columns for `name` and `population`.
2. Using `subset()`, extract a data frame called `states_gradincome_high` containing all columns, and all rows where the `income` is greater than the median of `income` or where `hs_grad` is greater than the median of `hs_grad`. (There are 35 such states listed in the table.) Next, do the same extraction with `[<row_selector>, <column_selector>]` syntax.
3. Create a table called `states_by_region_income`, where the rows are ordered first by region and second by income. (The `order()` function can take multiple parameters as in `order(vec1, vec2)`; they are considered in turn in determining the ordering.)
4. Use `order()`, `colnames()`, and `[<row_selector>,<column_selector>]` syntax to create a data frame `states_cols_ordered`, where the columns are alphabetically ordered (i.e., `hs_grad` first, then `income`, then `murder`, then `name`, and so on).
5. The `nrow()` function returns the number of rows in a data frame, while `rev()` reverses a vector and `seq(a,b)` returns a vector of integers from `a` to `b` (inclusive). Use these to produce `states_by_income_rev`, which is the states data frame but with rows appearing in *reverse* order of income (highest incomes on top).
6. Try converting the `states` data frame to a matrix by running it through the `as.matrix()` function. What happens to the data, and why do you think that is?

Chapter 32

Character and Categorical Data

Scientific data sets, especially those to be analyzed statistically, commonly contain "categorical" entries. If each row in a data frame represents a single measurement, then one column might represent whether the measured value was from a "male" or "female," or from the "control" group or "treatment" group. Sometimes these categories, while not numeric, have an intrinsic order, like "low," "medium," and "high" dosages.

Sadly, more often than not, these entries are not encoded for easy analysis. Consider the tab-separated file **`expr_long_coded.txt`**, where each line represents an (normalized) expression reading for a gene (specified by the ID column) in a given sample group. This experiment tested the effects of a chemical treatment on an agricultural plant species. The sample group encodes information about what genotype was tested (either `C6` or `L4`), the treatment applied to the plants (either `control` or `chemical`), the tissue type measured (either `A`, `B`, or `C` for leaf, stem, or root), and numbers for statistical replicates (`1`, `2`, or `3`).

```
id      annotation      expression      sample
LQ00X000020     Hypothetical protein    5.024142433     C6_control_A1
LQ00X000020     Hypothetical protein    4.646697026     C6_control_A3
LQ00X000020     Hypothetical protein    4.986591902     C6_control_B1
LQ00X000020     Hypothetical protein    5.164291761     C6_control_B2
LQ00X000020     Hypothetical protein    5.348809843     C6_control_B3
LQ00X000020     Hypothetical protein    5.519392911     C6_control_C1
LQ00X000020     Hypothetical protein    5.305463561     C6_control_C2
LQ00X000020     Hypothetical protein    5.535279863     C6_control_C3
LQ00X000020     Hypothetical protein    4.831259167     C6_chemical_A1
LQ00X000020     Hypothetical protein    4.784184435     C6_chemical_A3
LQ00X000020     Hypothetical protein    5.577740866     C6_chemical_B1
```

Initially, we'll read the table into a data frame. For this set of data, we'll likely want to work with the categorical information independently, for example, by extracting only values for the chemical treatment. This would be much easier if the data frame had individual columns for `genotype`, `treatment`, `tissue`, and `replicate` as opposed to a single, all-encompassing `sample` column.

A basic installation of R includes a number of functions for working with character vectors, but the `stringr` package (available via `install.packes("stringr")` on the interactive console) collects many of these into a set of nicely named functions with common options. For an overview, see `help(package = "stringr")`, but in this chapter we'll cover a few of the most important functions from that package.

```
library(stringr)

expr_long <- read.table("expr_long_coded.txt",
                        header = TRUE,
                        sep = "\t",
                        stringsAsFactors = FALSE)
```

Splitting and Binding Columns

The `str_split_fixed()` function from the `stringr` package operates on each element of a character vector, splitting it into pieces based on a pattern. With this function, we can split each element of the `expr_long$sample` vector into three pieces based on the pattern `"_"`. The "pattern" could be a regular expression, using the same syntax as used by Python (and similar to that used by sed).

```
sample_split <- str_split_fixed(expr_long$sample, "_", 3)
print(head(sample_split))
```

The value returned by the `str_split_fixed()` function is a matrix: like vectors, matrices can only contain a single data type (in fact, they are vectors with attributes specifying the number of rows and columns), but like data frames they can be accessed with `[<row_selector>, <column_selector>]` syntax. They may also have row and column names.

```
     [,1] [,2]      [,3]
[1,] "C6" "control" "A1"
[2,] "C6" "control" "A3"
[3,] "C6" "control" "B1"
[4,] "C6" "control" "B2"
[5,] "C6" "control" "B3"
[6,] "C6" "control" "C1"
```

Anyway, we'll likely want to convert the matrix into a data frame using the `data.frame()` function, and assign some reasonable column names to the result.

```
sample_split_df <- data.frame(sample_split)
colnames(sample_split_df) <- c("genotype", "treatment", "tissuerep")
```

At this point, we have a data frame `expr_long` as well as `sample_split_df`. These two have the same number of rows in a corresponding order, but with different columns. To get these into a single data frame, we can use the `cbind()` function, which binds such data frames by their columns, and only works if they contain the same number of rows.

```
expr_long_split <- cbind(expr_long, sample_split_df)
```

A quick `print(head(expr_long_split))` lets us know if we're headed in the right direction.

```
           id          annotation expression         sample genotype treatment
1 LQ00X000020 Hypothetical protein   5.024142 C6_control_A1       C6   control
2 LQ00X000020 Hypothetical protein   4.646697 C6_control_A3       C6   control
3 LQ00X000020 Hypothetical protein   4.986592 C6_control_B1       C6   control
4 LQ00X000020 Hypothetical protein   5.164292 C6_control_B2       C6   control
5 LQ00X000020 Hypothetical protein   5.348810 C6_control_B3       C6   control
6 LQ00X000020 Hypothetical protein   5.519393 C6_control_C1       C6   control
  tissuerep
1        A1
2        A3
3        B1
4        B2
5        B3
6        C1
```

At this point, the number of columns in the data frame has grown large, so `print()` has elected to wrap the final column around in the printed output.

Detecting and %in%

We still don't have separate columns for tissue and replicate, but we do have this information encoded together in a tissuerep column. Because these values are encoded without a pattern to obviously split on, str_split_fixed() may not be the most straightforward solution.

Although any solution assuming *a priori* knowledge of large data set contents is dangerous (as extraneous values have ways of creeping into data sets), a quick inspection of the data reveals that the tissue types are encoded as either A, B, or C, with apparently no other possibilities. Similarly, the replicate numbers are 1, 2, and 3.

A handy function in the stringr package detects the presence of a pattern in every entry of a character vector, returning a logical vector. For the column tissuerep containing "A1", "A3", "B1", "B2", "B3", "C1", ..., for example, str_detect(expr_long_split$tissuerep, "A") would return the logical vector TRUE, TRUE, FALSE, FALSE, FALSE, ... Thus we can start by creating a new tissue column, initially filled with NA values.

```
expr_long_split$tissue <- NA
```

Then we'll use selective replacement to fill this column with the value "A" where the tissuerep column has an "A" as identified by str_detect(). Similarly for "B" and "C".

```
expr_long_split$tissue[str_detect(expr_long_split$tissuerep, "A")] <- "A"
expr_long_split$tissue[str_detect(expr_long_split$tissuerep, "B")] <- "B"
expr_long_split$tissue[str_detect(expr_long_split$tissuerep, "C")] <- "C"
```

In chapter 34, "Reshaping and Joining Data Frames," we'll also consider more advanced methods for this sort of pattern-based column splitting. As well, although we're working with columns of data frames, it's important to remember that they are still vectors (existing as columns), and that the functions we are demonstrating primarily operate on and return vectors.

If our assumption, that "A", "B", and "C" were the only possible tissue types, was correct, there should be no NA values left in the tissue column. We should verify this assumption by attempting to print all rows where the tissue column *is* NA (using the is.na() function, which returns a logical vector).

```
print(expr_long_split[is.na(expr_long_split$tissue), ]) # should print 0 rows
```

In this case, a data frame with zero rows is printed. There is a possibility that tissue types like `"AA"` have been recoded as simple `"A"` values using this technique—to avoid this outcome, we could use a more restrictive regular expression in the `str_detect()`, such as `"^A\d$"`, which will only match elements that start with a single `"A"` followed by a single digit. See chapter 11, "Patterns (Regular Expressions)," and chapter 21, "Bioinformatics Knick-knacks and Regular Expressions," for more information on regular-expression patterns.

A similar set of commands can be used to fill a new replicate column.

```
expr_long_split$rep <- NA

expr_long_split$rep[str_detect(expr_long_split$tissuerep, "1")] <- "1"
expr_long_split$rep[str_detect(expr_long_split$tissuerep, "2")] <- "2"
expr_long_split$rep[str_detect(expr_long_split$tissuerep, "3")] <- "3"

print(expr_long_split[is.na(expr_long_split$rep), ]) # should print 0 rows
```

Again we search for leftover `NA` values, and find that this time there *are* some rows where the `rep` column is reported as `NA`, apparently because a few entries in the table have a replicate number of `0`.

```
              id                                             annotation
260  LQ00X000370                Putative Beta-hexosaminidase subunit B2
590  LQ00X001530                       Putative Peptide transporter PTR1
855  LQ00X002040              Probable adenylate kinase 2, chloroplastic
1827 LQ00X004320 Putative Serine/threonine-protein kinase BRI1-like 2
2135 LQ00X004780                             Phenylalanine ammonia-lyase
     expression          sample genotype treatment tissuerep tissue  rep
260   11.731868  C6_control_B0       C6   control        B0      B <NA>
590   12.982069 C6_chemical_C0       C6  chemical        C0      C <NA>
855    5.327353  L4_control_C0       L4   control        C0      C <NA>
1827   7.596899  C6_control_B0       C6   control        B0      B <NA>
2135   6.165839  L4_control_C0       L4   control        C0      C <NA>
```

There are a few ways we could handle this. We could determine what these five samples' replicate numbers *should* be; perhaps they were miscoded somehow. Second, we could add `"0"` as a separate replicate possibility (so a few groups were represented by four replicates, rather than three). Alternatively, we could remove these mystery entries.

Finally, we could remove all measurements for these gene IDs, including the other replicates. For this data set, we'll opt for the latter, as the existence of these "mystery" measurements throws into doubt the accuracy of the other measurements, at least for this set of five IDs.

To do this, we'll first extract a vector of the "bad" gene IDs, using logical selection on the `id` column based on `is.na()` on the `rep` column.

```
bad_ids <- expr_long_split$id[is.na(expr_long_split$rep)]
print(bad_ids)
```

```
[1] "LQ00X000370" "LQ00X001530" "LQ00X002040" "LQ00X004320" "LQ00X004780"
```

Now, for each element of the `id` column, which ones are equal to one of the elements in the `bad_ids` vector? Fortunately, R provides a `%in%` operator for this many-versus-many sort of comparison. Given two vectors, `%in%` returns a logical vector indicating which elements of the left vector match one of the elements in the right. For example, `c(3, 2, 5, 1) %in% c(1, 2)` returns the logical vector `FALSE, TRUE, FALSE, TRUE`. This operation requires comparing each of the elements in the left vector against each of the elements of the right vector, so the number of comparisons is roughly the length of the first *times* the length of the second. If both are very large, such an operation could take quite some time to finish.

Nevertheless, we can use the `%in%` operator along with logical selection to remove all rows containing a "bad" gene ID.

```
bad_rows <- expr_long_split$id %in% bad_ids       # logical vector

expr_long_split <- expr_long_split[!bad_rows, ] # logical selection
```

At this point, we could again check for `NA` values in the `rep` column to ensure the data have been cleaned up appropriately. If we wanted, we could also check `length(bad_rows[bad_rows])` to see how many bad rows were identified and removed. (Do you see why?)

Pasting

While above we discussed splitting contents of character vectors into multiple vectors, occasionally we want to do the opposite: join the contents of character vectors together into a single character vector, element by element. The `str_c()` function from the `stringr` library accomplishes this task.

```
first <- c("Tom", "Bob")
last <- c("Waits", "Dylan")
full <- str_c(first, last, sep = "_")
print(full)                                     # [1] "Tom_Waits" "Bob_Dylan"
```

The `str_c()` function is also useful for printing nicely formatted sentences for debugging.

```
print(str_c("Number of names: ", length(full)))  # [1] "Number of names: 2"
```

The Base-R function equivalent to `str_c()` is `paste()`, but while the default separator for `str_c()` is an empty string, `""`, the default separator for `paste()` is a single space, `" "`. The equivalent Base-R function for `str_detect()` is `grepl()`, and the closest equivalent to `str_split_fixed()` in Base-R is `strsplit()`. As mentioned previously, however, using these and other `stringr` functions for this type of character-vector manipulation is recommended.

Factors

By now, factors have been mentioned a few times in passing, mostly in the context of using `stringsAsFactors = FALSE`, to prevent character vectors from being converted into factor types when data frames are created. Factors are a type of data relatively unique to R, and provide an alternative for storing categorical data compared with simple character vectors.

A good method for understanding factors might be to understand one of the historical reasons for their development, even if the reason is no longer relevant today. How much space would the `treatment` column of the experimental data frame above require to store in memory, if the storage was done naively? Usually, a single character like `"c"` can be stored in a single byte (8 bits, depending on the encoding), so an entry like `"chemical"` would require 8 bytes, and `"control"` would require 7. Given that there are ~360,000 entries in the full table, the total space required would be ~0.36 megabytes. Obviously, the amount of space would be greater for a larger table, and decades ago even a few megabytes of data could represent a challenge.

But that's for a naive encoding of the data. An alternative could be to encode `"chemical"` and `"control"` as simple integers `1` and `2` (4 bytes can encode integers from -2.1 to 2.1 billion), as well as a separate lookup table mapping the integer `1` to `"chemical"` and `2` to `"control"`. This would be a space savings of about two times, or more if the terms were longer. This type of storage and mapping mechanism is exactly what factors provide.[101]

We can convert a character vector (or factor) into a factor using the `factor()` function, and as usual the `head()` function can be used to extract the first few elements.

101. Although character vectors are also efficiently stored in modern-day versions of R, factors still have unique uses.

```
treatment_factor <- factor(expr_long_split$treatment)
print(head(treatment_factor, n = 10))
```

When printed, factors display their `levels` as well as the individual data elements encoded to levels. Notice that the quote marks usually associated with character vectors are not shown.

```
 [1] control  control  control  control  control  control  control  control 
 [9] chemical chemical
Levels: chemical control
```

It is illustrating to attempt to use the `str()` and `class()` and `attr()` functions to dig into how factors are stored. Are they lists, like the results of the `t.test()` function, or something else? Unfortunately, they are relatively immune to the `str()` function; `str(treatment_factor)` reports:

```
 Factor w/ 2 levels "chemical","control": 2 2 2 2 2 2 2 2 1 1 ...
```

This result illustrates that the data appear to be coded as integers. If we were to run `print(class(treatment_factor))`, we would discover its class is `"factor"`.

As it turns out, the class of a data type is stored as an attribute.

```
print(attr(treatment_factor, "class"))           # [1] "factor"
```

Above, we learned that we could remove an attribute by setting it to `NULL`. Let's set the `"class"` attribute to `NULL`, and then run `str()` on it.

```
attr(treatment_factor, "class") <- NULL
str(treatment_factor)
```

```
 atomic [1:366016] 2 2 2 2 2 2 2 2 1 1 ...
 - attr(*, "levels")= chr [1:2] "chemical" "control"
```

Aha! This operation reveals a factor's true nature: an integer vector, with an attribute of `"levels"` storing a character vector of labels, and an attribute for `"class"` that specifies the class of the vector as a factor. The data itself are stored as either `1` or `2`, but the levels attribute has `"chemical"` as its first element (and hence an integer of `1` encodes `"chemical"`) and `"control"` as its second (so `2` encodes `"control"`).

This special "class" attribute controls how functions like str() and print() operate on an object, and if we want to change it, this is better done by using the class() accessor function rather than the attr() function as above. Let's change the class back to factor.

```
class(treatment_factor) <- "factor"
```

Renaming Factor Levels

Because levels are stored as an attribute of the data, we can easily change the names of the levels by modifying the attribute. We can do this with the attr() function, but as usual, a specific accessor function called levels() is preferred.

```
attr(treatment_factor, "levels") <- c("Chemical", "Water")
# better:
levels(treatment_factor) <- c("Chemical", "Water")
```

Why is the levels() function preferred over using attr()? Because when using attr(), there would be nothing to stop us from doing something irresponsible, like setting the levels to identical values, as in c("Water", "Water"). The levels() function will check for this and other absurdities.

What the levels() function can't check for, however, is the semantic meaning of the levels themselves. It would not be a good idea to mix up the names, so that "Chemical" would actually be referring to plants treated with water, and vice versa:

```
levels(treatment_factor) <- c("Water", "Chemical")  # Ack!
```

The reason this is a bad idea is that using levels() only modifies the "levels" attribute but does nothing to the underlying integer data, breaking the mapping.

Reordering Factor Levels

Although we motivated factors on the basis of memory savings, in modern versions of R, even character vectors are stored internally using a sophisticated strategy, and modern computers generally have larger stores of RAM besides. Still, there is another motivation for factors: the fact that the levels may have a meaningful order. Some statistical tests might like to compare certain subsets of a data frame as defined by a factor; for example, numeric values associated with factor levels `"low"` might be compared to those labeled `"medium"`, and those in turn should be compared to values labeled `"high"`. But, given these labels, it makes no sense to compare the `"low"` readings directly to the `"high"` readings. Factors provide a way to specify that the data are categorical, but also that `"low" < "medium" < "high"`.

Thus we might like to specify or change the order of the levels within a factor, to say, for example, that the `"Water"` treatment is somehow less than the `"Chemical"` treatment. But we can't do this by just renaming the levels.

The most straightforward way to specify the order of a character vector or factor is to convert it to a factor with `factor()` (even if it already is a factor) and specify the order of the levels with the optional `levels =` parameter. Usually, if a specific order is being supplied, we'll want to specify `ordered = TRUE` as well. The levels specified by the `levels =` parameter must match the existing entries. To simultaneously rename the levels, the `labels =` parameter can be used as well.

```
treatment_factor <- factor(expr_long_split$treatment,
                           levels = c("control", "chemical"), # existing levels
                           labels = c("Water", "Chemical"),   # new names
                           ordered = TRUE)                    # specify ordered

print(head(treatment_factor))
```

Now, `"Water"` is used in place of `"control"`, and the factor knows that `"Water" < "Chemical"`. If we wished to have `"Chemical" < "Water"`, we would have needed to use `levels = c("chemical", "control")` and `labels = c("Chemical", "Water")` in the call to `factor()`.

```
[1] Water Water Water Water Water Water
Levels: Water < Chemical
```

Disregarding the `labels =` argument (used only when we want to rename levels while reordering), because the `levels =` argument takes a character vector of the unique entries in the input vector, these could be precomputed to hold the levels in a given order. Perhaps we'd like to order the tissue types in reverse alphabetical order, for example:

```
tissues <- unique(expr_long_split$tissue)                   # B C A
tissues_sorted <- tissues[order(tissues)]                   # A B C
tissues_reversed <- rev(tissues_sorted)                     # C B A

tissues_factor <- factor(expr_long_split$tissue,
                         levels = tissues_reversed,
                         ordered = TRUE)

print(head(tissues_factor))
```

```
[1] A A B B B C
Levels: C < B < A
```

Rather than assigning to a separate `tissues_factor` variable, we could replace the data frame column with the ordered vector by assigning to `expr_long_split$tissue`.

We often wish to order the levels of a factor according to some other data. In our example, we might want the "first" tissue type to be the one with the smallest mean expression, and the last to be the one with the highest mean expression. A specialized function, `reorder()`, makes this sort of ordering quick and relatively painless. It takes three important parameters (among other optional ones):

1. The factor or character vector to convert to a factor with reordered levels.
2. A (generally numeric) vector of the same length to use as reordering data.
3. A function to use in determining what to do with argument 2.

Here's a quick canonical example. Suppose we have two vectors (or columns in a data frame), one of sampled fish species ("bass," "salmon," or "trout") and another of corresponding weights. Notice that the salmon are generally heavy, the trout are light, and the bass are in between.

```
species <- c("salmon", "trout", "salmon", "bass", "bass")
weights <- c(10.2, 3.5, 12.6, 6.2, 6.7)
```

If we were to convert the species vector into a factor with `factor(species)`, we would get the default alphabetical ordering: bass, salmon, trout. If we'd prefer to organize the levels according to the mean of the group weights, we can use `reorder()`:

```
species_factor <- reorder(species, weights, mean, order = TRUE)
```

With this assignment, `species_factor` will be an ordered factor with `trout < bass < salmon`. This small line of code does quite a lot, actually. It runs the `mean()` function on each group of weights defined by the different species labels, sorts the results by those means, and uses the corresponding group ordering to set the factor levels. What is even more impressive is that we could have just as easily used `median` instead of mean, or any other function that operates on a numeric vector to produce a numeric summary. This idea of specifying functions as parameters in other functions is one of the powerful "functional" approaches taken by R, and we'll be seeing more of it.

Final Notes about Factors

In many ways, factors work much like character vectors and vice versa; `%in%` and `==` can be used to compare elements of factors just as they can with character vectors (e.g., `tissues_factor == "A"` returns the expected logical vector).

Factors enforce that all elements are treated as one of the levels named in the `levels` attribute, or `NA` otherwise. A factor encoding 1 and 2 as `male` and `female`, for example, will treat any other underlying integer as `NA`. To get a factor to accept novel levels, the levels attribute must first be modified with the `levels()` function.

Finally, because factors work much like character vectors, but don't print their quotes, it can be difficult to tell them apart from other types when printed. This goes for simple character vectors when they are part of data frames. Consider the following printout of a data frame:

```
   id tissue count group
 1 AG4     A     1     2
 2 T35     C     2     1
 3 CX1     A     3     2
 4 L56     B     2     1
```

Because quotation marks are left off when printing data frames, it's impossible to tell from this simple output that the `id` column is a character vector, the `tissue` column is a factor, the `count` column is an integer vector, and the `group` column is a factor.[102] Using `class()` on individual columns of a data frame can be useful for determining what types they actually are.

Exercises

1. In the annotation file **`PZ.annot.txt`**, each sequence ID (column 1) may be associated with multiple gene ontology (GO) "numbers" (column 2) and a number of different "terms" (column 3). Many IDs are associated with multiple GO numbers, and there is nothing to stop a particular number or term from being associated with multiple IDs.

   ```
   ...
   PZ7180000023260_APN      GO:0005515      btb poz domain containing protein
   PZ7180000035568_APN      GO:0005515      btb poz domain containing protein
   PZ7180000020052_APQ      GO:0055114      isocitrate dehydrogenase (nad+)
   PZ7180000020052_APQ      GO:0006099      isocitrate dehydrogenase (nad+)
   PZ7180000020052_APQ      GO:0004449      isocitrate dehydrogenase (nad+)
   ...
   ```

   While most of the sequence IDs have an underscore suffix, not all do. Start by reading in this file (columns are tab separated) and then extracting a `suffix_only` data frame containing just those rows where the sequence ID contains an underscore. Similarly, extract a `no_suffix` data frame for rows where sequence IDs do *not* contain an underscore.

   Next, add to the `suffix_only` data frame columns for `base_id` and `suffix`, where base IDs are the parts before the underscore and suffices are the parts after the underscore (e.g., `base_id` is `"PZ7180000023260"` and `suffix` is `"APN"` for the ID `"PZ7180000023260_APN"`).

   Finally, produce versions of these two data frames where the `GO:` prefix has been removed from all entries of the second column.

2. The line `s <- sample(c("0", "1", "2", "3", "4"), size = 100, replace = TRUE)` generates a character vector of 100 random `"0"`s, `"1"`s, `"2"`s, `"3"`s, and `"4"s.` Suppose that `"0"` means "Strongly Disagree," `"1"` means "Disagree," `"2"` means "Neutral," `"3"` means "Agree,"

102. Factors created from integer vectors are a special kind of headache. Consider a line like `h <- factor(c(4, 1, 5, 6, 4))`; because factors are treated like character types, this would be converted so it is based on `c("4", "1", "5", "6", "4")`, where the underlying mapping and storage have an almost arbitrary relationship to the elements. Try `print(as.numeric(h))` to see the trouble one can get into when mixing up these types, as well as `class(h) <- NULL` followed by `str(h)` to see the underlying mapping and why this occurs.

and "4" means "Strongly Agree." Convert s into an ordered factor with levels Strongly Disagree < Disagree < Neutral < Agree < Strongly Agree.

3. Like vectors, data frames (both rows and columns) can be selected by index number (numeric vector), logical vector, or name (character vector). Suppose that grade_current is a character vector generated by grade_current <- sample(c("A", "B", "C", "D", "E"), size = 100, replace = TRUE), and gpa is a numeric vector, as in gpa <- runif(100, min = 0.0, max = 4.0). Further, we add these as columns to a data frame, grades <- data.frame(current_grade, gpa, stringsAsFactors = FALSE).

   We are interested in pulling out all rows that have "A", "B", or "C" in the current_grade column. Describe, in detail, what each of the three potential solutions does:

```
abc_grades <- grades[grades$current_grade %in% c("A", "B", "C"), ]
abc_grades <- grades[grades$current_grade == c("A", "B", "C"), ]
abc_grades <- grades[c("A", "B", "C"), ]
```

   How does R interpret each one (i.e., what will R try to do for each), and what would the result be? Which one(s) is (are) correct? Which will report errors? Are the following three lines any different from the above three in what R tries to do?

```
abc_grades <- subset(grades, current_grade %in% c("A", "B", "C"))
abc_grades <- subset(grades, current_grade == c("A", "B", "C"))
abc_grades <- subset(grades, c("A", "B", "C"))
```

Chapter 33

Split, Apply, Combine

Although we touched briefly on writing functions, the discussion has so far been largely about the various types of data used by R and the syntax and functions for working with them. These include vectors of various types, lists, data frames, factors, indexing, replacement, and vector recycling.

At some point won't we discuss analyzing a meaningful data set? Yes, and that point is now. We'll continue to work with the gene expression data from chapter 32, "Character and Categorical Data," after it's been processed to include the appropriate individual columns using the functions in the `stringr` library. Here's the code from the last chapter which pre-processed the input data:

```
library(stringr)

expr_long <- read.table("expr_long_coded.txt",
                        header = TRUE,
                        sep = "\t",
                        stringsAsFactors = FALSE)


# Split sample column into 3-column matrix on _'s
sample_split <- str_split_fixed(expr_long$sample, "_", 3)

# Turn the matrix into a data frame with appropriate column names
# and combine it with the original data frame into a larger set
sample_split_df <- data.frame(sample_split)
colnames(sample_split_df) <- c("genotype", "treatment", "tissuerep")
expr_long_split <- cbind(expr_long, sample_split_df)

# Create an individual tissue column
expr_long_split$tissue <- NA
expr_long_split$tissue[str_detect(expr_long_split$tissuerep, "A")] <- "A"
expr_long_split$tissue[str_detect(expr_long_split$tissuerep, "B")] <- "B"
expr_long_split$tissue[str_detect(expr_long_split$tissuerep, "C")] <- "C"

# We've already checked for NAs
#print(expr_long_split[is.na(expr_long_split$tissue), ]) # should print 0 rows

# Create a new rep column
expr_long_split$rep <- NA
expr_long_split$rep[str_detect(expr_long_split$tissuerep, "1")] <- "1"
expr_long_split$rep[str_detect(expr_long_split$tissuerep, "2")] <- "2"
expr_long_split$rep[str_detect(expr_long_split$tissuerep, "3")] <- "3"

# We've already checked for NAs, but a few were left
#print(expr_long_split[is.na(expr_long_split$rep), ]) # should print 0 rows

# So we remove all rows with such "bad" IDs
bad_ids <- expr_long_split$id[is.na(expr_long_split$rep)]
bad_rows <- expr_long_split$id %in% bad_ids       # logical vector
expr_long_split <- expr_long_split[!bad_rows, ]  # logical selection
```

As we continue, we'll likely want to rerun our analysis script in different ways, without having to rerun all of the code above, as it takes a few minutes to run this preprocessing script. One good strategy would be to write this "cleaned" data frame out to a text file using `write.table()`, and read it back in to an analysis-only script with `read.table()`. Just for variety, though, let's use the `save()` function, which stores any R object (such as a list, vector, or data frame) to a file in a compressed binary format.

```
save(expr_long_split, file = "expr_long_split.Rdata")
```

The traditional file extension for such an R-object file is `.Rdata`. Such data files are often shared among R programmers.

The first few lines of the saved data frame are all readings for the same gene. There are ~360,000 lines in the full data frame, with readings for over 11,000 genes. (The script above, which processes the **expr_long_coded.txt** file, can be found in the file **expr_preprocess.R**.) The figure below shows the results of `print(head(expr_long_split))`:

```
           id            annotation expression         sample genotype treatment
1 LQ00X000020 Hypothetical protein    5.024142 C6_control_A1       C6    control
2 LQ00X000020 Hypothetical protein    4.646697 C6_control_A3       C6    control
3 LQ00X000020 Hypothetical protein    4.986592 C6_control_B1       C6    control
4 LQ00X000020 Hypothetical protein    5.164292 C6_control_B2       C6    control
5 LQ00X000020 Hypothetical protein    5.348810 C6_control_B3       C6    control
6 LQ00X000020 Hypothetical protein    5.519393 C6_control_C1       C6    control
  tissuerep tissue rep
1        A1      A   1
2        A3      A   3
3        B1      B   1
4        B2      B   2
5        B3      B   3
6        C1      C   1
```

Let's start a new analysis script that loads this data frame before we begin our analysis of it. For this script we may also need the `stringr` library, and we're going to include the `dplyr` library as well (which we assume has been separately installed with `install.packages("dplyr")` in the interactive console).

```
library(stringr)
library(dplyr)

## This script analyzes the data stored in the data frame
## expr_long_split.Rdata, containing gene expression data

load("expr_long_split.Rdata")
expr <- expr_long_split
```

The `load()` function creates a variable with the same name that was used in `save()`, and for convenience we've created a copy of the data frame with a shorter variable name, `expr`. While these sorts of R data files are convenient for R users, the benefit of a `write.table()` approach would be that other users and languages would have more direct access to the data (e.g., in Python or even Microsoft Excel).

This data set represents a multifactor analysis of gene expression[103] in two genotypes of a plant, where a control treatment (water) and a chemical treatment (pesticide) have been applied to each. Further, three tissue types were tested (A, B, and C, for leaf, stem, and root, respectively) and two or three replicates were tested for each combination for statistical power. These data represent microarray readings and have already been normalized and so are ready for analysis.[104] (These data are also anonymized for this publication at the request of the researchers.)

For now, we'll focus on the genotype and treatment variables. Ignoring other variables, this gives us around 24 expression readings for each gene.

This sort of design is quite common in gene expression studies. It is unsurprising that gene expression will differ between two genotypes, and that gene expression will differ between two treatments. But genotypes and treatments frequently react in interesting ways. In this case, perhaps the L4 genotype is known to grow larger than the C6 genotype, and the chemical treatment is known to be harmful to the plants. Interestingly, it seems that the chemical treatment is less harmful to the L4 genotype.

Gene: `LQ00X00020`

Genotype	water	chemical
L4	15.2 14.2 15.5 13.1 10.9 14.1	7.65 6.58 6.32 7.91 5.04 6.94
C6	5.02 4.64 4.98 5.16 5.34 5.51	12.3 11.4 10.9 9.92 13.2 12.1

Treatment

Expressions

103. Here we'll be using the colloquial term "gene" to refer to a genetic locus producing messenger RNA and its expression to be represented by the amount of mRNA (of the various isoforms) present in the cell at the sampling time.

104. Microarrays are an older technology for measuring gene expression that have been largely supplanted by direct RNA-seq methods. For both technologies, the overall levels of expression vary between samples (e.g., owing to measurement efficiency) and must be modified so that values from different samples are comparable; microarray data are often $\log_2$ transformed as well in preparation for statistical analysis. Both steps have been performed for these data; we've avoided discussion of these preparatory steps as they are specific to the technology used and an area of active research for RNA-seq. Generally, RNA-seq analyses involve work at both the command line for data preprocessing and using functions from specialized R packages.

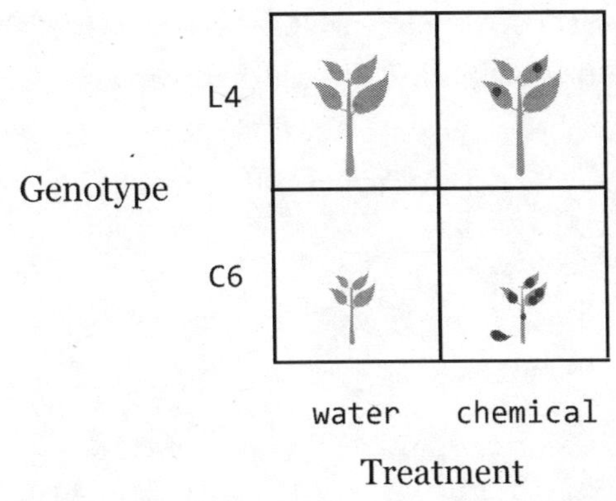

The natural question is: which genes might be contributing to the increased durability of L4 under chemical exposure? We can't simply look for the differences in mean expression between genotypes (with, say, `t.test()`), as that would capture many genes involved in plant size. Nor can we look for those with mean expression differences that are large between treatments, because we know many genes will react to a chemical treatment no matter what. Even both signals together (left side of the image below) are not that interesting, as they represent genes that are affected by the chemical and are involved with plant size, but may have nothing at all to do with unique features of L4 in the chemical treatment. What we are really interested in are genes with a *differential* response to the chemical between two the two genotypes (right side of the image below).

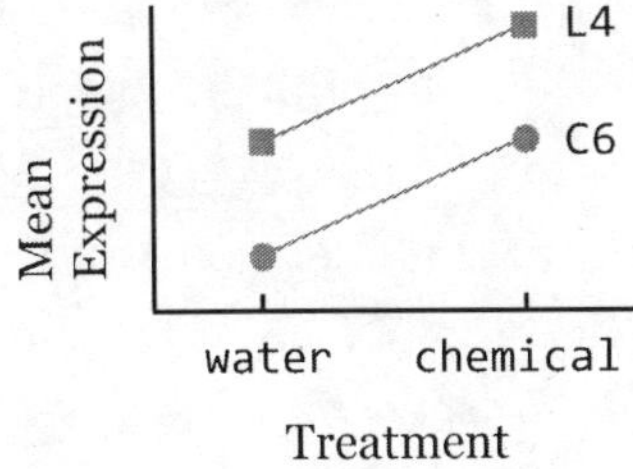

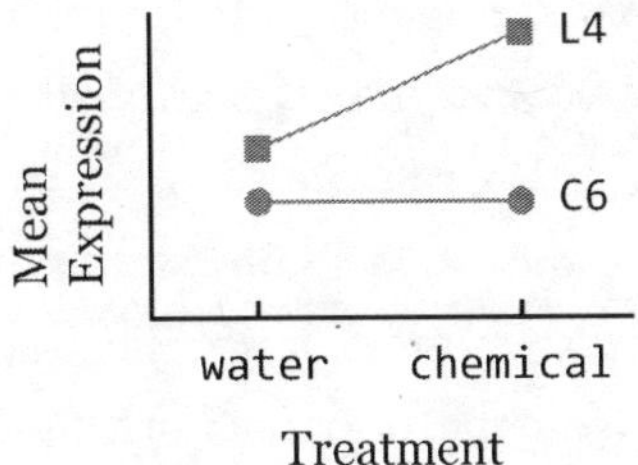

Does that mean that we can look for genes where the mean difference is large in one genotype, but not the other? Or within one treatment, but not that other? No, because many of these interesting scenarios would be missed:

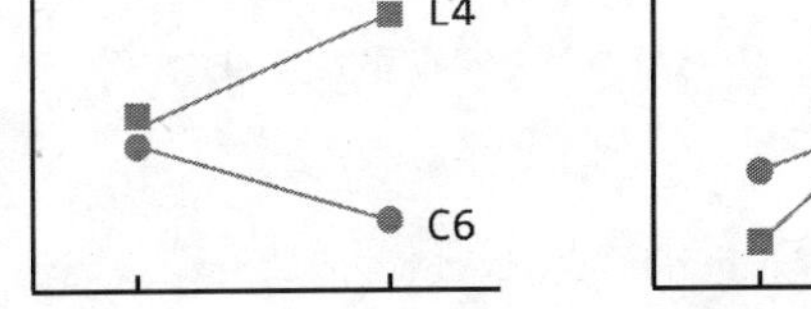

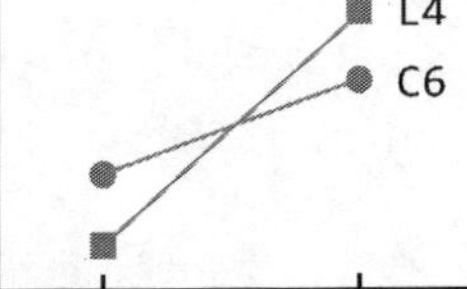

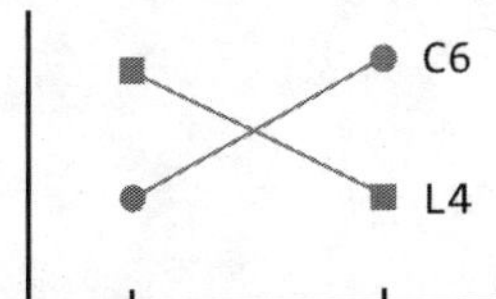

For each gene's expression measurements, we need a classic ANOVA (analysis of variance) model. After accounting for the variance explained by the genotype difference (to which we can assign a *p* value indicating whether there is a difference) as well as the variance explained by the treatment (which will also have a *p* value), is the leftover variance—for the interaction/differential response—significant?

A Single Gene Test, Formulas

To get started, we'll run an ANOVA for just a single gene.[105] First, we'll need to isolate a sub-data frame consisting of readings for only a single ID. We can use the `unique()` function to get a vector of unique IDs, and then the `%in%` operator to select rows matching only the first of those.

```
uniq_ids <- unique(expr$id)
expr1 <- expr[expr$id %in% uniq_ids[1], ]
```

This data frame contains rows for a single gene and columns for expression, genotype, treatment, and others (we could verify this with a simple `print()` if needed). We wish to run an ANOVA for these variables, based on a linear model predicting expression values from genotype, treatment, and the interaction of these terms.

The next step is to build a simple linear model using the `lm()` function, which takes as its only required parameter a "formula" describing how the predictor values should be related to the predicted values. Note the format of the formula: `response_vector ~ predictor_vector1 + predictor_vector2 + predictor_vector1 : predictor_vector2`, indicating that we wish to determine how to predict values in the response vector using elements from the predictor vectors.

```
lm1 <- lm(expr1$expression ~ expr1$genotype +                    # genotype
                             expr1$treatment +                   # treatment
                             expr1$genotype : expr1$treatment)   # interaction
```

If the vectors or factors of interest exist within a data frame (as is the case here), then the `lm()` function can work with column names only and a `data =` argument.

```
lm1 <- lm(expression ~ genotype + treatment + genotype:treatment, data = expr1)
```

105. There are many interesting ways we could analyze these data. For the sake of focusing on R rather than statistical methods, we'll stick to a relatively simple per gene 2x2 ANOVA model.

We can then print the linear model returned with `print(lm1)`, as well as its structure with `str(lm1)`, to see that it contains quite a bit of information as a named list:

```
Call:
lm(formula = expression ~ genotype + treatment + genotype:treatment,
    data = expr1)

Coefficients:
                (Intercept)                     genotypeL4
                    5.38363                       -0.08384
           treatmentcontrol  genotypeL4:treatmentcontrol
                   -0.19230                        0.16046

List of 13
 $ coefficients : Named num [1:4] 5.3836 -0.0838 -0.1923 0.1605
  ..- attr(*, "names")= chr [1:4] "(Intercept)" "genotypeL4" "treatmentcontrol" "
 $ residuals    : Named num [1:32] -0.167 -0.545 -0.205 -0.027 0.157 ...
  ..- attr(*, "names")= chr [1:32] "1" "2" "3" "4" ...
 $ effects      : Named num [1:32] -29.9003 -0.0102 0.317 0.2269 0.2552 ...
...
```

Before we continue, what is this `expression ~ genotype + treatment + genotype : treatment` "formula"? As usual, we can investigate a bit more by assigning it to a variable and using `class()` and `str()`.

```
f <- expression ~ genotype + treatment + genotype : treatment
str(f)
```

```
Class 'formula' length 3 expression ~ genotype + treatment + genotype:treatment
  ..- attr(*, ".Environment")=<environment: R_GlobalEnv>
```

The output indicates that the class is of type `"formula"` and that it has an `".Environment"` attribute. It's a bit opaque, but a formula type in R is a container for a character vector of variable names and a syntax for relating them to each other. The environment attribute specifies where those variables should be searched for by default, though functions are free to ignore that (as in the case of `data =` for `lm()`, which specifies that the variables should be considered column names in the given data frame). Formulas don't even need to specify variables that actually exist. Consider the following formula and the `all.vars()` function, which inspects a formula and returns a character vector of the unique variable names appearing in the formula.

```
f <- alpha ~ beta + beta : gamma
print(all.vars(f))                                  # [1] "alpha" "beta" "gamma"
```

Anyway, let's return to running the statistical test for the single gene we've isolated. The `lm1` result contains the model predicting expressions from the other terms (where this "model" is a list with a particular set of elements). To get the associated *p* values, we'll need to run it through R's `anova()` function.

```
anova1 <- anova(lm1)
print("Printed Result: ")
print(anova1)

print("Structure:")
str(anova1)
```

Printing the result shows that the *p* values (labeled `"Pr(>F)"`) for the genotype, treatment, and interaction terms are 0.97, 0.41, and 0.56, respectively. If we want to extract the *p* values individually, we'll need to first inspect its structure with `str()`, revealing that the result is both a list and a data frame—unsurprising because a data frame is a type of list. The three *p* values are stored in the `"Pr(>F)"` name of the list.

```
[1] "Printed Result: "
Analysis of Variance Table

Response: expression
                   Df Sum Sq  Mean Sq F value Pr(>F)
genotype            1 0.0001 0.000104  0.0007 0.9790
treatment           1 0.1005 0.100465  0.6770 0.4176
genotype:treatment  1 0.0515 0.051498  0.3470 0.5605
Residuals          28 4.1550 0.148395
[1] "Structure:"
Classes 'anova' and 'data.frame': 4 obs. of  5 variables:
 $ Df      : int  1 1 1 28
 $ Sum Sq  : num  0.000104 0.100465 0.051498 4.155049
 $ Mean Sq: num  0.000104 0.100465 0.051498 0.148395
 $ F value: num  0.000704 0.677016 0.347033 NA
 $ Pr(>F)  : num  0.979 0.418 0.561 NA
 - attr(*, "heading")= chr  "Analysis of Variance Table\n" "Response: expression"
```

We can thus extract the *p* values vector as `pvals1 <- anova1$"Pr(>F)"`; notice that we must use the quotations to select from this list by name because of the special characters (`(`, `>`, and `)`) in the name. For the sake of argument, let's store these three values in a data frame with a single row, and column names `"genotype"`, `"treatment"`, and `"genotype:treatment"` to indicate what the values represent.

```
pvals1 <- anova1$"Pr(>F)"
pvals_df1 <- data.frame(pvals1[1], pvals1[2], pvals1[3])
colnames(pvals_df1) <- c("genotype", "treatment", "genotype:treatment")

print(pvals_df1)
```

Output:

```
   genotype treatment genotype:treatment
1 0.9790277 0.4175689          0.5605206
```

This isn't bad, but ideally we wouldn't be "hard coding" the names of these *p* values into the column names. The information is, after all, represented in the printed output of the `print(anova1)` call above. Because `anova1` is a data frame as well as a list, these three names are actually the row names of the data frame. (Sometimes useful data hide in strange places!)

```
                                                   anova1$"Pr(>F)"

                   Df Sum Sq  Mean Sq F value Pr(>F)
genotype            1 0.0001 0.000104  0.0007 0.9790
treatment           1 0.1005 0.100465  0.6770 0.4176
genotype:treatment  1 0.0515 0.051498  0.3470 0.5605
Residuals          28 4.1550 0.148395

 rownames(anova1)
```

So, given any result of a call to `anova()`, no matter the formula used in the linear model, we should be able to computationally produce a single-row *p*-values data frame. But we've got to be clever. We can't just run `pvals_df1 <- data.frame(pvals1)`, as this will result in a single-column data frame with three rows, rather than a single-row data frame with three columns. Rather, we'll first convert the *p*-values vector into a list with `as.list()`, the elements of which will become the columns of the data frame because data frames are a type of list. From there, we can assign the `rownames()` of `anova1` to the `colnames()` of the `pvals_df1` data frame.[106]

106. For almost all data type conversions, R has a default interpretation (e.g., vectors are used as columns of data frames). As helpful as this is, it sometimes means nonstandard conversions require more creativity. Having a good understanding of the basic data types is thus useful in day-to-day analysis.

```
pvals1 <- anova1$"Pr(>F)"                   # vector of p-values
pvals_list1 <- as.list(pvals1)              # list of p-values
pvals_df1 <- data.frame(pvals_list1)        # single-row data frame
colnames(pvals_df1) <- rownames(anova1)  # column names from rownames(anova1)

print(pvals_df1)
```

The programmatically generated output is as follows:

```
   genotype treatment genotype:treatment Residuals
1 0.9790277 0.4175689          0.5605206        NA
```

This time, there's an extra column for the residuals, but that's of little significance.

This whole process—taking a sub–data frame representing expression values for a single gene and producing a single-row data frame of *p* values—is a good candidate for encapsulating with a function. After all, there are well-defined inputs (the sub–data frame of data for a single gene) and well-defined outputs (the *p*-values data frame), and we're going to want to run it several thousand times, once for each gene ID.

```
## Given a data frame with columns for expression,
## genotype, and treatement, runs a linear model
## and returns a single-row data frame with p-values in columns
sub_df_to_pvals_df <- function(sub_df) {
  lm1 <- lm(expression ~ genotype + treatment + genotype:treatment,
            data = sub_df)

  anova1 <- anova(lm1)
  pvals1 <- anova1$"Pr(>F)"
  pvals_list1 <- as.list(pvals1)
  pvals_df1 <- data.frame(pvals_list1)
  colnames(pvals_df1) <- rownames(anova1)

  return(pvals_df1)
}
```

To get the result for our `expr1` sub–data frame, we'd simply need to call something like `pvals_df1 <- sub_df_to_pvals_df(exp1)`. The next big question is: how are we going to run this function not for a single gene, but for all 11,000 in the data set? Before we can answer that question, we must learn more about the amazing world of functions in R.

Exercises

1. Generate a data frame with a large amount of data (say, one with two numeric columns, each holding 1,000,000 entries). Write the data frame to a text file with `write.table()` and save the data frame to an R data file with `save()`. Check the size of the resulting files. Is one smaller than the other? By how much?

2. The example we've covered—generating a small data frame of results (p values) from a frame of data (expression values) by using a function—is fairly complex owing to the nature of the experimental design and the test we've chosen. Still, the basic pattern is quite common and very useful.

   A default installation of R includes a number of example data sets. Consider the `CO2` data frame, which describes CO_2 uptake rates of plants in different treatments (`"chilled"` and `"nonchilled"`; see `help(CO2)` for more detailed information). Here's the output of `print(head(CO2))`:

   ```
     Plant   Type  Treatment conc uptake
   1   Qn1 Quebec nonchilled   95   16.0
   2   Qn1 Quebec nonchilled  175   30.4
   3   Qn1 Quebec nonchilled  250   34.8
   4   Qn1 Quebec nonchilled  350   37.2
   5   Qn1 Quebec nonchilled  500   35.3
   6   Qn1 Quebec nonchilled  675   39.2
   ```

   The conc column lists different ambient CO_2 concentrations under which the experiment was performed. Ultimately, we will want to test, for each concentration level, whether the uptake rate is different in chilled versus nonchilled conditions. We'll use a simple t test for this.

   Start by writing a function that takes a data frame with these five columns as a parameter and returns a single-row, single-column data frame containing a p value (reported by `t.test()`) for chilled uptake rates versus nonchilled uptake rates. (Your function will need to extract two vectors of update rates to provide to `t.test()`, one for chilled values and one for nonchilled.)

   Next, extract a sub–data frame containing rows where `conc` values are `1000`, and run your function on it. Do the same for a sub–data frame containing `conc` value of `675`.

Attack of the Functions!

While R has a dedicated function for computing the interquartile range (the range of data values between the 25th percentile and the 75th), let's write our own so that we can explore functions more generally. We'll call it `ipercentile_range`, to note that it computes an interpercentile range (where the percentiles just happen to be quartiles, for now).

```
ipercentile_range <- function(x) {
  lower_value <- quantile(x, 0.25)
  upper_value <- quantile(x, 0.75)
  return(upper_value - lower_value)
}
```

Next, let's make a list containing some samples of random data.

```
samples <- list()
samples$s1 <- rnorm(100, mean = 10, sd = 4)
samples$s2 <- rnorm(100, mean = 5, sd = 2)
samples$s3 <- rnorm(100, mean = 20, sd = 8)
```

To compute the three interquartile ranges, we could call the function three times, once per element of the list, and store the results in another list.

```
iqr1 <- ipercentile_range(samples$s1)                # 5.44
iqr2 <- ipercentile_range(samples$s2)                # 2.22
iqr3 <- ipercentile_range(samples$s3)                # 10.15
sample_iqrs <- list(iqr1, iqr2, iqr3)                # list of 5.44 2.22 10.15
```

Not too bad, but we can do better. Notice the declaration of the function—we've used the assignment operator `<-` to assign to `ipercentile_range`, just as we do with other variables. In fact, `ipercentile_range` is a variable! And just as with other data, we can check its class and print it.

```
print(class(ipercentile_range))          # [1] "function"
print(ipercentile_range)
```

When it is printed, we see the source code for the function:

```
function (x)
{
    lower_value <- quantile(x, 0.25)
    upper_value <- quantile(x, 0.75)
    return(upper_value - lower_value)
}
```

In R, functions are a type of data just like any other. Thus R is an example of a "functional" programming language.[107] One of the important implications is that functions can be passed as parameters to other functions; after all, they are just a type of data. This is a pretty tricky idea. Let's explore it a bit with the `lapply()` function. This function takes two parameters: first, a list, and second, a function to apply to each element of the list. The return value is a list representing the collated outputs of each function application. So, instead of using the four lines above to apply `ipercentile_range()` to each element of `samples`, we can say:

```
sample_irqs <- lapply(samples, ipercentile_range)  # list of 5.44 2.22 10.15
```

The resulting list will have the same three values as above, and the names of the elements in the output list will be inherited from the input list (so `sample_irqs$s1` will hold the first range or 5.44, for this random sample anyway).

This is an example of the powerful "split-apply-combine" strategy for data analysis. In general, this strategy involves splitting a data set into pieces, applying an operation or function to each, and then combining the results into a single output set in some way.

107. Not all programming languages are "functional" in this way, and some are only partially functional. Python is an example of a mostly functional language, in that functions defined with the `def` keyword are also data just like any other. This book does not highlight the functional features of Python, however, preferring instead to focus on Python's object-oriented nature. Here, R's functional nature is the focus, and objects in R will be discussed only briefly later.

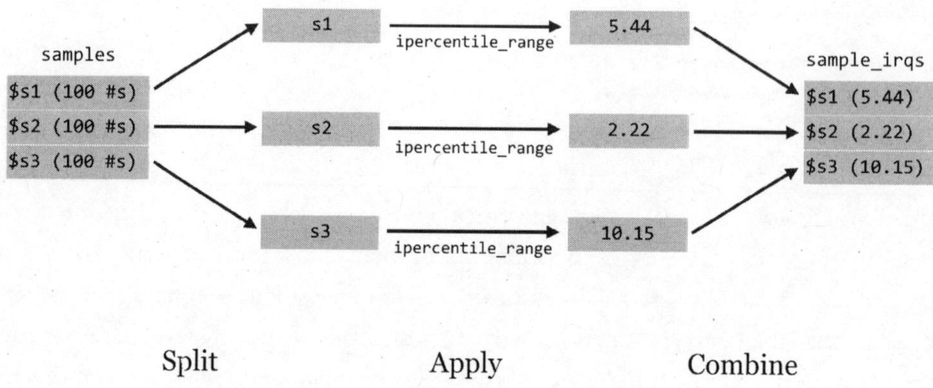

When the input and output are both lists, this operation is also known as a "map." In fact, this paradigm underlies some parallel processing supercomputer frameworks like Google's MapReduce and the open-source version of Hadoop (although these programs aren't written in R).

Because lists are such a flexible data type, `lapply()` is quite useful. Still, in some cases—like this one—we'd prefer the output to be a vector instead of a list. This is common enough to warrant a conversion function called `unlist()` that extracts all of the elements and subelements of a list to produce a vector.

```
irqs_vec <- unlist(sample_irqs)
print(irqs_vec)                              # [1] 5.44 2.22 10.15
```

Be warned: because lists are more flexible than vectors, if the list given to `unlist()` is not simple, the elements may be present in the vector in an odd order and they will be coerced into a common data type.

The `lapply()` function can also take as input a vector instead of a list, and it is but one of the many "apply" functions in R. Other examples include `apply()`, which applies a function to each row or column of a matrix,[108] and `sapply()` (for "simple apply"), which detects whether the return type should be a vector or matrix depending on the function applied. It's easy to get confused with this diversity, so to start, it's best to focus on `lapply()` given its relatively straightforward (but useful) nature.

108. Because the `apply()` function operates on matrices, it will silently convert any data frame given to it into a matrix. Because matrices can hold only a single type (whereas data frames can hold different types in different columns), using `apply()` on a data frame will first force an autoconversion so that all columns are the same type. As such, it is almost never correct to use `apply()` on a data frame with columns of different types.

Collecting Parameters with ...

The help page for `lapply()` is interesting. It specifies that the function takes three parameters: `X` (the list to apply over), `FUN` (the function to apply), and `...`, which the help page details as "optional arguments to `FUN`."

```
Usage:

     lapply(X, FUN, ...)
```

The `...` parameter is common in R, if a bit mysterious at times. This "parameter" collects any number of named parameters supplied to the function in a single unit. This collection can then be used by the function; often they are passed on to functions called by the function.

To illustrate this parameter, let's modify our `ipercentile_range()` function to take two optional parameters, a `lower` and `upper` percentile from which the range should be computed. While the interquartile range is defined as the range in the data from the 25th to the 75th percentile, our function will be able to compute the range from any lower or upper percentile.

```
ipercentile_range <- function(x, lower = 0.25, upper = 0.75) {
  lower_value <- quantile(x, lower)
  upper_value <- quantile(x, upper)
  return(upper_value - lower_value)
}
```

Now we can call our function as `ipercentile_range(sample$s1)` to get the default interquartile range of the first sample, or as `ipercentile_range(samples$s2, lower = 0.05, upper = 0.95)`, to get the middle 90th percentile range. Similarly, we have the ability to supply the optional parameters through the `...` of `lapply()`, which will in turn supply them to every call of the `ipercentile_range` function during the apply step.

```
inter_90s <- lapply(samples, ipercentile_range, lower = 0.05, upper = 0.95)
print(unlist(inter_90s))                                 # [1] 12.12 6.30 22.08
```

This syntax might seem a bit awkward, but something like it is necessary if we hope to pass functions that take multiple parameters *as* parameters to other functions. This also should help illuminate one of the commonly misunderstood features of R.

```
inter_90s <- lapply(samples, ipercentile_range, lower = 0.05, upper = 0.95)
```

X FUN(,) ...

each element

Functions Are Everywhere

Although mostly hidden from view, most (if not all) operations in R are actually function calls. Sometimes function names might contain nonalphanumeric characters. In these cases, we can enclose the function name in backticks, as in `` iqr1 <- `ipercentile_range`(samples$s1) ``. (In chapter 27, "Variables and Data," we learned that any variable name can be enclosed in backticks this way, though it's rarely recommended.)

What about something like the lowly addition operator, `+`? It works much like a function: given two inputs (left-hand-side and right-hand-side vectors), it returns an output (the element-by-element sum).

```
s <- c(1, 2) + c(6, 3)                         # 7 5
```

In fact, `+` really *is* a function that takes two parameters, but to call it that way we need to use backticks.

```
s <- `+`(c(1, 2), c(6, 3))                      # 7 5
```

This might be considered somewhat amazing—the "human-readable" version is converted into a corresponding function call behind the scenes by the interpreter. The latter, functional representation is easier for the interpreter to make use of than the human-friendly representation. This is an example of syntactic sugar: adjustments to a language that make it "sweeter" for human programmers. R contains quite a lot of syntactic sugar. Even the assignment operator `<-` is a function call! The line `a <- c(1, 2, 3)` is actually syntactic sugar for `` `<-`("a", c(1, 2, 3)) ``.

Because functions are just a type of data and function names are variables, we can easily reassign new functions to old names. Perhaps a colleague has left her R session open, so we decide to redefine `+` to mean "to the power of":

```
# Redefines + to mean power! Mwahaha
`+` <- function(a, b) {
  value <- a ^ b
  return(value)
}

print(3 + 7)                                  # [1] 2187
```

Ok, maybe that's a bit too evil. Another feature of R is that function names that begin and end with `%` and take two parameters signals that the function can be used as an infix function via syntactic sugar.

```
`%equalish%` <- function(a, b, epsilon = 0.00001) {
  result <- abs(a - b) < epsilon
  return(result)
}

answer <- c(4.00000001, 6) %equalish% c(4, 2)
print(answer)                                 # [1] TRUE FALSE
```

Sugary Data Frame Split-Apply-Combine

Although the basic units of data in R are vectors, and lists are used to store heterogeneous data, data frames are important as a storage mechanism for tabular data. Because `lapply()` can apply a function to each element of a list, it can also apply a function to each column of a data frame. This fact can be used, for example, to extract all numeric columns from a data frame with an application of `lapply()` and `is.numeric()` (the latter returns `TRUE` when its input is a numeric type; we'll leave the details as an exercise).

For most tabular data, though, we aren't interested in independently applying a function to each column. Rather, we more often wish to apply a function to different sets of rows grouped by one or more of the columns. (This is exactly what we want to do with the `sub_df_to_pvals()` function we wrote for our gene expression analysis.) The `dplyr` package (`install.packages("dplyr")`) provides this ability and is both powerful and easy to use, once we get used to its specialized syntactic sugar.

Initially, we'll cover two functions in the `dplyr` package, `group_by()` and `do()`: `group_by()` adds metadata (as attributes) to a data frame indicating which categorical columns define groups of rows and returns such a "grouped" data frame, while `do()` applies a given function to each group of a grouped data frame, and returns a data frame of results with grouping information included.

To illustrate, we'll first create a simple data frame on which to work, representing samples of fish from one of two different lakes: Green Lake or Detroit Lake.

```
species <- c("bass", "bass", "trout", "bass", "trout", "bass")
lake <- c("detroit", "green", "green", "detroit", "detroit", "green")
weight <- c(10.4, 9.6, 4.2, 12.2, 6.5, 8.4)

fish <- data.frame(species, lake, weight)
print(fish)
```

The nicely formatted printout:

```
  species    lake weight
1    bass detroit   10.4
2    bass   green    9.6
3   trout   green    4.2
4    bass detroit   12.2
5   trout detroit    6.5
6    bass   green    8.4
```

First, we'll group the data frame by the species column using `group_by()`. This function takes a data frame (or column-named matrix or similar) and a listing of column names (without quotes) that define the groups as additional parameters. As with using the `data =` parameter for `lm()`, the function looks within the data frame for the columns specified.

```
fish_by_species <- group_by(fish, species)
print(fish_by_species)
```

Is this data frame different from the original? Yes, but mostly in the attributes/metadata. When printed, we get some information on the "source" (where the data are stored—`dplyr` can access data from remote databases) and the groups:

```
Source: local data frame [6 x 3]
Groups: species

  species    lake weight
1    bass detroit   10.4
2    bass   green    9.6
...
```

Handily, unlike regular data frames, "grouped" data frames print only the first few rows and columns, even if they are many thousands of rows long. Sometimes this is an undesirable feature—fortunately, running `data.frame()` on a grouped data frame returns an ungrouped version. The `class()` of a grouped data frame returns `"data.frame"` as well as `"tbl"`, `"tbl_df"`, and `"grouped_df"`. Because a grouped data frame is also a regular data frame (and is also a list!), we can still do all of the fancy indexing operations on them covered in previous chapters.

The `do()` function applies a function to each group of rows of a grouped data frame using a split-apply-combine strategy. The function applied must take as its first parameter a data frame, and it must return a data frame. Let's write a function that, given a data frame or sub-data frame (group) of the fish data, returns a single-row data frame with columns for `mean_weight` and `sd_weight`. Both the `mean()` and `sd()` functions can take an `na.rm` argument to strip out any potential `NA` values in their inputs, so perhaps our function should take a similar optional parameter.

```
mean_sd_weight <- function(sub_df, remove_nas = FALSE) {
  weights <- sub_df$weight
  meanw <- mean(weights, na.rm = remove_nas)
  sdw <- sd(weights, na.rm = remove_nas)
  ret_df <- data.frame(mean_weight = meanw, sd_weight = sdw)
  return(ret_df)
}
```

This function takes as input a data frame with a column for `weight` and returns a data frame with summary statistics. We can run it on the original data frame:

```
stats <- mean_sd_weight(fish)
print(stats)
```

```
  mean_weight sd_weight
1        8.55   2.86339
```

Alternatively, we can call `do()` on the grouped data frame, telling it to run `mean_sd_weight()` on each group (sub-data frame). The syntax for `do()` differs slightly from that of `lapply()` in that we specify a `.` for the positional argument representing the sub-data frame.

```
stats_by_species <- do(fish_by_species, mean_sd_weight(.))
print(stats_by_species)
```

The result of the `do()` is another grouped data frame, made up of the rows returned by the applied function. Notice that the grouping columns have been added, even though we didn't specify them in the `ret_df` inside the function.

```
Source: local data frame [2 x 3]
Groups: species

  species mean_weight sd_weight
1    bass       10.15  1.594783
2   trout        5.35  1.626346
```

When developing functions that work with `do()`, you might run into an error like `Error: Results are not data frames at positions: 1, 2`. This error message indicates that the function is not returning a data frame type, which is required for `do()`. To diagnose problems like this, you can add `print()` statements inside of the function to inspect the contents of variables as the function is applied.

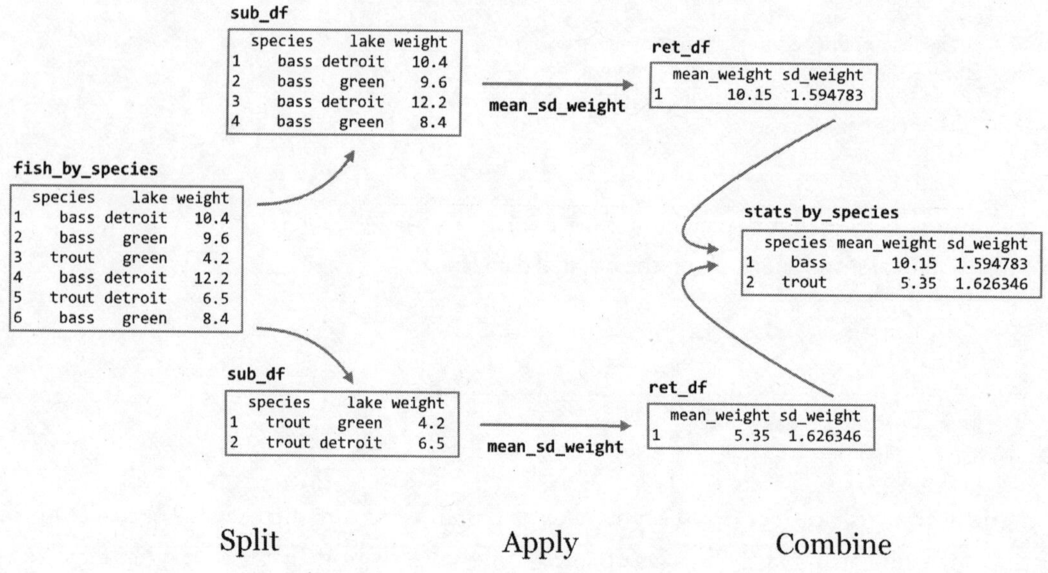

We can also group data frames by multiple columns, resulting in a single group per combination of column entries.[109]

```
fish_by_species_lake <- group_by(fish, species, lake)
stats_by_species_lake <- do(fish_by_species_lake, mean_sd_weight(.))
print(stats_by_species_lake)
```

```
  species    lake mean_weight sd_weight
1    bass detroit        11.3 1.2727922
2    bass   green         9.0 0.8485281
3   trout detroit         6.5        NA
4   trout   green         4.2        NA
```

The `NA` values for some standard deviations are the result of calling `sd()` on a vector of length one (because there was only one trout measurement per lake).

Although the applied function must take a data frame and return a data frame, there are no restrictions on the nature of the returned data frame. Here our function returns a single-row data frame, but it could return multiple rows that would be stitched together in the combine step. As an example, here's a function that, given a data frame, computes the `mean()` of the `weight` column, and subtracts that mean from all of the entries, returning the modified data frame (so-called "mean normalization" of the data).

```
mean_normalize_weight <- function(sub_df, remove_nas = FALSE) {
  ret_df <- sub_df
  meanw <- mean(sub_df$weight)
  ret_df$weight <- sub_df$weight - meanw
  return(ret_df)
}
```

And then we can easily mean-normalize the data on a *per group* basis!

```
mean_norm_by_species <- do(fish_by_species, mean_normalize_weight(.))
print(mean_norm_by_species)
```

109. Because `group_by()` and similar functions take "bare" or "unquoted" column names, they can be difficult to call given a character vector. For functions in the `dplyr` and `tidyr` packages (covered later), versions with `_` as part of the function name support this. For example, `fish_by_species_lake <- group_by(fish, species, lake)` can be replaced with `fish_by_species_lake <- group_by_(fish, c("species", "lake"))`.

```
  species    lake weight
1    bass detroit   0.25
2    bass   green  -0.55
3    bass detroit   2.05
4    bass   green  -1.75
5   trout   green  -1.15
6   trout detroit   1.15
```

In the above output, `-1.15` and `1.15` are the deviations from the mean of the trout group, and the others are deviations from the mean for the bass group.

More Sugar, Optional Parameters, Summarize

Something to note about the call to `do()` is that it differs syntactically from the call to `lapply()`. In `do()`, we specify not only the function to apply to each group, but also *how* that function will be called, using `.` to denote the input group data frame. This is somewhat clearer when we want to specify optional arguments to the applied function. In this case, we may want to specify that `NA` values should be removed by setting `remove_nas = TRUE` in each call to `mean_sd_weight()`.

```
stats_by_species <- do(fish_by_species, mean_sd_weight(., remove_nas = TRUE))
```

Speaking of syntactic sugar, the `magrittr` package (which is installed and loaded along with `dplyr`, though written by a different author) provides an interesting infix function, `%>%`. Consider the common pattern above; after the creation of a data frame (`fish`), we run it through a function to create an intermediary result (`fish_by_species` from `group_by()`), run that through another function to get another result (`stats_by_species` from `mean_sd_weight()`), and so on. While this process is clear, we could avoid the creation of multiple lines of code if we were willing to nest some of the function calls.

```
stats_by_species <- do(group_by(fish, species), mean_sd_weight(.))
```

This line of code certainly suffers from some readability problems. The infix `%>%` function supplies its left-hand side to its right-hand side (located by `.` in the right-hand side), returning the result of the call to the right-hand function. Thus we can rewrite the above as follows.

```
stats_by_species <- fish %>% group_by(., species) %>% do(., mean_sd_weight(.))
```

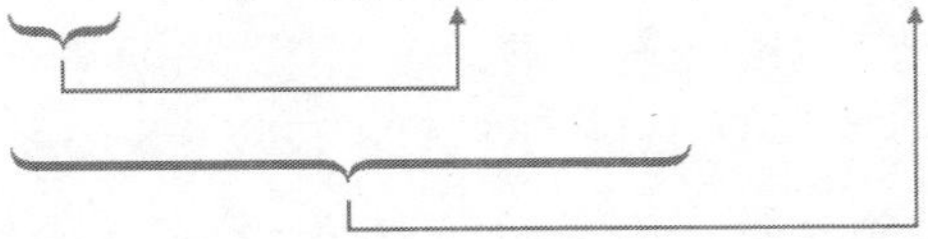

Notice the similarity between this and method chaining in Python and stdin/stdout pipes on the command line. (It's also possible to omit the `.` if it would be the first parameter, as in `fish %>% group_by(species) %>% do(mean_sd_weight(.))`.)

We demonstrated that functions applied by `do()` can return multiple-row data frames (in the mean-normalization example), but our `mean_sd_weight()` function only returns a single-row data frame with columns made of simple summary statistics. For this sort of simplified need, the `dplyr` package provides a specialized function called `summarize()`, taking a grouped data frame and other parameters indicating how such group-wise summary statistics should be computed. This example produces the same output as above, without the need for the intermediary `mean_sd_weight()` function.

```
fish_by_species <- group_by(fish, species)
stats_by_species <- summarize(fish_by_species,
                              mean_weight = mean(weight),
                              sd_weight = sd(weight))
```

The `dplyr` package provides quite a few other features and functions to explore, but even the simple combination of `group_by()` and `do()` represent a powerful paradigm. Because R is such a flexible language, the sort of advanced syntactic sugar used by `dplyr` is likely to become more common. Although some effort is required to understand these domain-specific languages (DSLs) and how they relate to "normal" R, the time spent is often worth it.

Exercises

1. The primary purpose of `lapply()` is to run a function on each element of a list and collate the results into a list. With some creativity, it can also be used to run an identical function call a large number of times. In this exercise we'll look at how *p* values are distributed under the "null" model—when two random samples from the same distribution are compared.

Start by writing a function `null_pval()` that generates two random samples with `rnorm(100, mean = 0, sd = 1)`, compares them with `t.test()`, and returns the *p* value. The function should take a single parameter, say, `x`, but not actually make any use of it.

Next, generate a list of 10,000 numbers with `10k_nums_list <- as.list(seq(1,10000))`, and call `10k_pvals_list <- lapply(10k_nums, null_pval)`. Turn this into a vector with `10k_pvals_vec <- unlist(10k_pvals_list)` and inspect the distribution with `hist(10k_pvals_vec)`.

What does this test reveal? What is the code you wrote doing, and why does it work? Why does the function need to take an `x` parameter that isn't even used? What happens if you change one of the random samples to use `rnorm(100, mean = 0.1, sd = 1)`?

2. If `df` is a data frame, using `[]` indexing treats it like a list of elements (columns). Write a function called `numeric_cols()` that takes a data frame as a parameter, and returns a version of the data frame with only the numeric columns kept. You may want to make use of `lapply()`, `unlist()`, `as.numeric()`, and the fact that lists (and data frames when treated like lists) can be indexed by logical vector.

   As an example, if `df1 <- data.frame(id = c("PRQ", "XL2", "BB4"), val = c(23, 45.6, 62))`, then `print(numeric_cols(df1))` should print a data frame with only the `val` column. If `df2 <- data.frame(srn = c(461, 514), name = c("Mel", "Ben"), age = c(27, 24))`, then `print(numeric_cols(df2))` should print a data frame with only `srn` and `age` columns.

3. Write a function called `subset_rows()` that takes two parameters: first, a data frame `df`, and second, an integer `n`. The function should return a data frame consisting of `n` random rows from `df` (you may find the `sample()` function of use).

   The `iris` data frame is a commonly-referenced built-in R data set, describing measurements of petals for a variety of iris species. Here's the output of `print(head(iris))`:

```
  Sepal.Length Sepal.Width Petal.Length Petal.Width Species
1          5.1         3.5          1.4         0.2  setosa
2          4.9         3.0          1.4         0.2  setosa
3          4.7         3.2          1.3         0.2  setosa
4          4.6         3.1          1.5         0.2  setosa
5          5.0         3.6          1.4         0.2  setosa
...
```

Use group_by() to group the iris data frame by the Species column, and then do() along with your subset_rows() function to generate a data frame consisting of 10 random rows per species.

4. Interestingly, there's nothing stopping a function called by do() (or lapply()) from itself calling do() (and/or lapply()). Write a set of functions that compares every species in the iris data frame to every other species, reporting the mean difference in Petal.Width. (In this printout the difference in means compares SpeciesA to SpeciesB.)

```
     Species    SpeciesA    SpeciesB diff_mean_petal_width
1      setosa      setosa      setosa                 0.00
2 versicolor versicolor      setosa                 1.08
3  virginica  virginica      setosa                 1.78
4      setosa      setosa versicolor                -1.08
...
```

To accomplish this, you will want to write a function compare_petal_widths() that takes two sub-data frames, one containing data for species A (sub_dfa) and the other for species B (sub_dfb), and returns a data frame containing the name of the A species, the name of the B species, and the difference of mean Petal.Width. You can test your function by manually extracting data frames representing setosa and versicolor.

Next, write a function one_vs_all_by_species() that again takes two parameters; the first will be a sub-data frame representing a single species (sub_df), but the second will be a data frame with data for all species (all_df). This function should group all_df by species, and use do() to call compare_petal_widths() on each group sending along sub_df as a secondary parameter. It should return the result.

Finally, a function all_vs_all_by_species() can take a single data frame df, group it by species, and call one_vs_all_by_species() on each group, sending along df as a secondary parameter, returning the result. In the end, all that is needed is to call this function on iris.

Gene Expression, Finished

With the discussion of split-apply-combine and dplyr under our belts, let's return to the task of creating and analyzing a linear model for each ID in the gene expression data set. As a reminder, we had left off having read in the "cleaned" data set, extracting a sub-data frame representing a single

ID, and writing a function that takes such a sub-data frame and returns a single-row data frame of *p* values. (It should now be clear why we went through the trouble of ensuring our function takes a data frame as a parameter and returns one as well.)

```
library(stringr)
library(dplyr)

## This script analyzes the data stored in the data frame
## expr_long_split.Rdata, containing gene expression data

load("expr_long_split.Rdata")
expr <- expr_long_split

## Given a data frame with columns for expression,
## genotype, and treatement, runs a linear model
## and returns a single-row data frame with p-values in columns
sub_df_to_pvals_df <- function(sub_df) {
  lm1 <- lm(expression ~ genotype + treatment + genotype:treatment,
            data = sub_df)

  anova1 <- anova(lm1)
  pvals1 <- anova1$"Pr(>F)"
  pvals_list1 <- as.list(pvals1)
  pvals_df1 <- data.frame(pvals_list1)
  colnames(pvals_df1) <- rownames(anova1)

  return(pvals_df1)
}


uniq_ids <- unique(expr$id)
expr1 <- expr[expr$id %in% uniq_ids[1], ]

pvals_df1 <- sub_df_to_pvals_df(expr1)
```

Now, we can use `group_by()` on the `expr` data frame to group by the `id` column, and `do()` to apply the `sub_df_to_pvals_df()` function to each group. Rather than work on the entire data set, though, let's create a `expr10` to hold a data frame representing measurements for 10 IDs; if we are satisfied with the results, we can always instead analyze the full `expr` table (though the full data set takes only a couple of minutes to analyze).

```
ten_ids <- head(uniq_ids, n = 10)
expr10 <- expr[expr$id %in% ten_ids, ]

expr10_by_id <- group_by(expr10, id)
pvals_df <- do(expr10_by_id, sub_df_to_pvals_df(.))

print(pvals_df)
```

The result is a nicely organized table of *p* values for each gene in the data set:

```
Source: local data frame [10 x 5]
Groups: id

          id     genotype treatment genotype:treatment Residuals
1  LQ00X000020 0.9790276672 0.4175689          0.5605206        NA
2  LQ00X000200 0.9563376003 0.6023259          0.5273140        NA
3  LQ00X000210 0.3882136652 0.1738578          0.1057537        NA
4  LQ00X000280 0.5574234253 0.5687726          0.6549019        NA
5  LQ00X000290 0.0003209239 0.5995250          0.2806888        NA
...
```

There is one more important issue to consider for an analysis like this: *multiple test correction.* Suppose for a moment that none of the ~11,000 genes are differentially expressed in any way. Because *p* values are distributed evenly between zero and one under the null hypothesis (no difference), for the `genotype` column alone we could expect ~11,000 * 0.05 = 550 apparently (but erroneously) significant differences. Altering the scenario, suppose there were about 50 genes that truly were differentially expressed in the experiment: these would likely have small *p* values, but it is difficult to separate them from the other 550 apparently significant values.

There are a variety of solutions to this problem. First, we could consider a much smaller threshold for significance. What threshold should that be? Well, if we wanted the number of apparently—but not really—significant *p* values to be small (on average), say, 0.05, we could solve for 11,000 * α = 0.05, suggesting a cutoff of α = 0.000004545. This is known as Bonferroni correction. The trouble is, this is a conservative correction, and it's likely that even our 50 significant genes won't pass that small threshold.

The reason Bonferroni correction is so conservative is that we've specified that we'll only accept the average number of false positives to be 0.05, which is quite small. We might alternatively say we would accept 10 false positives (and solve for 11,000 * α = 10); if the number of kept results is 100, that's only a 10% false discovery rate (FDR), and so we can expect 90% of the kept results to be real (but we won't know which are which!). On the other hand, if the number of kept results ends up being 15, that's a 75% false discovery rate, indicating that we can expect most of the kept results to be false positives.

Instead of specifying the *number* of false positives we're willing to accept, we can instead specify the *rate* and deal with whatever number of kept results come out. There are several of these "FDR-controlling" methods available, and some make more assumptions about the data than others. For example, the methods of Benjamini and Hochberg (described in 1995 and known by the acronym "BH") and Benjamini, Hochberg, and Yekutieli (from 2001, known as "BY") differ in the amount of independence the many tests are assumed to have. The BY method assumes less independence between tests but generally results in more results for a given FDR rate. (Consult your local statistician or statistics textbook for details.)

The `p.adjust()` function allows us to run either the Bonferroni, BH, or BY correction methods. It takes two arguments: first, a vector of *p* values to adjust, and, second, a `method =` parameter that can be set to `"bonferroni"`, `"BH"`, or `"BY"`. (There are other possibilities, but these three are the most commonly used.) Returned is a vector of the same length of adjusted FDR values—selecting all entries with values less than *Q* is equivalent to setting an FDR rate cutoff of *Q*.

In the final analysis of all genes, we will produce a BY-adjusted column for the interaction *p* values, and then select from the data set only those rows where the FDR rate is less than 0.05. (The full analysis script is located in the file **`expr_analyze.R`**.)

```
## Run all data
expr_by_id <- group_by(expr, id)
pvals_df <- do(expr_by_id, sub_df_to_pvals_df(.))

## Add BY-adjusted column
pvals_df$interaction_BY <- p.adjust(pvals_df$"genotype:treatment")

## Extract with BY-adjusted FDR < 0.05
pvals_interaction_sig <- pvals_df[pvals_df$interaction_BY < 0.05, ]
print(pvals_interaction_sig)
```

Unfortunately (or fortunately, depending on how much data one hopes to further analyze), only one gene is identified as having a significant interaction after BY correction in this analysis. (If 100 were returned, we would expect 5 of them to be false positives due to the 0.05 FDR cutoff used.)

```
Source: local data frame [1 x 6]
Groups: id

           id   genotype    treatment genotype:treatment Residuals
1 LQ00X058430 0.01177861 7.113263e-07        4.24831e-06        NA
Variables not shown: interaction_BY (dbl)
```

For grouped data frames, `print()` will omit printing some columns if they don't fit in the display window. To see all the columns (including our new `interaction_BY` column), we can convert the grouped table back into a regular data frame and print the `head()` of it with `print(head(data.frame(pvals_interaction_sig)))`. For the curious, the `interaction_BY` value of this ID is 0.048.

Exercises

1. We spent a fair amount of work ensuring that the data frame generated and returned by `sub_df_to_pvals_df()` was programmatically generated. Take advantage of this fact by making the formula a secondary parameter of the function, that is, as `function(sub_df, form)`, so that within the function the linear model can be produced as `lm1 <- lm(form, data = sub_df)`.
   Next, run the analysis with `group_by()` and `do()`, but specify some other formula in the `do()` call (perhaps something like `expression ~ genotype + treatment + tissue`).

2. In a previous exercise in this chapter, we wrote a function that took a portion of the `CO2` data frame and returned a data frame with a *p* value for a comparison of CO_2 uptake values between chilled and nonchilled plants. Now, use `group_by()` and `do()` to run this test for each `conc` group, producing a data frame of *p* values. Use `p.adjust()` with Bonferroni correction to add an additional column of multiply-corrected values.

3. While functions called by `do()` must take a data frame as their first parameter and must return a data frame, they are free to perform any other actions as well, like generating a plot or writing a file. Use `group_by()`, `do()`, and `write.table()` to write the contents of the `CO2` data frame into seven text files, one for each `conc` level. Name them `conc_95.txt`, `conc_175.txt`, and so on. You may need `paste()` or `str_c()` (from the `stringr` library) to generate the file names.

Chapter 34

Reshaping and Joining Data Frames

Looking back at the split-apply-combine analysis for the gene expression data, it's clear that the organization of the data worked in our favor (after we split the `sample` column into `genotype`, `treatment`, and other columns, at least). In particular, each row of the data frame represented a single measurement, and each column represented a variable (values that vary across measurements). This allowed us to easily group measurements by the ID column. It also worked well when building the linear model, as `lm()` uses a formula describing the relationship of equal-length vectors or columns (e.g., `expr$expression ~ expr$treatment + expr$genotype`), which are directly accessible as columns of the data frame.

Data are often not organized in this "tidy" format. This data set, for example, was originally in a file called **`expr_wide.txt`** with the following format:

```
           id C6_chemical_A1 C6_chemical_A3 C6_chemical_B1 C6_chemical_B2 ...
LQ00X000020         4.831259       4.784184       5.577741       5.143742 ...
LQ00X000200        14.046963      14.045607      14.448088      14.927930 ...
LQ00X000210        12.546436      12.410848      12.566147      12.638267 ...
LQ00X000280        10.002939      10.162578       9.425349       9.774258 ...
LQ00X000290        11.930745      10.757338      11.769187      12.391130 ...
LQ00X000300         5.146014       5.386916       6.048761       5.263776 ...
...
```

Data formats like this are frequently more convenient for human readers than machines. Here, the `id` and `annotation` columns (the latter not shown) are variables as before, but the other column names each encode a variety of variables, and each row represents a large number of measurements. Data in this "shape" would be much more difficult to analyze with the tools we've covered so far.

Converting between this format and the preferred format (and back) is the primary goal of the tidyr package. As usual, this package can be installed in the interactive interpreter with install.packages("tidyr"). The older reshape2 package also handles this sort of data reorganization, and while the functions provided there are more flexible and powerful, tidyr covers the majority of needs while being easier to use.

Gathering for Tidiness

The gather() function in the tidyr package makes most untidy data frames tidier. The first parameter taken is the data frame to fix up, and the second and third are the "key" and "value" names for the newly created columns, respectively (without quotes). The remaining parameters specify the column names that need to be tidied (again, without quotes). Suppose we had a small, untidy data frame called expr_small, with columns for id, annotation, and columns for expression in the C6 and L4 genotypes.

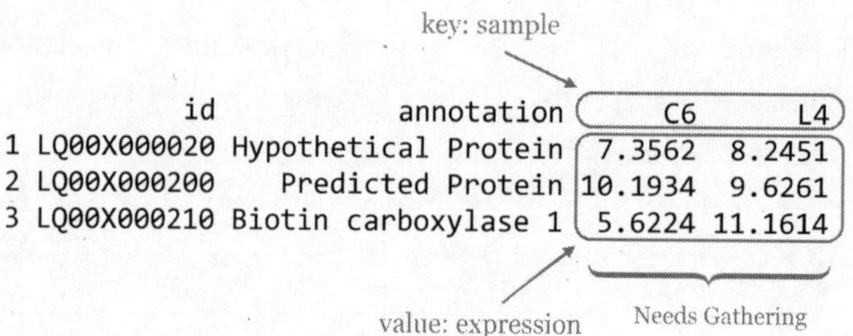

key: sample
id annotation C6 L4
1 LQ00X000020 Hypothetical Protein 7.3562 8.2451
2 LQ00X000200 Predicted Protein 10.1934 9.6261
3 LQ00X000210 Biotin carboxylase 1 5.6224 11.1614
value: expression
Needs Gathering

In this case, we would run the gather() function as follows, where sample and expression are the new column names to create, and C6 and L4 are the columns that need tidying. (Notice the lack of quotation marks on all column names; this is common to both the tidyr and dplyr packages' syntactic sugar.)

```
library(tidyr)

expr_gathered_small <- gather(expr_small, sample, expression, C6, L4)
print(expr_gathered_small)
```

```
           id            annotation sample expression
1 LQ00X000020 Hypothetical Protein     C6     7.3562
2 LQ00X000200    Predicted Protein     C6    10.1934
3 LQ00X000210 Biotin carboxylase 1     C6     5.6224
4 LQ00X000020 Hypothetical Protein     L4     8.2451
5 LQ00X000200    Predicted Protein     L4     9.6261
6 LQ00X000210 Biotin carboxylase 1     L4    11.1614
```

Notice that the data in the nongathered, nontidied columns (`id` and `annotation`) have been repeated as necessary. If no columns to tidy have been specified (`C6` and `L4` in this case), the `gather()` assumes that all columns need to be reorganized, resulting in only two output columns (`sample` and `expression`). This would be obviously incorrect in this case.

Listing all of the column names that need tidying presents a challenge when working with wide data frames like the full expression data set. To gather this table, we'd need to run `gather(expr_wide, sample, expression, C6_chemical_A1, C6_chemical_A3, C6_chemical_B1,` and so on, listing each of the 35 column names. Fortunately, `gather()` accepts an additional bit of syntactic sugar: using `-` and specifying the columns that don't need to be gathered. We could thus gather the full data set with

```
expr_gathered <- gather(expr_wide, sample, expression, -id, -annotation)
```

The `gather()` function takes a few other optional parameters, for example, for removing rows with `NA` values. See `help("gather")` in the interactive interpreter for more information.

Ungathering with spread()

While this organization of the data—with each row being an observation and each column being a variable—is usually most convenient for analysis in R, sharing data with others often isn't. People have an affinity for reading data tables where many measurements are listed in a single row, whereas "tidy" versions of data tables are often long with many repeated entries. Even when programmatically analyzing data tables, sometimes it helps to have multiple measurements in a single row. If we wanted to compute the difference between the `C6` and `L4` genotype expressions for each treatment condition, we might wish to have `C6_expression` and `L4_expression` columns, so that we can later use vectorized subtraction to compute a `C6_L4_difference` column.

The `spread()` function in the `tidyr` provides this, and in fact is the complement to the `gather()` function. The three important parameters are (1) the data frame to spread, (2) the column to use as the "key," and (3) the column to use as the "values." Consider the `expr_gathered_small` data frame from above.

key (entries ⇒ new columns)   value (entries ⇒ new row entries)

```
          id            annotation sample expression
1 LQ00X000020 Hypothetical Protein     C6     7.3562
2 LQ00X000200    Predicted Protein     C6    10.1934
3 LQ00X000210 Biotin carboxylase 1     C6     5.6224
4 LQ00X000020 Hypothetical Protein     L4     8.2451
5 LQ00X000200    Predicted Protein     L4     9.6261
6 LQ00X000210 Biotin carboxylase 1     L4    11.1614
```

Needs Spreading

Converting this data frame back into the "wide" version is as simple as:

```
expr_small <- spread(expr_gathered_small, sample, expression)
print(expr_small)
```

```
          id           annotation      C6      L4
1 LQ00X000020 Hypothetical Protein  7.3562  8.2451
2 LQ00X000200    Predicted Protein 10.1934  9.6261
3 LQ00X000210 Biotin carboxylase 1  5.6224 11.1614
```

Because the entries in the "key" column name become new column names, it would usually be a mistake to use a numeric column here. In particular, if we were to mix up the order and instead run `spread(expr_gathered_small, expression, sample)`, we'd end up with a column for each unique value in the `expression` column, which could easily number in the hundreds of thousands and would likely crash the interpreter.

In combination with `group_by()`, `do()`, and `summarize()` from the `dplyr` package, `gather()` and `spread()` can be used to aggregate and analyze tabular data in an almost limitless number of ways. Both the `dplyr` and `tidyr` packages include a number of other functions for working with data frames, including filtering rows or columns by selection criteria and organizing rows and columns.

Splitting Columns

In chapter 32, "Character and Categorical Data," we learned how to work with character vectors using functions like `str_split_fixed()` to split them into pieces based on a pattern, and `str_detect()` to produce a logical vector indicating which elements matched a pattern. The `tidyr` package also includes some specialized functions for these types of operations. Consider a small data frame `expr_sample` with columns for `id`, `expression`, and `sample`, like the precleaned data frame considered in previous chapters.

```
             id expression         sample
1 LQ00X000020   5.024142 C6_control_A1
2 LQ00X000020   4.646697 C6_control_A3
3 LQ00X000020   4.986592 C6_control_B1
4 LQ00X000020   5.164292 C6_control_B2
5 LQ00X000020   5.348810 C6_control_B3
6 LQ00X000020   5.519393 C6_control_C1
```

The `tidyr` function `separate()` can be used to quickly split a (character or factor) column into multiple columns based on a pattern. The first parameter is the data frame to work on, the second is the column to split within that data frame, the third specifies a character vector of newly split column names, and the fourth optional `sep =` parameter specifies the pattern (regular expression) to split on.

```
expr_sample_separated <- separate(expr_sample,
                                  sample,
                                  c("genotype", "treatment", "tissuerep"),
                                  sep = "_")
print(expr_sample_separated)
```

```
             id expression genotype treatment tissuerep
1 LQ00X000020   5.024142       C6   control        A1
2 LQ00X000020   4.646697       C6   control        A3
3 LQ00X000020   4.986592       C6   control        B1
4 LQ00X000020   5.164292       C6   control        B2
5 LQ00X000020   5.348810       C6   control        B3
6 LQ00X000020   5.519393       C6   control        C1
```

Similarly, the `extract()` function splits a column into multiple columns based on a pattern (regular expression), but the pattern is more general and requires an understanding of regular expressions and back-referencing using `()` capture groups. Here, we'll use the regular expression pattern "([A-Z])([0-9])" to match any single capital letter followed by a single digit, each of which get captured by a pair of parentheses. These values will become the entries for the newly created columns.

```
expr_sample_extracted <- extract(expr_sample_separated,
                                 tissuerep,
                                 c("tissue", "rep"),
                                 regex = "([A-Z])([0-9])")
print(expr_sample_extracted)

           id expression genotype treatment tissue rep
1 LQ0X000020   5.024142       C6   control      A   1
2 LQ0X000020   4.646697       C6   control      A   3
3 LQ0X000020   4.986592       C6   control      B   1
4 LQ0X000020   5.164292       C6   control      B   2
5 LQ0X000020   5.348810       C6   control      B   3
6 LQ0X000020   5.519393       C6   control      C   1
```

Although we covered regular expressions in earlier chapters, for entries like `C6_control_b3` where we assume the encoding is well-described, we could use a regular expression like `"(C6|L4)_(control|chemical)_(A|B|C)(1|2|3)"`.

While these functions are convenient for working with columns of data frames, an understanding of `str_split_fixed()` and `str_detect()` is nevertheless useful for working with character data in general.

Joining/Merging Data Frames, cbind() and rbind()

Even after data frames have been reshaped and otherwise massaged to make analyses easy, occasionally similar or related data are present in two different data frames. Perhaps the annotations for a set of gene IDs are present in one data frame, while the *p* values from statistical results are present in another. Usually, such tables have one or more columns in common, perhaps an `id` column.

Sometimes, each entry in one of the tables has a corresponding entry in the other, and vice versa. More often, some entries are shared, but others are not. Here's an example of two small data frames where this is the case, called `heights` and `ages`.

heights

```
  first     last height
1   Joe Withmore    5.5
2  Mary  O'Leary    5.2
3 Brent   Liston    6.1
4 Karen Streeter    5.3
```

ages

```
  first     last age
1 Brent   Liston  27
2   Joe    Jones  19
3 Karen Streeter  22
4 Chris Peterson  34
```

The `merge()` function (which comes with the basic installation of R) can quickly join these two data frames into one. By default, it finds all columns that have common names, and uses all entries that match in all of those columns. Here's the result of `merge(heights, ages)`:

```
  first     last height age
1 Brent   Liston    6.1  27
2 Karen Streeter    5.3  22
```

This is much easier to use than the command line join program: it can join on multiple columns, and the rows do not need to be in any common order. If we like, we can optionally specify a `by =` parameter, to specify which column names to join by as a character vector of column names. Here's `merge(heights, ages, by = c("first"))`:

```
  first   last.x height   last.y age
1 Brent   Liston    6.1   Liston  27
2   Joe Withmore    5.5    Jones  19
3 Karen Streeter    5.3 Streeter  22
```

Because we specified that the joining should only happen by the first column, `merge()` assumed that the two columns named `last` could contain different data and thus should be represented by two different columns in the output.

By default, `merge()` produces an "inner join," meaning that rows are present in the output only if entries are present for both the left (`heights`) and right (`ages`) inputs. We can specify `all = TRUE` to perform a full "outer join." Here's `merge(heights, ages, all = TRUE)`.

```
  first     last height age
1 Brent   Liston    6.1  27
2   Joe Withmore    5.5  NA
3   Joe    Jones     NA  19
4 Karen Streeter    5.3  22
5  Mary  O'Leary    5.2  NA
6 Chris Peterson     NA  34
```

In this example, `NA` values have been placed for entries that are unspecified. From here, rows with `NA` entries in either the `height` or `age` column can be removed with row-based selection and `is.na()`, or a "left outer join" or "right outer join," can be performed with `all.x = TRUE` or `all.y = TRUE`, respectively.

In chapter 32, we also looked at `cbind()` after splitting character vectors into multicolumn data frames. This function binds two data frames into a single one on a column basis. It won't work if the two data frames don't have the same number of rows, but it will work if the two data frames have column names that are identical, in which case the output data frame might confusingly have multiple columns of the same name. This function will also leave the data frame rows in their original order, so be careful that the order of rows is consistent before binding. Generally, using `merge()` to join data frames by column is preferred to `cbind()`, even if it means ensuring some identifier column is always present to serve as a binding column.

The `rbind()` function combines two data frames that may have different numbers of rows but have the same number of columns. Further, the column names of the two data frames must be identical. If the types of data are different, then after being combined with `rbind()`, columns of different types will be converted to the most general type using the same rules when mixing types within vectors.

Using `rbind()` requires that the data from each input vector be copied to produce the output data frame, even if the variable name is to be reused as in `df <- rbind(df, df2)`. Wherever possible, data frames should be generated with a split-apply-combine strategy (such as with `group_by()` and `do()`) or a reshaping technique, rather than with many repeated applications of `rbind()`.

Exercises

1. As discussed in exercises in Chapter 33, Split, Apply, Combine, the built-in `CO2` data frame contains measurements of CO_2 uptake rates for different plants in different locations under different ambient CO_2 concentrations.

   Use the `spread()` function in the `tidyr` library to produce a `CO2_spread` data frame that looks like so:

```
    Plant        Type  Treatment   95  175  250  350  500  675 1000
1    Qn1       Quebec nonchilled 16.0 30.4 34.8 37.2 35.3 39.2 39.7
2    Qn2       Quebec nonchilled 13.6 27.3 37.1 41.8 40.6 41.4 44.3
3    Qn3       Quebec nonchilled 16.2 32.4 40.3 42.1 42.9 43.9 45.5
4    Qc1       Quebec    chilled 14.2 24.1 30.3 34.6 32.5 35.4 38.7
5    Qc3       Quebec    chilled 15.1 21.0 38.1 34.0 38.9 39.6 41.4
...
```

Next, undo this operation with a `gather()`, re-creating the `CO2` data frame as `CO2_recreated`.

2. Occasionally, we want to "reshape" a data frame while simultaneously computing summaries of the data. The `reshape2` package provides some sophisticated functions for this type of computation (specifically `melt()` and `cast()`), but we can also accomplish these sorts of tasks with `group_by()` and `do()` (or `summarize()`) from the `dplyr` package in combination with `gather()` and `spread()` from `tidyr`.

   From the `CO2` data frame, generate a data frame like the following, where the last two columns report mean uptake for each `Type/conc` combination:

```
          Type conc nonchilled_mean_uptake chilled_mean_uptake
1       Quebec   95               15.26667            12.86667
2       Quebec  175               30.03333            24.13333
3       Quebec  250               37.40000            34.46667
4       Quebec  350               40.36667            35.80000
5       Quebec  500               39.60000            36.66667
6       Quebec  675               41.50000            37.50000
7       Quebec 1000               43.16667            40.83333
8  Mississippi   95               11.30000             9.60000
9  Mississippi  175               20.20000            14.76667
10 Mississippi  250               27.53333            16.10000
...
```

   You'll likely want to start by computing appropriate group-wise means from the original `CO2` data.

3. The built-in data frames `beaver1` and `beaver2` describe body temperature and activity observations of two beavers. Merge these two data frames into a single one that contains all the data and looks like so:

```
  day time  temp activ    name
1 307  930 36.58     0 beaver2
2 307  940 36.73     0 beaver2
3 307  950 36.93     0 beaver2
4 307 1000 37.15     0 beaver2
5 307 1010 37.23     0 beaver2
...
```

   Notice the column for `name`—be sure to include this column so it is clear to which beaver each measurement corresponds!

Chapter 35

Procedural Programming

For many (perhaps most) languages, control-flow structures like for-loops and if-statements are fundamental building blocks for writing useful programs. By contrast, in R, we can accomplish a lot by making use of vectorized operations, vector recycling, logical indexing, split-apply-combine, and so on. In general, R heavily emphasizes vectorized and functional operations, and these approaches should be used when possible; they are also usually optimized for speed. Still, R provides the standard repertoire of loops and conditional structures (which form the basis of the "procedural programming" paradigm) that can be used when other methods are clumsy or difficult to design. We've already seen functions, of course, which are incredibly important in the vast majority of programming languages.

Conditional Looping with While-Loops

A while-loop executes a block of code over and over again, testing a given condition (which should be or return a logical vector) before each execution. In R, blocks are delineated with a pair of curly brackets. Standard practice is to place the first opening bracket on the line, defining the loop and the closing bracket on a line by itself, with the enclosing block indented by two spaces.

```
count <- 1

while(count < 4) {
  print("Count is: ")
  print(count)
  count <- count + 1
}

print("Done!")
```

In the execution of the above, when the `while(count < 4)` line is reached, `count < 4` is run and returns the single-element logical vector `TRUE`, resulting in the block being run and printing `"Count is: "` and `1`, and then incrementing `count` by 1. Then the loop returns to the check; `count` being 2 is still less than 4, so the printing happens and `count` is incremented again to 3. The loop starts over, the check is executed, output is printed, and `count` is incremented to 4. The loop returns back to the top, but this time `count < 4` results in `FALSE`, so the block is skipped, and finally execution moves on to print `"Done!"`.

```
[1] "Count is: "
[1] 1
[1] "Count is: "
[1] 2
[1] "Count is: "
[1] 3
[1] "Done!"
```

Because the check happens at the start, if `count` were to start at some larger number like `5`, then the loop block would be skipped entirely and only `"Done!"` would be printed.

Because no "naked data" exist in R and vectors are the most basic unit, the logical check inside the `while()` works with a logical vector. In the above, that vector just happened to have only a single element. What if the comparison returned a longer vector? In that case, only the first element of the logical vector will be considered by the `while()`, and a warning will be issued to the effect of `condition has length > 1`. This can have some strange consequences. Consider if instead of `count <- 1`, we had `count <- c(1, 100)`. Because of the vectorized nature of addition, the output would be:

```
[1] "Count is: "
[1]   1 100
[1] "Count is: "
[1]   2 101
[1] "Count is: "
[1]   3 102
Warning messages:
1: In while (count < 4) { :
  the condition has length > 1 and only the first element will be used
2: In while (count < 4) { :
  the condition has length > 1 and only the first element will be used
3: In while (count < 4) { :
  the condition has length > 1 and only the first element will be used
4: In while (count < 4) { :
  the condition has length > 1 and only the first element will be used
[1] "Done!"
```

Two handy functions can provide a measure of safety from such errors when used with simple conditionals: `any()` and `all()`. Given a logical vector, these functions return a single-element logical vector indicating whether any, or all, of the elements in the input vector are `TRUE`. Thus our `while`-loop conditional above might better be coded as `while(any(count < 4))`. (Can you tell the difference between this and `while(any(count) < 4)`?)

Generating Truncated Random Data, Part I

The unique statistical features of R can be combined with control-flow structures like while-loops in interesting ways. R excels at generating random data from a given distribution; for example, `rnorm(1000, mean = 20, sd = 10)` returns a 100-element numeric vector with values sampled from a normal distribution with mean 20 and standard deviation 10.

```
sample <- rnorm(1000, mean = 20, sd = 10)
hist(sample)
```

Although we'll cover plotting in more detail later, the `hist()` function allows us to produce a basic histogram from a numeric vector. (Doing so requires a graphical interface like Rstudio for easily displaying the output. See chapter 37, "Plotting Data and `ggplot2`," for details on plotting in R.)

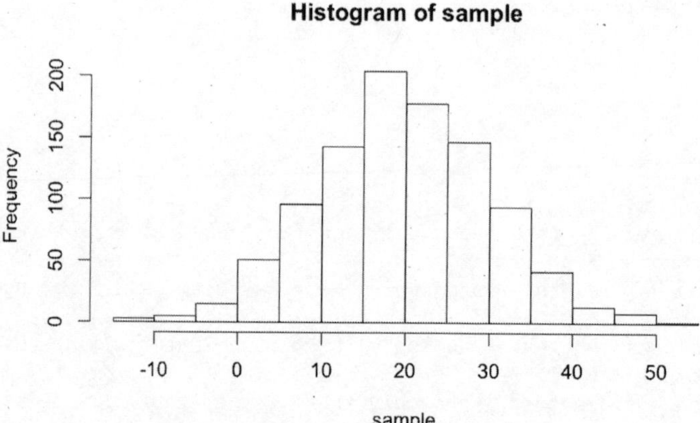

What if we wanted to sample numbers originally from this distribution, but limited to the range 0 to 30? One way to do this is by "resampling" any values that fall outside of this range. This will effectively truncate the distribution at these values, producing a new distribution with altered mean and standard deviation. (These sorts of "truncated by resampling" distributions are sometimes needed for simulation purposes.)

Ideally, we'd like a function that encapsulates this idea, so we can say `sample <- rnorm_trunc(0, 30, 1000, mean = 20, sd = 10)`. We'll start by defining our function, and inside the function we'll first produce an initial sample.

```
rnorm_trunc <- function(lower, upper, count, mean, sd) {
  sample <- rnorm(count, mean = mean, sd = sd)

  return(sample)
}
```

Note the usage of `mean = mean` in the call to `rnorm()`. The right-hand side refers to the parameter passed to the `rnorm_trunc()` function, the left-hand side refers to the parameter taken by `rnorm()`, and the interpreter has no problem with this usage.

Now, we'll need to "fix" the sample so long as *any* of the values are less than `lower` or greater than `upper`. We'll check as many times as needed using a while-loop.

```
rnorm_trunc <- function(lower, upper, count, mean, sd) {
  sample <- rnorm(count, mean = mean, sd = sd)

  while(any(sample < lower | sample > upper)) {
    # fix the sample

  }

  return(sample)
}
```

If any values are outside the desired range, we don't want to just try an entirely new sample set, because the probability of generating just the right sample (within the range) is incredibly small, and we'd be looping for quite a while. Rather, we'll only resample those values that need it, by first generating a logical vector of "bad" elements. After that, we can generate a resample of the needed size and use selective replacement to replace the bad elements.

```
rnorm_trunc <- function(lower, upper, count, mean, sd) {
  sample <- rnorm(count, mean = mean, sd = sd)

  while(any(sample < lower | sample > upper)) {
    # fix the sample
    bad <- sample < lower | sample > upper # logical
    bad_values <- sample[bad]
    count_bad <- length(bad_values)

    new_vals <- rnorm(count_bad, mean = mean, sd = sd)
    sample[bad] <- new_vals
  }

  return(sample)
}
```

Let's try it:

```
sample_trunc <- rnorm_trunc(0, 30, 1000, mean = 20, sd = 10)
hist(sample_trunc)
```

The plotted histogram reflects the truncated nature of the data set:

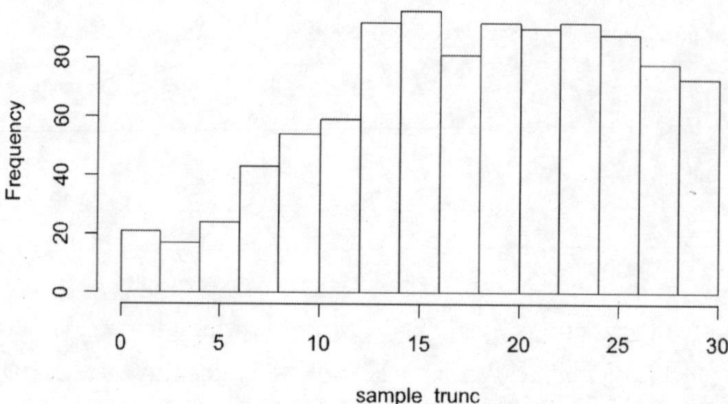

If-Statements

An if-statement executes a block of code based on a conditional (like a while-loop, but the block can only be executed once). Like the check for `while`, only the first element of a logical vector is checked, so using `any()` and `all()` is suggested for safety unless we are sure the comparison will result in a single-element logical vector.

```
# a single random number between 1 and 50
rand <- runif(1, min = 1, max = 50)

if(rand < 10) {
  print("The random number is less than 10")
} else if(rand < 20) {
  print("The random number is 10 or higher")
  print("But it's also less than 20")
} else if(rand < 30) {
  print("The random number is 20 or higher")
  print("And it's less than 30")
} else {
  print("The random number is 30 or higher")
}
```

Like if-statements in Python, each conditional is checked in turn. As soon as one evaluates to `TRUE`, that block is executed and all the rest are skipped. Only one block of an if/else chain will be executed, and perhaps none will be. The `if`-controlled block is required to start the chain, but one or more `else if`-controlled blocks and the final `else`-controlled block are optional.

In R, if we want to include one or more `else if` blocks or a single `else` block, the `else if()` or `else` keywords must appear on the same line as the previous closing curly bracket. This differs slightly from other languages and is a result of the way the R interpreter parses the code.

Generating Truncated Random Data, Part II

One of the issues with our `rnorm_trunc()` function is that if the desired range is small, it might still require many resampling efforts to produce a result. For example, calling `sample_trunc <- rnorm_trunc(15, 15.01, 1000, 20, 10)` will take a long time to finish, because it is rare to randomly generate values between 15 and 15.01 from the given distribution. Instead, what we might like is for the function to give up after some number of resamplings (say, 100,000) and return a vector containing `NA`, indicating a failed computation. Later, we can check the returned result with `is.na()`.

```
rnorm_trunc <- function(lower, upper, count, mean, sd) {
  sample <- rnorm(count, mean = mean, sd = sd)
  looped <- 0

  while(any(sample < lower | sample > upper)) {
    # fix the sample
    bad <- sample < lower | sample > upper # logical
    bad_values <- sample[bad]
    count_bad <- length(bad_values)

    new_vals <- rnorm(count_bad, mean = mean, sd = sd)
    sample[bad] <- new_vals

    if(looped > 100000) {
      return(NA)
    }
    looped <- looped + 1
  }
  return(sample)
}

sample_trunc <- rnorm_trunc(0, 30, 1000, mean = 20, sd = 10)
if(!is.na(sample_trunc)) {
  hist(sample_trunc)
}
```

The example so far illustrates a few different things:

1. NA can act as a placeholder in our own functions, much like mean() will return NA if any of the input elements are themselves NA.
2. If-statements may use else if and else, but they don't have to.
3. Functions may return from more than one point; when this happens, the function execution stops even if it's inside of a loop.[110]
4. Blocks of code, such as those used by while-loops, if-statements, and functions, may be nested. It's important to increase the indentation level for all lines within each nested block; otherwise, the code would be difficult to read.

For-Loops

For-loops are the last control-flow structure we'll study for R. Much like while-loops, they repeat a block of code. For-loops, however, repeat a block of code once for each element of a vector (or list), setting a given variable name to each element of the vector (or list) in turn.

```
ids <- c("CYP6B", "CATB", "AGP4")

for(id_el in ids) {
  print("id_el is now: ")
  print(id_el)
}

print("Done!")
```

```
[1] "id_el is now: "
[1] "CYP6B"
[1] "id_el is now: "
[1] "CATB"
[1] "id_el is now: "
[1] "AGP4"
[1] "Done!"
```

110. Some programmers believe that functions should have only one return point, and it should always go at the end of a function. This can be accomplished by initializing the variable to return at the start of the function, modifying it as needed in the body, and finally returning at the end. This pattern is more difficult to implement in this type of function, though, where execution of the function needs to be broken out of from a loop.

While for-loops are important in many other languages, they are not as crucial in R. The reason is that many operations are vectorized, and functions like `lapply()` and `do()` provide for repeated computation in an entirely different way.

There's a bit of lore holding that for-loops are slower in R than in other languages. One way to determine whether this is true (and to what extent) is to write a quick loop that iterates one million times, perhaps printing on every 1,000th iteration. (We accomplish this by keeping a loop counter and using the modulus operator `%%` to check whether the remainder of `counter` divided by `1000` is `0`.)

```
counter <- 1
for(i in seq(1, 1000000)) {
  if(counter%%1000 == 0) {
    print("Counter is now")
    print(counter)
  }
  counter <- counter + 1
}
```

When we run this operation we find that, while not quite instantaneous, this loop doesn't take long at all. Here's a quick comparison measuring the time to execute similar code for a few different languages (without printing, which itself takes some time) on a 2013 MacBook Pro.

Language	Loop 1 Million	Loop 100 Million
R (version 3.1)	0.39 seconds	~30 seconds
Python (version 3.2)	0.16 seconds	~12 seconds
Perl (version 5.16)	0.14 seconds	~13 seconds
C (g++ version 4.2.1)	0.0006 seconds	0.2 seconds

While R is the slowest of the bunch, it is "merely" twice as slow as the other interpreted languages, Python and Perl.

The trouble with R is not that for-loops are slow, but that one of the most common uses for them—iteratively lengthening a vector with `c()` or a data frame with `rbind()`—is very slow. If we wanted to illustrate this problem, we could modify the loop above such that, instead of adding one to counter, we increase the length of the counter vector by one with counter <- c(counter, 1). We'll also modify the printing to reflect the interest in the length of the counter vector rather than its contents.

```
counter <- 1
for(i in seq(1, 1000000)) {
  if(length(counter)%%1000 == 0) {
    print("Length of counter is now")
    print(length(counter))
  }
  counter <- c(counter, 1)
}
```

In this case, we find that the loop starts just as fast as before, but as the counter vector gets longer, the time between printouts grows. And it continues to grow and grow, because the time it takes to do a single iteration of the loop is depending on the length of the counter vector (which is growing).

To understand why this is the case, we have to inspect the line counter <- c(counter, 1), which appends a new element to the counter vector. Let's consider a related set of lines, wherein two vectors are concatenated to produce a third.

```
first_names <- c("Joseph", "Robert")
last_names <- c("Anderson", "Wilson")
names <- c(first_names, last_names)

print(names)          # [1] "Joseph" "Robert" "Anderson" "Wilson"
print(first_names)    # [1] "Joseph" "Robert"
print(last_names)     # [1] "Anderson" "Wilson"
```

In the above, names will be the expected four-element vector, but the first_names and last_names have not been removed or altered by this operation—they persist and can still be printed or otherwise accessed later. In order to remove or alter items, the R interpreter copies information from each smaller vector into a new vector that is associated with the variable names.

Now, we can return to counter <- c(counter, 1). The right-hand side copies information from both inputs to produce a new vector; this is then assigned to the variable counter. It makes little difference to the interpreter that the variable name is being reused and the original vector will no longer be accessible: the c() function (almost) always creates a new vector in RAM. The amount of time it takes to do this copying thus grows along with the length of the counter vector.

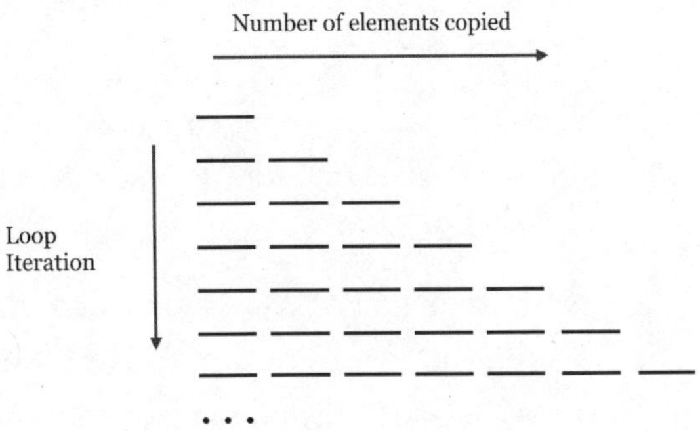

Number of elements copied

Loop Iteration

. . .

The total amount of time taken to append n elements to a vector in a for-loop this way is roughly $n^2/2$, meaning the time taken to grow a list of n elements grows quadratically in its final length! This problem is exacerbated when using `rbind()` inside of a for-loop to grow a data frame row by row (as in something like `df <- rbind(df, c(val1, val2, val3))`), as data frame columns are usually vectors, making `rbind()` a repeated application of `c()`.

One solution to this problem in R is to "preallocate" a vector (or data frame) of the appropriate size, and use replacement to assign to the elements in order using a carefully constructed loop.

```
# pre-allocate a counter vector of length 1 million
counter <- rep(0, 1000000)
for(i in seq(1, 1000000)) {
  counter[i] <- 1
}
```

(Here we're simply placing values of `1` into the vector, but more sophisticated examples are certainly possible.) This code runs much faster but has the downside of requiring that the programmer know in advance how large the data set will need to be.[III]

Does this mean we should never use loops in R? Certainly not! Sometimes looping is a natural fit for a problem, especially when it doesn't involve dynamically growing a vector, list, or data frame.

Exercises

1. Using a for-loop and a preallocated vector, generate a vector of the first 100 Fibonacci

III. Another way to lengthen a vector or list in R is to assign to the index just off of the end of it, as in `counter[length(counter) + 1] <- val`. One might be tempted to use such a construction in a for-loop, thinking that the result will be the `counter` vector being modified "in place" and avoiding the expensive copying. One would be wrong, at least as of the time of this writing, because this line is actually syntactic sugar for a function call and assignment: ``counter <- `[<-`(counter, length(counter) + 1, val)``, which suffers the same problem as the `c()` function call.

numbers, 1, 1, 2, 3, 5, 8, and so on (the first two Fibonacci numbers are 1; otherwise, each is the sum of the previous two).

2. A line like `result <- readline(prompt = "What is your name?")` will prompt the user to enter their name, and store the result as a character vector in `result`. Using this, a while-loop, and an if-statement, we can create a simple number-guessing game.

   Start by setting `input <- 0` and `rand` to a random integer between 1 and 100 with `rand <- sample(seq(1,100), size = 1)`. Next, while `input != rand`: Read a guess from the user and convert it to an integer, storing the result in `input`. If `input < rand`, print `"Higher!"`, otherwise if `input > rand`, print `"Lower!"`, and otherwise report `"You got it!"`.

3. A paired student's *t*-test assesses whether two vectors reveal a significant mean element-wise difference. For example, we might have a list of student scores before a training session and after.

```
scores <- c(50, 35, 56, 36, 90)
scores_after <- c(70, 45, 50, 42, 91)
improvement_test <- t.test(scores, scores_after, paired = TRUE)
print(improvement_test)
```

   But the paired *t*-test should only be used when the differences (which in this case could be computed with `scores_after - scores`) are normally distributed. If they aren't, a better test is the Wilcoxon signed-rank test:

```
improvement_test <- wilcox.test(scores, scores_after, paired = TRUE)
print(improvement_test)
```

   (While the *t*-test checks to determine whether the mean difference is significantly different from `0`, the Wilcoxon signed-rank test checks to determine whether the median difference is significantly different from `0`.)

   The process of determining whether data are normally distributed is not an easy one. But a function known as the Shapiro test is available in R, and it tests the null hypothesis that a numeric vector is not normally distributed. Thus the *p* value is small when data are not normal. The downside to a Shapiro test is that it and similar tests tend to be oversensitive to non-normality when given large samples. The `shapiro.test()` function can be explored by running `print(shapiro.test(rnorm(100, mean = 10, sd = 4)))` and `print(shapiro.test(rexp(100, rate = 2.0)))`.

Write a function called `wilcox_or_ttest()` that takes two equal-length numeric vectors as parameters and returns a *p* value. If the `shapiro.test()` reports a *p* value of less than 0.05 on the difference of the vectors, the returned *p* value should be the result of a Wilcoxon rank-signed test. Otherwise, the returned *p* value should be the result of a `t.test()`. The function should also print information about which test is being run. Test your function with random data generated from `rexp()` and `rnorm()`.

A Functional Extension

Having reviewed the procedural control-flow structures `if`, `while`, and `for`, let's expand on our truncated random sampling example to explore more "functional" features of R and how they might be used. The function we designed, `rnorm_trunc()`, returns a random sample that is normally distributed but limited to a given range via resampling. The original sampling distribution is specified by the `mean =` and `sd =` parameters, which are passed to the call to `rnorm()` within `rnorm_trunc()`.

```
rnorm_trunc <- function(lower, upper, count, mean, sd) {
  sample <- rnorm(count, mean = mean, sd = sd)
  looped <- 0

  while(any(sample < lower | sample > upper)) {
    # fix the sample
    bad <- sample < lower | sample > upper # logical
    bad_values <- sample[bad]
    count_bad <- length(bad_values)

    new_vals <- rnorm(count_bad, mean = mean, sd = sd)
    sample[bad] <- new_vals

    if(looped > 100000) {
      return(NA)
    }
    looped <- looped + 1
  }
  return(sample)
}
```

What if we wanted to do the same thing, but for `rexp()`, which samples from an exponential distribution taking a `rate =` parameter?

```
sample <- rexp(1000, rate = 1.5)
hist(sample)
```

The distribution normally ranges from 0 to infinity, but we might want to resample to, say, 1 to 4.

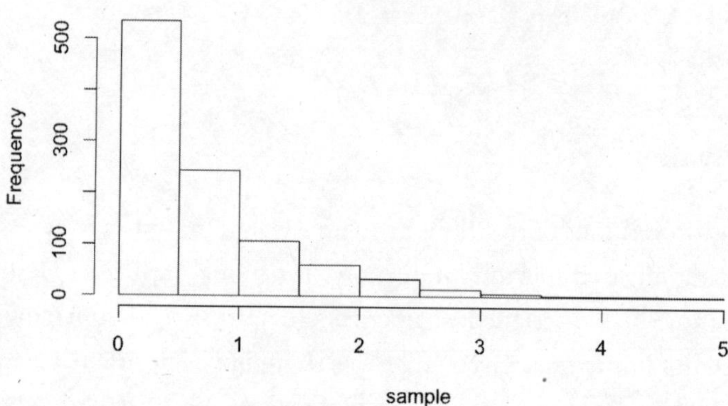

One possibility would be to write an `rexp_trunc()` function that operates similarly to the `rnorm_trunc()` function, with changes specific for sampling from the exponential distribution.

```
rexp_trunc <- function(lower, upper, count, rate) {
  sample <- rexp(count, rate = rate)
  looped <- 0

  while(any(sample < lower | sample > upper)) {
    # fix the sample
    bad <- sample < lower | sample > upper # logical
    bad_values <- sample[bad]
    count_bad <- length(bad_values)

    new_vals <- rexp(count_bad, rate = rate)
    sample[bad] <- new_vals

    if(looped > 100000) {
      return(NA)
    }
    looped <- looped + 1
  }
  return(sample)
}
```

The two functions `rnorm_trunc()` and `rexp_trunc()` are incredibly similar—they differ only in the sampling function used and the parameters passed to them. Can we write a single function to do both jobs? We can, if we remember two important facts we've learned about functions and parameters in R.

1. Functions like `rnorm()` and `rexp()` are a type of data like any other and so can be passed as parameters to other functions (as in the split-apply-combine strategy).
2. The special parameter `...` "collects" parameters so that functions can take arbitrary parameters.

Here, we'll use `...` to collect an arbitrary set of parameters and pass them on to internal function calls. When defining a function to take `...`, it is usually specified last. So, we'll write a function called `sample_trunc()` that takes five parameters:

1. The lower limit, `lower`.
2. The upper limit, `upper`.
3. The sample size to generate, `count`.
4. The function to call to generate samples, `sample_func`.
5. Additional parameters to pass on to `sample_func`, `...`.

```
sample_trunc <- function(lower, upper, count, sample_func, ...) {
  sample <- sample_func(count, ...)
  looped <- 0

  while(any(sample < lower | sample > upper)) {
    # fix the sample
    bad <- sample < lower | sample > upper # logical
    bad_values <- sample[bad]
    count_bad <- length(bad_values)

    new_vals <- sample_func(count_bad, ...)
    sample[bad] <- new_vals

    if(looped > 100000) {
      return(NA)
    }
    looped <- looped + 1
  }
  return(sample)
}
```

We can call our `sample_trunc()` function using any number of sampling functions. We've seen `rnorm()`, which takes `mean =` and `sd =` parameters, and `rexp()`, which takes a `rate =` parameter, but there are many others, like `dpois()`, which generates Poisson distributions and takes a `lambda =` parameter.

```
sample1 <- sample_trunc(0, 30, 1000, rnorm, mean = 20, sd = 10)
sample2 <- sample_trunc(1, 4, 1000, rexp, rate = 1.5)
sample3 <- sample_trunc(3, 6, 1000, rpois, lambda = 2)

if(!is.na(sample1)) {
  hist(sample1)
}

if(!is.na(sample2)) {
  hist(sample2)
}

if(!is.na(sample3)) {
  hist(sample3)
}
```

In the first example above, `mean = 20, sd = 10` is collated into `...` in the call to `sample_trunc()`, as is `rate = 1.5` in the second example and `lambda = 2` in the third example.

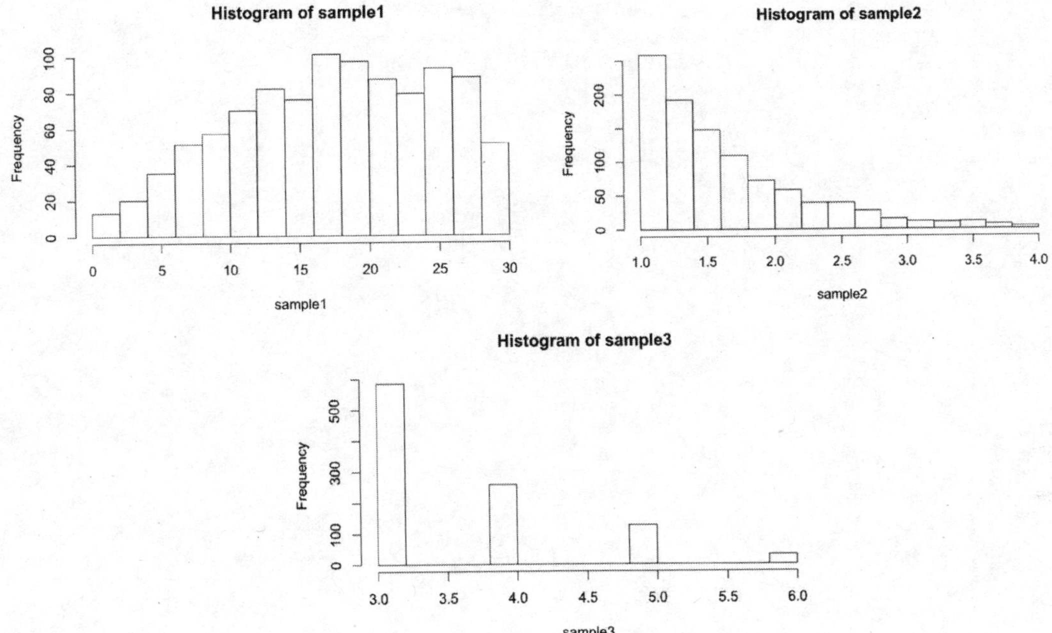

Exercises

1. As discussed in a previous exercise, the `t.test()` and `wilcox.test()` functions both take two numeric vectors as their first parameters, and return a list with a `$p.value` entry. Write a function `pval_from_test()` that takes four parameters: the first two should be two numeric vectors (as in the `t.test()` and `wilcox.test()` functions), the third should be a test function (either `t.test` or `wilcox.test`), and the fourth should be any optional parameters to pass on (`...`). It should return the *p* value of the test run.

   We should then be able to run the test like so:

```
sample1 <- rnorm(20, mean = 10, sd = 3)
sample2 <- rnorm(20, mean = 20, sd = 3)

pval1 <- pval_from_test(sample1, sample2, t.test)
pval2 <- pval_from_test(sample1, sample2, wilcox.test)
pval3 <- pval_from_test(sample1, sample2, t.test, paired = TRUE)
pval4 <- pval_from_test(sample1, sample2, wilcox.test, paired = TRUE)
```

   The four values `pval1`, `pval2`, `pval3`, and `pval4` should contain simple *p* values.

Chapter 36

Objects and Classes in R

In the chapters covering Python, we spent a fair amount of time discussing objects and their blueprints, known as classes. Generally speaking, an object is a collection of data along with functions (called "methods" in this context) designed specifically to work on that data. Classes comprise the definitions of those methods and data.

As it turns out, while functions are the major focus in R, objects are also an important part of the language. (By no means are any of these concepts mutually exclusive.) While class definitions are nicely encapsulated in Python, in R, the pieces are more distributed, at least for the oldest and most commonly used "S3" system we'll discuss here.[112] With this in mind, it is best to examine some existing objects and methods before attempting to design our own. Let's consider the creation of a small, linear model for some sample data.

```
treatments <- c("w", "w", "p", "p")
heights <- c(4.2, 5.4, 2.1, 3.2)
lm_result <- lm(heights ~ treatments)
anova_result <- anova(lm_result)
```

In chapter 30, "Lists and Attributes," we learned that functions like `lm()` and `anova()` generally return a list (or a data frame, which is a type of list), and we can inspect the structure with `str()`.

112. Modern versions of R have not one, not two, but *three* different systems for creating and working with objects. We'll be discussing only the oldest and still most heavily used, known as S3. The other two are called S4 and Reference Classes, the latter of which is most similar to the class/object system used by Python. For more information on these and other object systems (and many other advanced R topics), see Norman Matloff, *The Art of R Programming* (San Francisco: No Starch Press, 2011), and Hadley Wickham, *Advanced R* (London: Chapman and Hall/CRC, 2014).

```
print("Structure of lm_result:")
str(lm_result)
print("Structure of anova_result:")
str(anova_result)
```

Here's a sampling of the output lines for each call (there are quite a few pieces of data contained in the `lm_result` list):

```
[1] "Structure of lm_result:"
List of 13
 $ coefficients : Named num [1:2] 2.65 2.15
  ..- attr(*, "names")= chr [1:2] "(Intercept)" "treatmentsw"
...
  .. .. .. ..- attr(*, "names")= chr [1:2] "heights" "treatments"
 - attr(*, "class")= chr "lm"
[1] "Structure of anova_result:"
Classes 'anova' and 'data.frame': 2 obs. of  5 variables:
 $ Df      : int  1 2
 $ Sum Sq : num  4.62 1.33
 $ Mean Sq: num  4.622 0.663
 $ F value: num  6.98 NA
 $ Pr(>F) : num  0.118 NA
 - attr(*, "heading")= chr  "Analysis of Variance Table\n" "Response: heights"
```

If these two results are so similar—both types of lists—then why are the outputs so different when we call `print(lm_result)`

```
Call:
lm(formula = heights ~ treatments)

Coefficients:
(Intercept)  treatmentsw
       2.65         2.15
```

and `print(anova_result)`?

```
Analysis of Variance Table

Response: heights
           Df Sum Sq Mean Sq F value Pr(>F)
treatments  1 4.6225  4.6225  6.9774 0.1184
Residuals   2 1.3250  0.6625
```

How these printouts are produced is dictated by the `"class"` attribute of these lists, `"lm"` for `lm_result` and "anova" for `anova_result`. If we were to remove this attribute, we would get a default printed output similar to the result of `str()`. There are several ways to modify or remove the class attribute of a piece of data: using the `attr()` accessor function with `attr(lm_result, "class") <- NULL`, setting it using the more preferred `class()` accessor, as in `class(lm_result) <- NULL`, or using the even more specialized `unclass()` function, as in `lm_result <- unclass(lm_result)`. In any case, running `print(lm_result)` after one of these three options will result in `str()`-like default printout.

Now, how does R produce different output based on this class attribute? When we call `print(lm_result)`, the interpreter notices that the `"class"` attribute is set to `"lm"`, and searches for another function with a different name to actually run: `print.lm()`. Similarly, `print(anova_result)` calls `print.anova()` on the basis of the class of the input. These specialized functions assume that the input list will have certain elements and produce an output specific to that data. We can see this by trying to confuse R by setting the class attribute incorrectly with `class(anova_result) <- "lm"` and then `print(anova_result)`:

```
Call:
NULL

No coefficients
```

Notice that the class names are part of the function names. This is R's way of creating methods, stating that objects with class `"x"` should be printed with `print.x()`; this is known as *dispatching* and the general `print()` function is known as a *generic function*, whose purpose is to dispatch to an appropriate method (class-specific function) based on the class attribute of the input.

In summary, when we call `print(result)` in R, because `print()` is a generic function, the interpreter checks the `"class"` attribute of `result`; suppose the class is `"x"`. If a `print.x()` exists, that function will be called; otherwise, the print will fall back to `print.default()`, which produces output similar to `str()`.

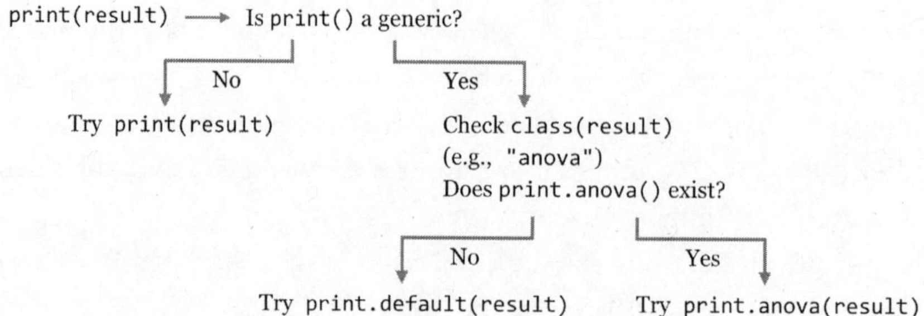

There are many different `"print."` methods; we can see them with `methods("print")`.

```
  [1] print.acf*
  [2] print.anova*
  [3] print.aov*
  [4] print.aovlist*
  [5] print.ar*
  [6] print.Arima*
...
```

Similarly, there are a variety of `".lm"` methods specializing in dealing with data that have a `"class"` attribute of `"lm"`. We can view these with `methods(class = "lm")`.

```
 [1] add1.lm*          alias.lm*            anova.lm*       case.names.lm*
 [5] confint.lm        cooks.distance.lm* deviance.lm*    dfbeta.lm*
 [9] dfbetas.lm*       drop1.lm*            dummy.coef.lm   effects.lm*
[13] extractAIC.lm*    family.lm*           formula.lm*     hatvalues.lm*
[17] influence.lm*     kappa.lm             labels.lm*      logLik.lm*
[21] model.frame.lm*   model.matrix.lm      nobs.lm*        plot.lm*
[25] predict.lm        print.lm*            proj.lm*        qr.lm*
[29] residuals.lm      rstandard.lm*        rstudent.lm*    simulate.lm*
[33] summary.lm        variable.names.lm* vcov.lm*

   Non-visible functions are asterisked
```

The message about nonvisible functions being asterisked indicates that, while these functions exist, we can't call them directly as in `print.lm(lm_result)`; we must use the generic `print()`. Many functions that we've dealt with are actually generics, including `length()`, `mean()`, `hist()`, and even `str()`.

So, in its own way R, is also quite "object oriented." A list (or other type, like a vector or data frame) with a given class attribute constitutes an object, and the various specialized methods are part of the class definition.

Creating Our Own Classes

Creating novel object types and methods is not something beginning R programmers are likely to do often. Still, an example will uncover more of the inner workings of R and might well be useful.

First, we'll need some type of data that we wish to represent with an object. For illustrative purposes, we'll use the data returned by the `nrorm_trunc()` function defined in chapter 35, "Structural Programming." Rather than producing a vector of samples, we might also want to store with that vector the original sampling mean and standard deviation (because the truncated data will have a different actual mean and standard deviation). We might also wish to store in this object the requested upper and lower limits. Because all of these pieces of data are of different types, it makes sense to store them in a list.

```
truncated_normal_sample <- function(lower, upper, count, mean, sd) {
  obj <- list()
  obj$sample <- rnorm_trunc(lower, upper, count, mean, sd)
  obj$lower <- lower
  obj$upper <- upper
  obj$original_mean <- mean
  obj$original_sd <- sd
  class(obj) <- "truncated_normal_sample"
  return(obj)
}
```

The function above returns a list with the various elements, including the sample itself. It also sets the class attribute of the list to `truncated_normal_sample`—by convention, this class attribute is the same as the name of the function. Such a function that creates and returns an object with a defined class is called a *constructor*.

Now, we can create an instance of a `"truncated_normal_sample"` object and print it.

```
trsamp <- truncated_normal_sample(0, 30, 25, 20, 10)
print(trsamp)
```

Because there is no `print.truncated_normal_sample()` function, however, the generic `print()` dispatches to `print.default()`, and the output is not pleasant.

```
$sample
 [1] 18.570116 28.058587 18.988116 23.923216 21.362590 27.071478  9.017976
 [8] 13.627638 14.859320  2.023175  1.335886 24.238769 19.714821  8.070529
[15]  5.084282 13.383528 22.007093  8.546593 11.442850 17.214193 17.037683
[22] 13.786232  7.832597 16.553821 17.877895

$lower
[1] 0

$upper
[1] 30

$original_mean
[1] 20

$original_sd
[1] 10

attr(,"class")
[1] "truncated_normal_sample"
```

If we want to stylize the printout, we need to create the customized method. We might also want to create a customized `mean()` function that returns the mean of the stored sample.

```
print.truncated_normal_sample <- function(obj) {
  print("Truncated normal sample, limited to:")
  print(c(obj$lower, obj$upper))
  print("Original sampling mean and sd:")
  print(c(obj$original_mean, obj$original_sd))
  print("First 10 elements:")
  print(head(obj$sample, n = 10))
}

mean.truncated_normal_sample <- function(obj) {
  answer = mean(obj$sample)
  return(answer)
}

print("Printing trsamp")
print(trsamp)
print("Calling mean(trsamp)")
print(mean(trsamp))
```

The output:

```
[1] "Printing trsamp"
[1] "Truncated normal sample, limited to:"
[1]  0 30
[1] "Original sampling mean and sd:"
[1] 20 10
[1] "First 10 elements:"
 [1] 15.21184 26.31239 11.38106 19.10591 24.98551 14.89133 23.83144 15.57210
 [9] 29.12310 20.32704
[1] "Calling mean(trsamp)"
[1] 19.20806
```

This customized print function is rather crude; more sophisticated printing techniques (like `cat()` and `paste()`) could be used to produce friendlier output.

So far, we've defined a custom `mean.truncated_normal_sample()` method, which returns the mean of the sample when we call the generic function `mean()`. This works because the generic function `mean()` already exists in R. What if we wanted to call a generic called `originalmean()`, which returns the object's `original_mean`? In this case, we need to create our own specialized method as well as the generic function that dispatches to that method. Here's how that looks:

```
# generic function, will dispatch based on class of obj
originalmean <- function(obj) {
  UseMethod("originalmean", obj)
}

# method, dispatched to for objects of class "truncated_normal_sample"
originalmean.truncated_normal_sample <- function(obj) {
  answer = obj$original_mean
  return(answer)
}

print(originalmean(trsamp))              # [1] 20
```

These functions—the constructor, specialized methods, and generic functions that don't already exist in R—need to be defined only once, but they can be called as many times as we like. In fact, packages in R that are installed using `install.packages()` are often just such a collection of functions, along with documentation and other materials.

Object-oriented programming is a large topic, and we've only scratched the surface. In particular, we haven't covered topics like polymorphism, where an object may have multiple classes listed in the `"class"` attribute. In R, the topic of polymorphism isn't difficult to describe in a technical sense, though making effective use of it is a challenge in software engineering. If an object has multiple classes, like `"anova"` and `"data.frame"`, and a generic like `print()` is called on it, the interpreter will first look for `print.anova()`, and if that fails, it will try `print.data.frame()`, and failing that will fall back on `print.default()`. This allows objects to capture "is a type of" relationships, so methods that work with data frames don't have to be rewritten for objects of class anova.

Exercises

1. Many functions in R are generic, including (as we'll explore in chapter 37, "Plotting Data and `ggplot2`") the `plot()` function, which produces graphical output. What are all of the different classes that can be plotted with the generic `plot()`? An example is `plot.lm()`; use `help("plot.lm")` to determine what is plotted when given an input with class attribute of `"lm"`.
2. What methods are available for data with a class attribute of `"matrix"`? (For example, is there a `plot.matrix()` or `lm.matrix()`? What others are there?)
3. Create your own class of some kind, complete with a constructor returning a list with its class attribute set, a specialized method for `print()`, and a new generic and associated method.
4. Explore using other resources the difference between R's S3 object system and its S4 object system.

Chapter 37

Plotting Data and ggplot2

R provides some of the most powerful and sophisticated data visualization tools of any program or programming language (though `gnuplot` mentioned in chapter 12, "Miscellanea," is also quite sophisticated, and Python is catching up with increasingly powerful libraries like `matplotlib`). A basic installation of R provides an entire set of tools for plotting, and there are many libraries available for installation that extend or supplement this core set. One of these is `ggplot2`, which differs from many other data visualization packages in that it is designed around a well-conceived "grammar" of graphics.[113]

The `ggplot2` package is not the most powerful or flexible—the graphics provided by default with R may take that title. Neither is `ggplot2` the easiest—simpler programs like Microsoft Excel are much easier to use. What `ggplot2` provides is a remarkable balance of power and ease of use. As a bonus, the results are usually professional looking with little tweaking, and the integration into R makes data visualization a natural extension of data analysis.

A Brief Introduction to Base-R Graphics

Although this chapter focuses on the `ggplot2` package, it is worth having at least passing familiarity with some of the basic plotting tools included with R. First, how plots are generated depends on whether we are running R through a graphical user interface (like RStudio) or on the command

113. The idea of a grammar of graphics was first introduced in 1967 by Jacques Bertin in his book *Semiologie Graphique* (Paris: Gauthier-Villars, 1967) and later expanded upon in 2005 by Leland Wilkinson et al. in *The Grammar of Graphics* (New York: Springer, 2005). Hadley Wickham, "A Layered Grammar of Graphics," *Journal of Computational and Graphical Statistics* 19 (2010): 3–28, originally described the R implementation.

line via the interactive R console or executable script. Although writing a noninteractive program for producing plots might seem counterintuitive, it is beneficial as a written record of how the plot was produced for future reference. Further, plotting is often the end result of a complex analysis, so it makes sense to think of graphical output much like any other program output that needs to be reproducible.

When working with a graphical interface like RStudio, plots are by default shown in a pop-up window or in a special plotting panel for review. Alternatively, or if we are producing plots via a remote command line login, each plot will be saved to a PDF file called `Rplots.pdf`. The name of this file can be changed by calling the `pdf()` function, giving a file name to write to. To finish writing the PDF file, a call to `dev.off()` is required (it takes no parameters).

The most basic plotting function (other than `hist()`, which we've already seen) is `plot()`. Like `hist()`, `plot()` is a generic function that determines what the plot should look like on the basis of class attributes of the data given to it. For example, given two numeric vectors of equal length, it produces a dotplot.

```
vecx <- rnorm(20, mean = 5, sd = 2)
vecy <- rnorm(20, mean = 10, sd = 1)

pdf("dotplot.pdf")  # not necessary if using a graphical interface
plot(vecx, vecy)
dev.off()           # not necessary if using a graphical interface
```

The contents of `dotplot.pdf`:

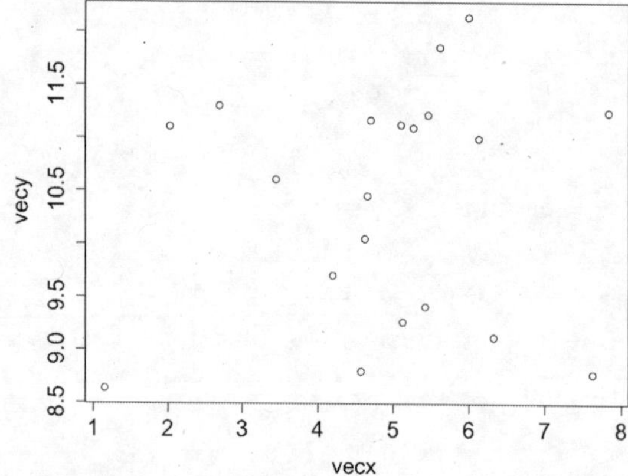

For the rest of this chapter, the `pdf()` and `dev.off()` calls are not specified in code examples.

We can give the `plot()` function a hint about the type of plot we want by using the `type =` parameter, setting it to `"p"` for points, `"l"` for lines, `"b"` for both, and so on. For basic vector plotting like the above, `plot()` respects the order in which the data appear. Here's the output of `plot(vecx, vecy, type = "l")`:

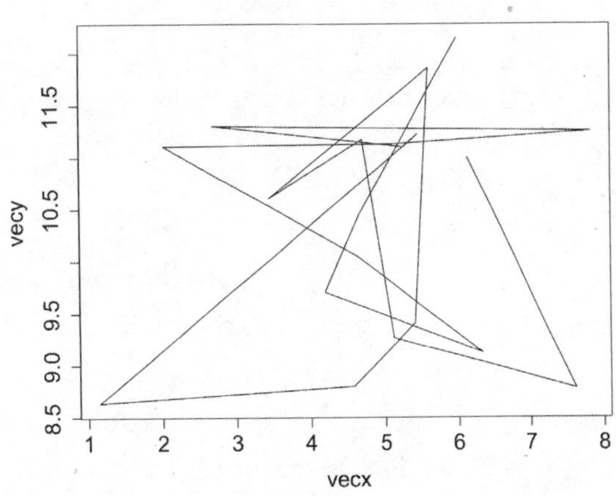

We would have had to sort one or both input vectors to get something more reasonable, if that makes sense for the data.

Other plotting functions like `hist()`, `curve()`, and `boxplot()` can be used to produce other plot types. Some plot types (though not all) can be added to previously plotted results as an additional layer by setting `add = TRUE`. For example, we can produce a dotplot of two random vectors, along with a histogram with normalized bar heights by using `hist()` with `probability = TRUE` and `add = TRUE`.

```
vecx <- rnorm(100, mean = 5, sd = 2)
vecy <- rnorm(100, mean = 0.5, sd = 0.2)
plot(vecx, vecy)
hist(vecx, probability = TRUE, add = TRUE)
```

A plot like this will only look reasonable if the axes ranges are appropriate for both layers, which we must ensure ourselves. We do this by specifying ranges with the `xlim =` and `ylim =` parameters in the call to `plot()`, specifying length-two vectors.

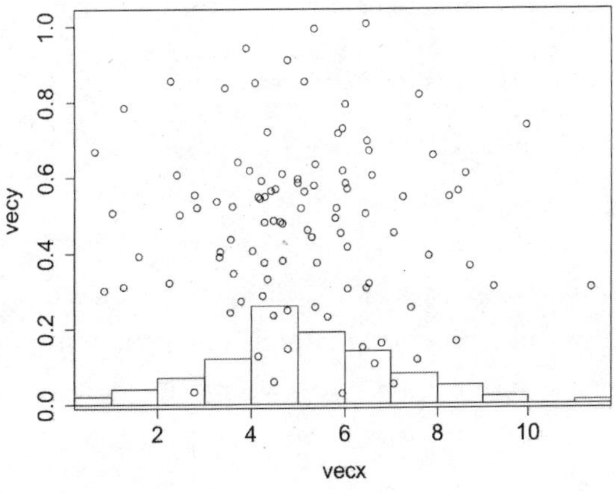

```
vecx <- rnorm(100, mean = 5, sd = 2)
vecy <- rnorm(100, mean = 0.5, sd = 0.2)
plot(vecx, vecy, xlim = c(-4, 14), ylim = c(0, 1))
hist(vecx, probability = TRUE, add = TRUE)
```

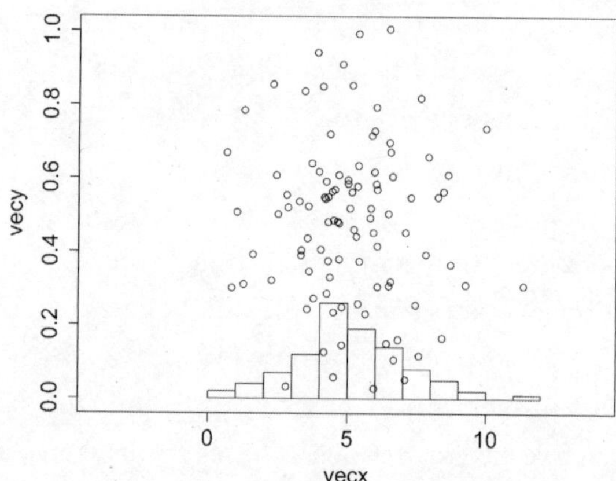

There are a number of hidden rules here. For example, `plot()` must be called before `hist()`, as `add = TRUE` isn't accepted by the `plot()` function. Although `hist()` accepts `xlim` and `ylim` parameters, they are ignored when `hist()` is used with `add = TRUE`, so they must be specified in the `plot()` call in this example. There are many individual plotting functions like `plot()` and `hist()`, and each takes dozens of parameters with names like `"las"`, `"cex"`, `"pch"`, and `"tck"` (these control the orientation of *y*-axis labels, font size, dot shapes, and tick-mark size, respectively). Unfortunately, the documentation of all of these functions and parameters oscillates between sparse and confusingly complex, though there are a number of books dedicated solely to the topic of plotting in R.

Despite its complexities, one of the premier benefits of using `plot()` is that, as a generic, the plotted output is often customized for the type of input. As an example, let's quickly create some linearly dependent data and run them through the `lm()` linear modeling function.

```
vecx <- rnorm(100, mean = 10, sd = 2)
vecy <- 2 * vecx + rnorm(100, mean = 1, sd = 1)
lm_result <- lm(vecy ~ vecx)
plot(vecx, vecy)
plot(lm_result)
```

If we give the `lm_result` (a list with class attribute `"lm"`) to `plot()`, the call will be dispatched to `plot.lm()`, producing a number of plots appropriate for linear model parameters and data. The first of the five plots below was produced by the call to `plot(vecx, vecy)`, while the remaining four are plots specific to linear models and were produced by the single call to `plot(lm_result)` as a multipage PDF file.

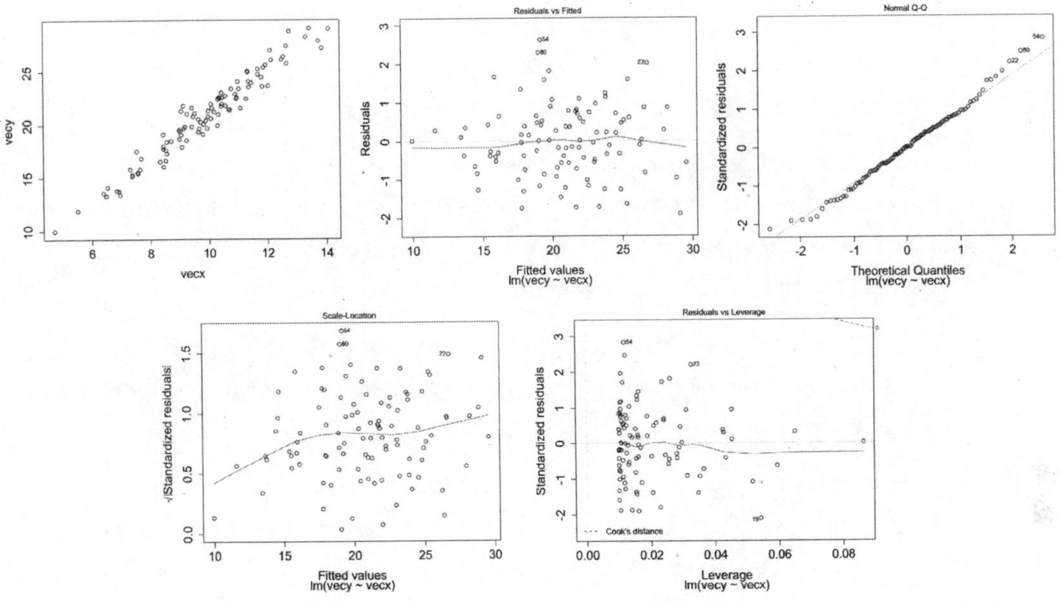

Overview of ggplot2 and Layers

As mentioned previously, the `ggplot2` package seeks to simplify the process of plotting while still providing a large amount of flexibility and power by implementing a "grammar" of graphical construction. Given this structure, we're going to have to start by learning some (more) specialized vocabulary, followed by some (more) specialized syntax. There are several ways of interacting with `ggplot2` of various complexity. We'll start with the most complex first, to understand the structure of the grammar, and then move on to simpler methods that are easier to use (once framed in terms of the grammar).

Unlike the generic `plot()` function, which can plot many different types of data (such as in the linear model example above), `ggplot2` specializes in plotting data stored in data frames.

A "plot" in `ggplot2` is made up of the following components:

1. One or more *layers*, each representing how some data should be plotted. Plot layers are the most important pieces, and they consist of five subparts:

    a. The `data` (a data frame), which contains the data to plot.

   b. A `stat`, which stands for "statistical mapping." By default, this is the `"identity"` stat (for no modification) but could be some other mapping that processes the `data` data frame and produces a new data set that is actually plotted.

   c. A `geom`, short for "geometric object," which specifies the shape to be drawn. For example, `"point"` specifies that data should be plotted with points (as in a dotplot), `"line"` specifies that data should be plotted with lines (as in a line plot), and `"bar"` specifies that data should be plotted with bars (as in a histogram).

   d. A `mapping` of "aesthetics," describing how properties of the `geom` (e.g., `x` and `y` position of points, or their `color`, `size`, and so on) relate to columns of the `stat`-transformed data. Each `geom` type cares about a specific set of aesthetics, though many are common.

   e. A `position` for the `geom`; usually, this is `"identity"` to suggest that the position of the `geom` should not be altered. Alternatives include `"jitter"`, which adds some random noise to the location (so overlapping `geom` shapes are more easily viewable).

2. One `scale` for each aesthetic used in layers. For example, each point's `x` value occurs on a horizontal scale that we might wish to limit to a given range, and each point's `color` value occurs on a color scale that we might wish to range from purple to orange, or black to blue.

3. A `facet` specification, describing how paneling data across multiple similar plots should occur.

4. A `coordinate` system to which to apply some scales. Usually, we won't modify these from the defaults. For example, the default for `x` and `y` is to use a `cartesian` coordinate system, though `polar` is an option.

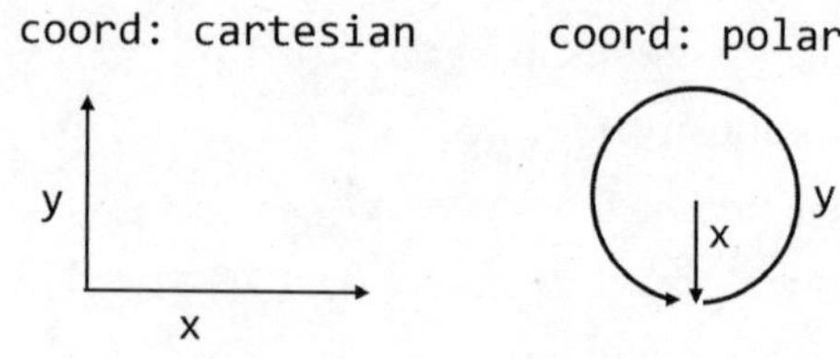

5. Some `theme` information, such as font sizes and rotations, axis ticks, aspect ratios, background colors, and the like.

6. A set of defaults. Defaults are an odd thing to list as a specific part of a plot, as these are the properties we *don't* specify, but `ggplot2` makes heavy use of them, so they are worth mentioning.

When installed with `install.packages("ggplot2")` (in the interactive console) and loaded with `library(ggplot2)`, the package comes with a data frame called `diamonds`. Each row of this data frame specifies some information about a single diamond; with about 54,000 rows and many types of columns (including numeric and categorical), it is an excellent data set with which to explore plotting.

```
library(ggplot2)
print(head(diamonds))
```

```
  carat       cut color clarity depth table price    x    y    z
1  0.23     Ideal     E     SI2  61.5    55   326 3.95 3.98 2.43
2  0.21   Premium     E     SI1  59.8    61   326 3.89 3.84 2.31
3  0.23      Good     E     VS1  56.9    65   327 4.05 4.07 2.31
4  0.29   Premium     I     VS2  62.4    58   334 4.20 4.23 2.63
5  0.31      Good     J     SI2  63.3    58   335 4.34 4.35 2.75
6  0.24 Very Good     J    VVS2  62.8    57   336 3.94 3.96 2.48
```

Let's start by exploring the most important concept in the list of definitions above: the layer and its five components. To create a layer, we can start by creating an "empty" gg object by calling `ggplot()` with no parameters. To this we'll add a layer with + and calling the `layer()` function, specifying the five components we want.[114] Because these plotting commands become fairly long, we break them up over multiple lines (ending broken lines with + or , to let the interpreter know the command isn't finished) and indent them to help indicate where different pieces are contributing.

```
p <- ggplot() +
     layer(data = diamonds,
           stat = "identity",
           geom = "point",
           mapping = aes(x = carat, y = price, color = cut),
           position = "identity")

plot(p)
```

114. Given the information in previous chapters on objects, you might have guessed that the type of object returned by `ggplot()` is a list with class attribute set, where the class is set to `"gg"` and `"ggplot"`. Further, a specialized `` `+.gg`() `` method is defined for objects of the `"gg"` class, and + is not just syntactic sugar for `` `+`() ``, but also a generic function that is dispatched!

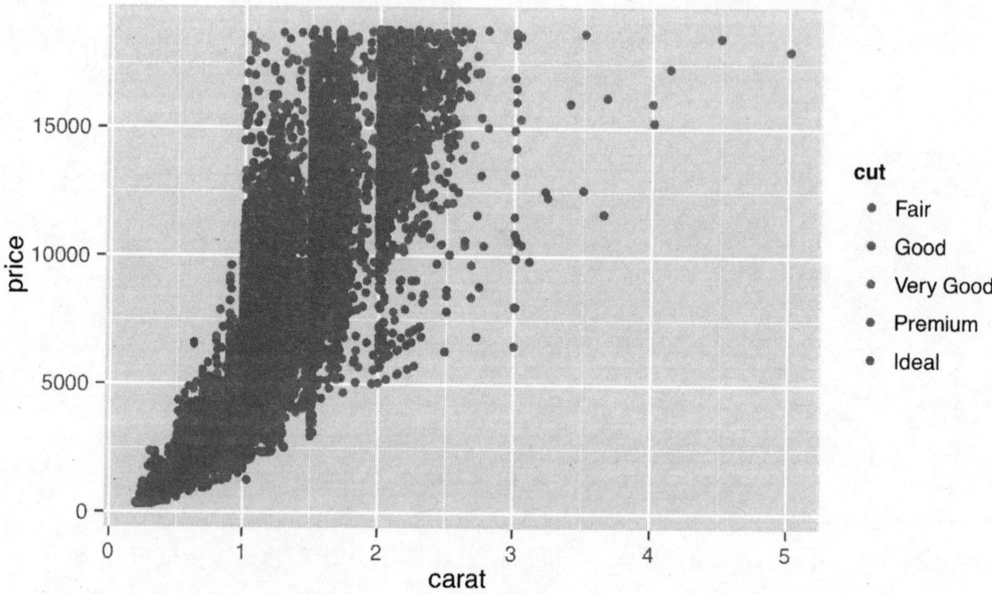

Here, we've specified each of the five layer components described above. For the `mapping` of aesthetics, there is an internal call to an `aes()` function that describes how aesthetics of the geoms (`x` and `y`, and `color` in this case) relate to columns of the `stat`-adjusted `data` (in this case, the output columns from the stat are identical to the input columns).[115] Finally, we note that `ggplot2` has seamlessly handled the categorical column of `cut`.

To save the result to a file or when not working in a graphical interface, we can use the `pdf()` function before the call to `plot()` followed by `dev.off()`, as we did for the Base-R graphics. Alternatively, we can use the specialized `ggsave()` function, which also allows us to specify the overall size of the plot (in inches at 300 dpi by default for PDFs).

```
ggsave("diamond_layer1.pdf", p, width = 7, height = 4)
```

Let's add a layer to our plot that will also plot points on the `x` and `y` axes, by `carat` and `price`. This additional layer, however, will use a `"smooth"` stat, and we won't color the points. (In recent

115. If you are reading this in black-and-white, then you'll have to trust that the colors differentiate the points. Later we'll discuss the importance of choosing color schemes that work in both grayscale and for colorblind readers, though we haven't covered the code concepts yet to do so.

versions of `ggplot2`, this layer example also requires a `params = list(method = "auto")` which sets the stat's smoothing method. Below we'll see how to write more compact code with this and other parameters set automatically.)

```
p <- ggplot() +
     layer(data = diamonds,
           stat = "identity",
           geom = "point",
           mapping = aes(x = carat, y = price, color = cut),
           position = "identity") +
     layer(data = diamonds,
           stat = "smooth",
           geom = "point",
           mapping = aes(x = carat, y = price),
           position = "identity"),
           params = list(method = "auto"))
```

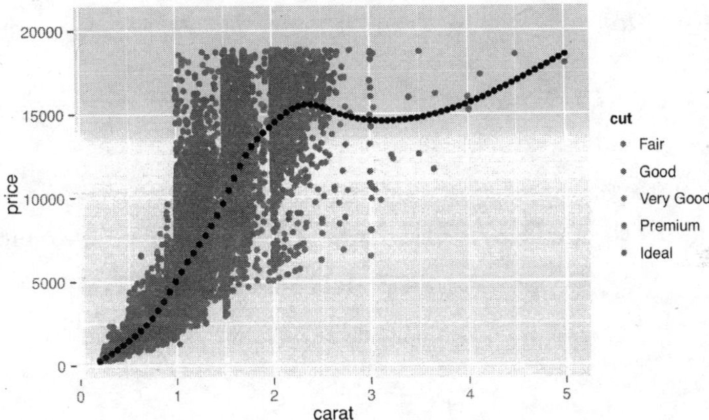

In this case, the original data have been transformed by the `"smooth"` stat, so `x = carat, y = price` now specifies the columns in the stat-transformed data frame. If we were to switch this layer's `geom` to `"line"`, we would get a plot like below on the left, and if we add a `color = cut` in the `aes()` call, we would get a plot like below on the right.

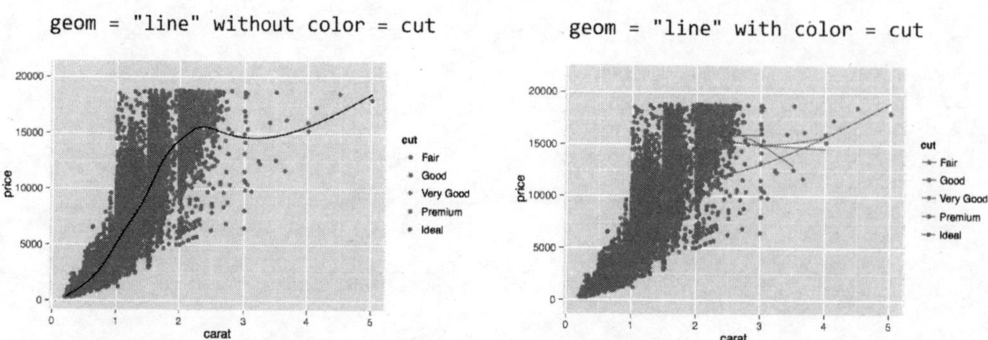

In the right plot above, multiple lines have been created; they are a bit difficult to see, but we'll see how to fix that in a bit. Note that the order of the layers matters: the second layer was plotted on *top* of the first.

This second layer illustrates one of the more confusing aspects of `ggplot2`, namely, that aesthetic mappings (properties of geoms) and stat mappings interact. In the second layer above, we specified `aes(x = carat, y = price)`, but we also specified the `"smooth"` stat. As a consequence, the underlying data representing `carat` and `price` were modified by the stat, and the stat knew which variables to smooth on the basis of this aesthetic mapping.

For a second example, let's look at the `"bin"` stat and the `"bar"` geom, which together create a histogram. The `"bin"` stat checks the `x` aesthetic mapping to determine which column to bin into discrete counts, and also creates some entirely new columns in the stat-transformed data, including one called `..count..`. The extra dots indicate to the user that the column produced by the stat is novel. The `"bar"` geom pays attention to the `x` aesthetic (to determine each bar's location on the horizontal axis) and the `y` aesthetic (to determine each bar's height).

```
p <- ggplot() +
     layer(data = diamonds,
           stat = "bin",
           geom = "bar",
           mapping = aes(x = carat, y = ..count..),
           position = "identity")
```

The result of plotting the above is shown below on the left. To complete the example, below on the right shows the same plot with `geom = "point"`. We can easily see that the stat-transformed data contain only about 30 rows, with columns for bin centers (`carat` mapped on `x`) and counts (`..count..`). The `"bin"` stat also generates a `..density..` column you can explore.

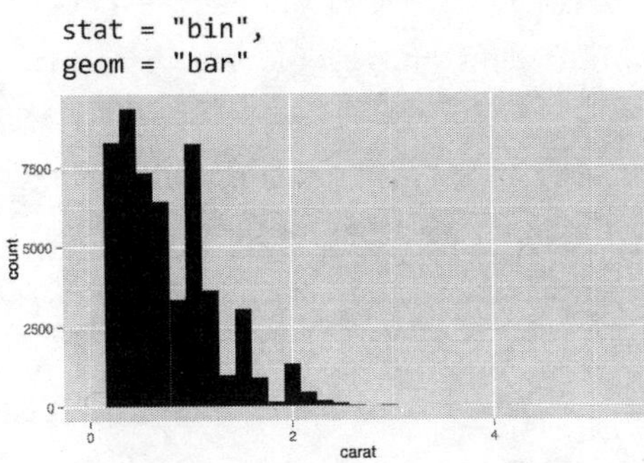

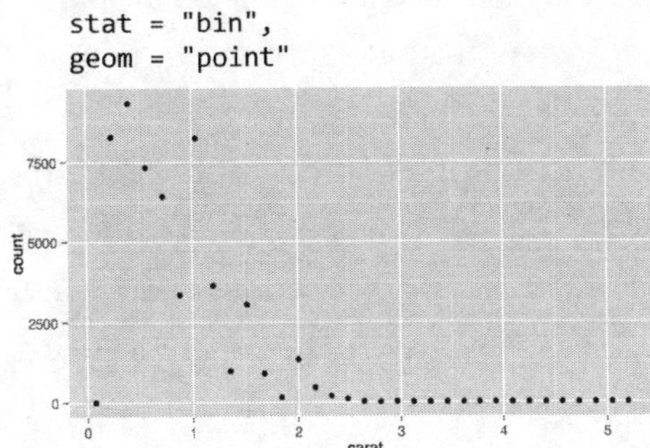

Smart Defaults, Specialized Layer Functions

Most of us are probably familiar with the standard plot types: histograms, fitted-line plots, scatterplots, and so on. Usually, the statistical mapping used will determine the common plot type and the geom to use. For example, a binning statistic likely goes with a bar geom to produce a histogram, and the point geom is most commonly paired with the identity statistic. (In fact, in older versions of `ggplot2`, if a stat or `geom` was unspecified in the call to `layer()`, a default was used on the basis of the other. Newer versions require both `geom =` and `stat =` to be specified when calling `layer()`.) Similarly, when a binning statistic is used, the `y` aesthetic mapping defaults to `..count..` if unspecified.

Because it is so common to specify a plot by specifying the statistical mapping to use, `ggplot2()` provides specialized layer functions that effectively move the specification of the stat to the function name and use the default geom (though it can still be changed). Similarly, we often wish to specify a plot by the geom type and accept the default stat for that geom (though, again, this can be changed by adding a `stat =` parameter.) Here are two more ways to plot the left histogram above:

```
p <- ggplot() + stat_bin(data = diamonds, mapping = aes(x = carat))
p <- ggplot() + geom_histogram(data = diamonds, mapping = aes(x = carat))
```

(There is a `geom_bar()` layer function, but its default is not the binning statistic; hence the alternative `geom_histogram()`.) To reproduce the plot above on the right, we could use `stat_bin(data = diamonds, mapping = aes(x = carat), geom = "point")`.

With so many defaults being set, the plotting commands can become quite small. These specialized layer functions represent the most commonly used methods for plotting in ggplot2, being both flexible and quick.

Another example of defaults being set is the geom_boxplot() layer function, which uses a "boxplot" geom (a box with whiskers) and a default "boxplot" stat. The boxplot geom recognizes a number of aesthetics for the various pieces that position the box and whiskers, including x, y, middle, upper, lower, ymin, and ymax. Fortunately, most of these required values are created by the boxplot stat and set accordingly (much like the y aesthetic defaults to ..count.. for histograms); only the x and y aesthetics are required to determine the others.

```
# default stat chosen by geom, many aesthetics autoset accordingly
p <- ggplot() + geom_boxplot(data = diamonds,
                             mapping = aes(x = color, y = price))
```

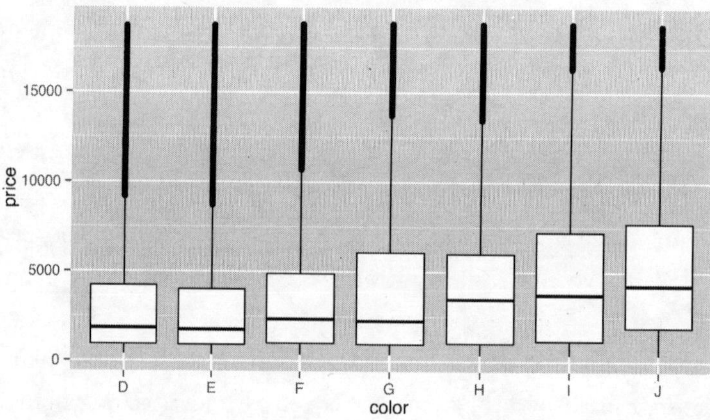

We've mapped a discrete variable to x and a continuous variable to y—a boxplot would make considerably less sense the other way around! There is also a corresponding stat_boxplot() layer function, which specifies the stat and uses the default corresponding geom (boxplot).

So far, we've been individually specifying the data = and mapping = parameters for each layer. It is common to see them set only once in the call to ggplot(), in which case they are inherited for all subsequent layers. We can also leave off the data = and mapping = names if we first specify the data.

```
# data and default mappings set in call to ggplot()
p <- ggplot(diamonds, aes(x = carat, y = price)) +
     geom_point(aes(color = cut)) +
     geom_smooth()
```

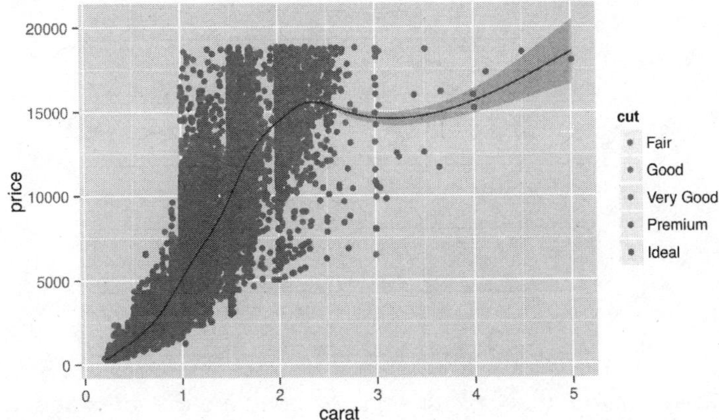

Most users of `ggplot2` prefer to utilize many of these "fall-through" settings and defaults, though we'll specify all data and mappings for each layer for clarity. (And yes, different layers can use different data frames for their `data =` parameter.)

There are many specialized layer functions for specific stats and geoms. Documentation for them and other features of `ggplot2` can be found at http://docs.ggplot2.org.

Exercises

1. Use the Base-R `plot()` function to produce a PDF plot that looks like the following:

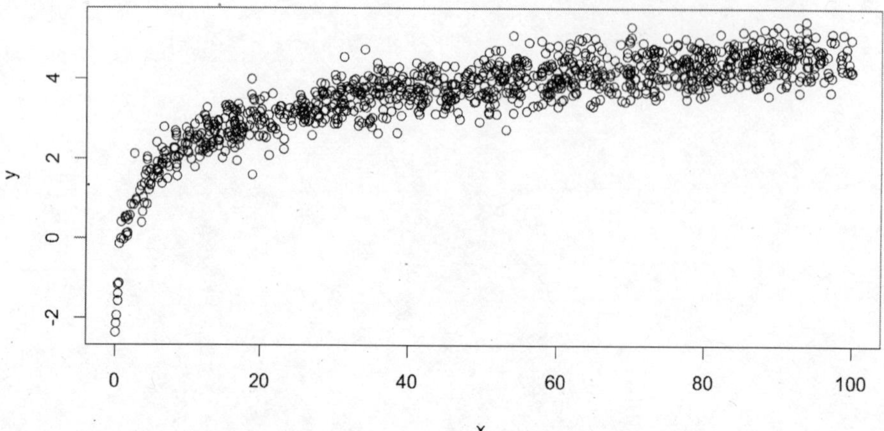

There are 1,000 *x* values randomly (and uniformly) distributed between 0 and 100, and the *y* values follow a curve defined by the `log()` of *x* plus some random noise.

2. R's `hist()` function is a convenient way to produce a quick histogram. Try running `hist(rnorm(1000, mean = 0, sd = 3))`. Read the help for `hist()` to determine how to increase the number of "bins."

3. Use the `layer()` function in `ggplot2` to produce a histogram of diamond prices. Next, try using `geom_histogram()` and `stat_bin()`. For these latter plots, try changing the number of bins from the default of 30.

4. Try creating a dotplot of diamonds (point geom, identity stat) using `layer()` with `x` mapped to `color` and `y` mapped to `price`. What happens when you change the `position` to `"jitter"`? Next, add a boxplot layer on top of this dotplot layer.

More Aesthetics and Mathematical Expressions

The `geom_point()` layer function deserves some special attention, not only because scatterplots are a particularly useful plot type for data exploration, but also because we can use it to illustrate more features of the `ggplot2` package. Unfortunately, scatterplots tend to suffer from "overplotting" issues when the number of data points is large, as in previous examples. For now, we'll get around this issue by generating a random subset of 1,000 diamonds for plotting, placing the sample in a data frame called dd.

```
dd <- diamonds[sample(seq(1, nrow(diamonds)), 1000), ]
```

First, the geom_point() layer accepts a number of aesthetics that might be useful in other situations as well.

```
p <- ggplot() +
     geom_point(data = dd,
                mapping = aes(x = carat,
                              y = depth,
                              color = price/carat,
                              size = table,
                              alpha = cut))
```

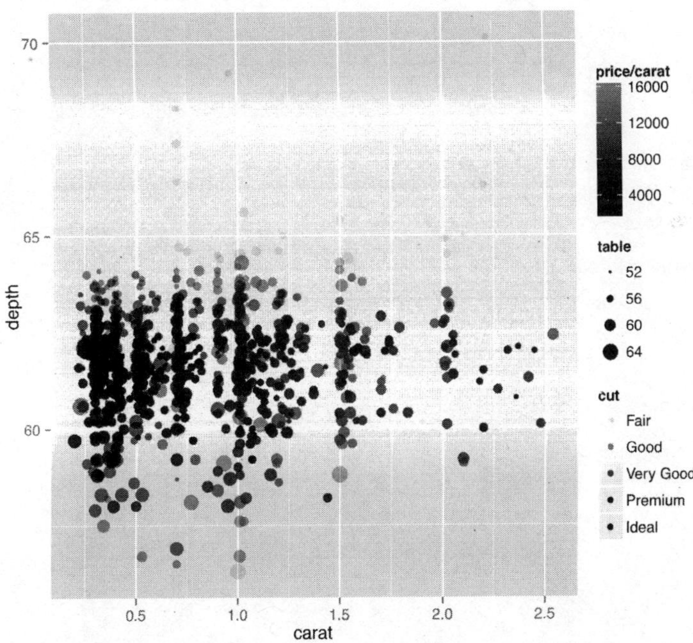

The result is probably too many aesthetics to map to values from the data, but sometimes for exploratory purposes (rather than publication), this isn't a bad thing. In this plot we can see a few interesting features of the data. (1) Diamonds are usually cut to carat sizes that are round numbers

(0.25, 0.5, 1.0, etc.). (2) The price per carat (mapped to color) is generally higher for larger diamonds. This means that not only are larger diamonds more expensive, but they are also more expensive on a per carat basis (they are rarer, after all). (3) Although it's a bit difficult to see in this plot, it appears that points with the best cut (mapped to alpha, or "transparency") are also of intermediate depth (mapped to the *y* axis). Or, at the very least, outliers in depth tend to have poorer "cut" values.

This plot also illustrates that aesthetics can generally be mapped to continuous or discrete data, and `ggplot2` will handle the data with grace. Further, we can specify the data as mathematical expressions of the columns in the input data frame (or even other data), as we did with `color = price/carat`. Doing so allows us to explore data sets in a powerful and flexible way. We could quickly discretize the price per carat by setting `color = price/carat > 5000`, for example, or we could reduce the range of these values by plotting with `color = log(price/carat)`. In the subsection on scales, however, we'll see that `ggplot2` contains some friendly features for plotting data on log-adjusted and similar scales.

Because this plot is far too busy, let's fix it by removing the alpha and size mapping, remembering the possibility of a relationship between cut and depth for future exploration. We'll also add a layer for a smoothed fit line.

```
p <- ggplot() +
     geom_point(data = dd,
                mapping = aes(x = carat,
                              y = depth,
                              color = price/carat)) +
     stat_smooth(data = dd,
                 mapping = aes(x = carat,
                               y = depth))
```

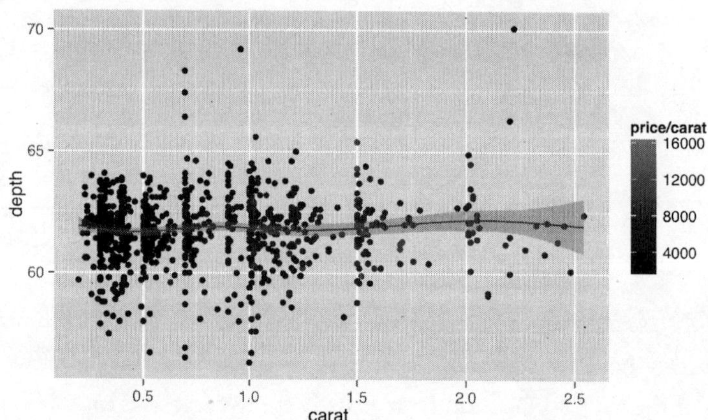

Layer functions like `layer()` and `geom_point()` accept additional options on top of `data`, `stat`, `geom`, and `mapping`. In this example, the first layer would look better if it were partially transparent, so that the top layer could stand out. But we don't want to map the alpha aesthetic to a property of the data—we just want it to set it to a constant for all the geoms in the layer. Thus we can set `alpha = 0.2` outside the `aes()` mapping as an option for the layer itself. Similarly, we might want our fit line to be red, and we note from the documentation that the `stat_smooth()` layer takes a `method` argument that can be set; `"auto"` is the default for a non-linear smooth and `"lm"` will produce a linear fit.

```
p <- ggplot() +
     geom_point(data = dd,
                mapping = aes(x = carat,
                              y = depth,
                              color = price/carat),
                alpha = 0.2) +
     stat_smooth(data = dd,
                 mapping = aes(x = carat,
                               y = depth),
                 color = "red",
                 method = "lm")
```

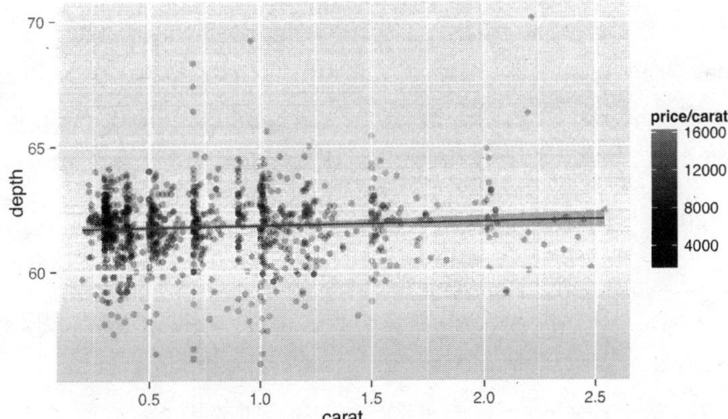

As a general note, `ggplot2` enforces that all layers share the same scales. For this reason, we would not want to plot in one layer `x = carat, y = depth` and in another `x = carat, y = price`; all of the `y` values will be forced into a single scale, and because depth ranges up to ~70 while prices range up to ~18,000, the depth values would be indiscernible. This also applies to color mappings, if multiple layers utilize them.

The `ggplot2` package enforces these rules for ease of understanding of plots that use multiple layers. For the same reasons, `ggplot2` doesn't support multiple *y* axes (i.e., a "left" *y* axis and "right" *y* axis), nor does it natively support three-dimensional plots that are difficult to read on a two-dimensional surface, like paper or a computer monitor.[116]

Faceting

Faceting is one of the techniques that can be fairly difficult to do in other graphics packages, but is easy in `ggplot2`. Facets implement the idea of "small multiples" championed by Edward Tufte.[117] Different subsets of the data are plotted in independent panels, but each panel uses the same axes (scales), allowing for easy comparisons across panels. In `ggplot2()`, we can add a facet specification by adding another function call to the "chain" of layer functions defining the plot. Faceting is almost always done on discrete columns of the input data frame, though below are some techniques for discretizing continuous columns.

The first facet-specification function is `facet_wrap()`. It takes one required parameter, three less often used but occasionally helpful parameters, and a few more that we won't discuss here.

1. The first parameter is a formula specifying the columns of the input data frame to facet by. For `facet_wrap()`, this formula is usually of the form `~ <column>`, with no "variable" specified to the left of the `~`. If we wish to facet on all combinations of multiple input columns, we can use `~ <column_1> + <column_2> + ...`.
2. The `nrow =` parameter controls the number of rows of panels that should appear in the output.
3. The `ncol =` parameter controls the number of columns of panels that should appear in the output. (The `nrow` times `ncol` parameters should be large enough to hold all of the needed panels. By default, `nrow` and `ncol` are determined automatically.)

116. Yet we can quickly explore multidimensional data by using additional aesthetics like color, size, shape, alpha, and so on. If multiple *y* axes or three-dimensional plots are required, some other R packages can provide these features, as can the `gnuplot` command line utility.

117. Many in the data visualization community consider Tufte's books, especially *The Visual Display of Quantitative Information* (Cheshire, CT: Graphics Press, 1983), to be staples. These volumes are beautiful in their own right as examples of information presentation.

4. The `scales =` parameter can be set to `"fixed"` (the default), which specifies that all axes should be similarly scaled across panels. Setting this parameter to `"free_x"` allows the *x* axes to vary across panels, `"free_y"` allows the *y* axes to vary, and `"free"` allows both scales to vary. Because facets are usually designed to allow comparisons between panels, settings other than `"fixed"` should be avoided in most situations.

Returning to the dotplots of the diamonds data, recall that there appeared to be some relationship between the "cut" and "depth" of diamonds, but in the previous plots, this relationship was difficult to see. Let's plot the same thing as above, but this time facet on the cut column.

```
p <- ggplot() +
     geom_point(data = dd,
                mapping = aes(x = carat,
                              y = depth,
                              color = price/carat),
                alpha = 0.2) +
     stat_smooth(data = dd,
                 mapping = aes(x = carat,
                               y = depth),
                 color = "red",
                 method = "lm") +
     facet_wrap(~ cut)
```

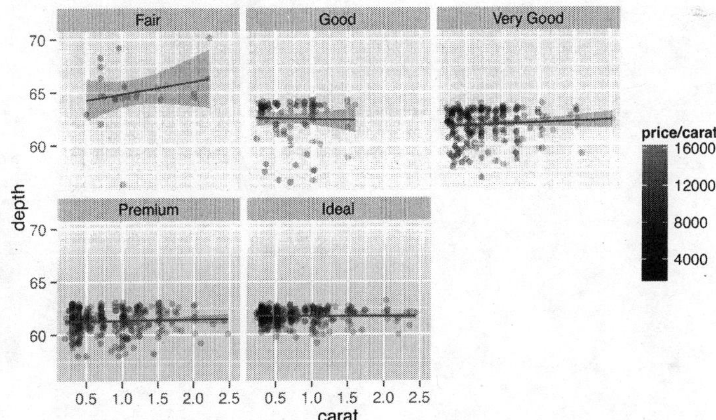

Faceting this way produces a separate panel for each cut type, all on the same axes, but anything plot-specific (such as the fit lines) are computed independently. This reveals that "ideal" cut diamonds tend to have a depth of around 62, whereas lower-quality cuts deviate from this value. To facet on cut and color, the formula would be `~ cut + color`, and a panel would be created for each cut/color combination.

The `facet_grid()` function works much like `facet_wrap()` but allows us to facet according to not one but two variables (or two combinations of multiple variables) simultaneously. One of the variables will vary across rows of panels, and the other will vary across columns. The first parameter is still a formula, but usually of the form `<column_1> ~ <column_2>`. Let's replace the `facet_wrap(~ cut)` with `facet_grid(cut ~ color)`:

```
p <- ggplot() +
     geom_point(data = dd,
                mapping = aes(x = carat,
                              y = depth,
                              color = price/carat),
                alpha = 0.2) +
     stat_smooth(data = dd,
                 mapping = aes(x = carat,
                               y = depth),
                 color = "red",
                 method = "lm") +
     facet_grid(cut ~ color)
```

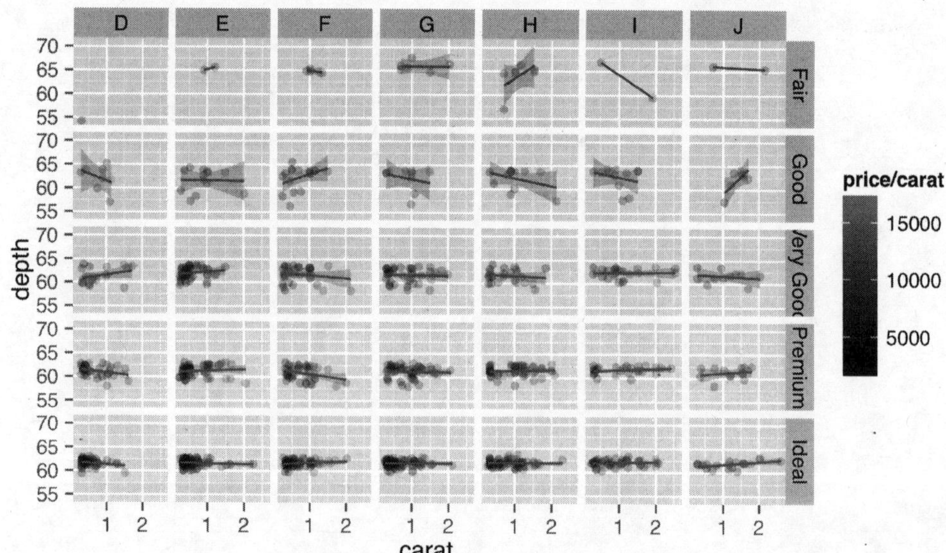

As the number of facets increases, readability can decrease, but having cut along the rows and color along the columns certainly helps reveal trends. In this facet specification, we see that "fair" diamonds are relatively rare in the data set, and the largest diamonds tend to be premium or ideal cuts in poor colors (I and J). Producing PDFs with larger dimensions in `ggsave()` can make plots with many features more readable.

Like `facet_wrap()`, `facet_grid()` can take a `scales =` parameter to allow panel *x* and *y* axes to vary, though again this isn't generally recommended, as it makes panels difficult to compare visually. The `facet_grid()` function includes another optional parameter, `margins =`, which when set to `TRUE` adds "aggregate" panels for each column and row of panels. Here's the plot with `facet_grid(cut ~ color, margins = TRUE)`:

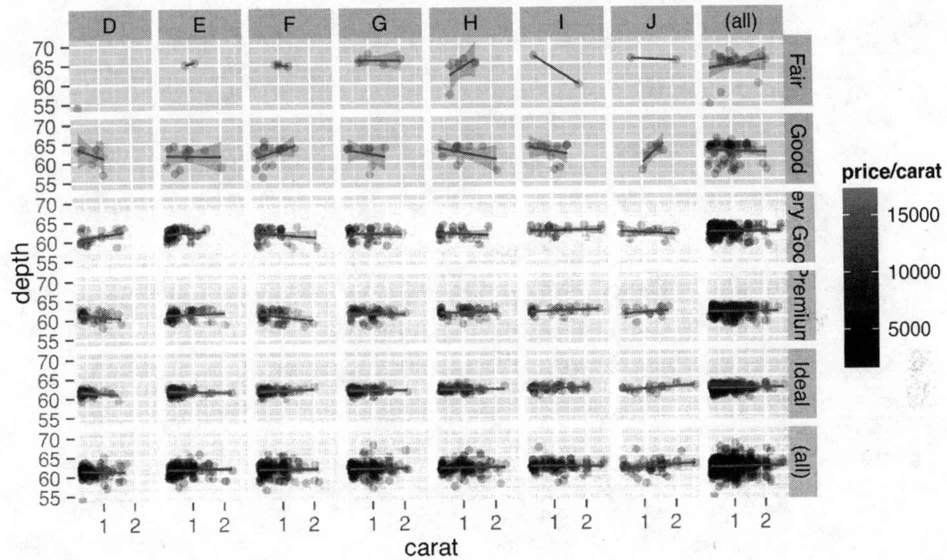

When plotting a grid of facets, we can also use a period in the specification to indicate that the grid should consist of only a single row or column of panels. While `facet_wrap()` can also produce only a single row or grid, the `margins = TRUE` option can be used with `facet_grid()` to produce a row or column while simultaneously producing an aggregate panel. Here's the same plot with `facet_grid(. ~ cut, margins = TRUE)`.

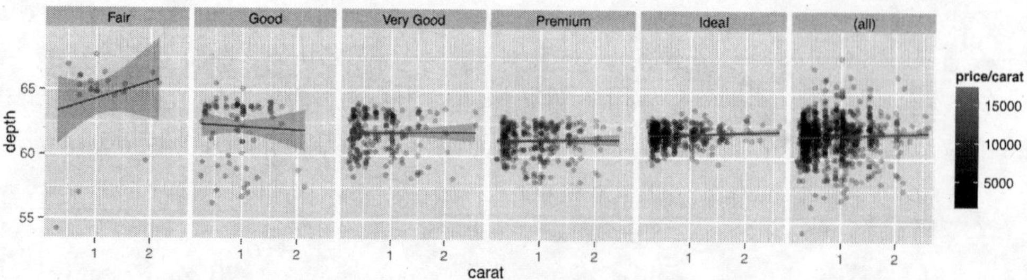

Facets are usually created for different values of categorical data. An attempt to facet over all the different values in a continuous column could result in millions of panels! (Or, more likely, a crashed instance of R.) Still, occasionally we do want to facet over continuous data by producing individual discretized "bins" for values. Consider if we wanted to facet on diamond price/carat, categorized as "high" (at least $10,000) or "low" (less than $10,000). Unfortunately, the formulas taken by `facet_wrap()` and `facet_grid()` do not allow for vectorized expressions, so `facet_wrap(~ price/carat < 10000)` won't work.

The solution is to first create a new column of the data frame before plotting, as in `dd$price_carat_low <- dd$price/dd$carat < 10000`, to create a new logical column, and then facet on the newly created column with `facet_wrap(~ price_carat_low)`.

R also includes a function called `cut()` specifically for discretizing continuous vectors into factors. The first parameter is the vector to discretize, and the second parameter is either a number of (equal-sized) bins to break the input into, or a vector of breakpoints for the bins.

```
dd$price_per_carat_category <- cut(dd$price/dd$carat, breaks = 6)
p <- ggplot() +
    geom_point(data = dd,
               mapping = aes(x = carat,
                             y = depth,
                             color = price/carat)) +
    facet_wrap(~ price_per_carat_category)
```

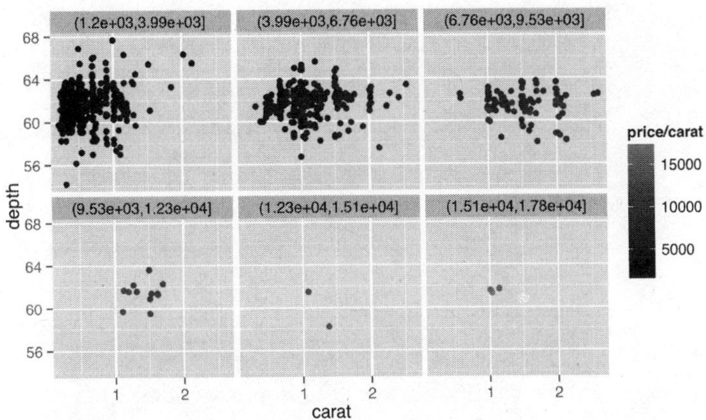

The `cut()` can take a `labels =` parameter as well, for cases when the default labels aren't sufficient.

Scales

Each aesthetic used in a plot is associated with a scale, whether it be the `x` and `y` or even `color` or `size`. For faceted plots, usually any scale adjustments are shared across facets, and for each layer in a plot, all scale properties must be shared as well. The different types of scales—continuous for `x` and `y`, color values for `color`, sizes for `size`—can all be modified to change the scale name, range, locations of breaks (or tick-marks) and labels of the breaks. There's more to say about scales than can be covered in this book, but we'll try to hit the high points.

Rather than continue plotting the diamonds data set, let's load up a data frame more biologically relevant. This table, stored in a file called **`contig_stats.txt`**, summarizes sequence statistics from a de novo genome assembly. Each row represents a "contig" from the assembly (a single sequence meant to represent a contiguous piece of the genome sequence produced from many overlapping sequenced pieces).

```
Contig:   ACGAGACGAGAGCGATACGATAGGACTAGACGACGGTAGACGGATACCAGATATTCGCAGAA
        { ACGAGACGAGAGCGATA          ACTAGACGACGGTAGAC   ATACCAGATATTCGCAGAA
Reads:  {           AGAGCGATACGATAGGAC          CGGTAGACGGATACC
        {                 CGATACGATAGGACTAG           AGACGGATACCAGATAT

Average_coverage: 1.93
  Consensus_length: 62
         gccontent: 0.5
```

Each contig has an Average_coverage value, representing the average number of sequence reads covering each base in the assembled contig. These values can vary widely, potentially because of duplicated regions of the genome that were erroneously assembled into a single contig. The Consensus_length column indicates the length of the contig (in base pairs), and the gccontent column shows the percentage of G or C bases of the contig.

```
ctg_stats <- read.table("contig_stats.txt",
                        header = TRUE,
                        stringsAsFactors = FALSE)

print(head(ctg_stats))
```

```
      ID Average_coverage Consensus_length gccontent
1 NODE_2       49011.9000              604  0.206954
2 NODE_3         319.4400              298  0.251678
3 NODE_4         194.6960             8141  0.148508
4 NODE_5         166.2010              338  0.248521
5 NODE_6       15913.8000             2184  0.459249
6 NODE_7          76.3737             5023  0.227752
```

Let's produce a dotplot for this data with Average_coverage on the x axis and Consensus_length on the y.

```
p <- ggplot() +
     geom_point(data = ctg_stats,
                mapping = aes(x = Average_coverage,
                              y = Consensus_length))
```

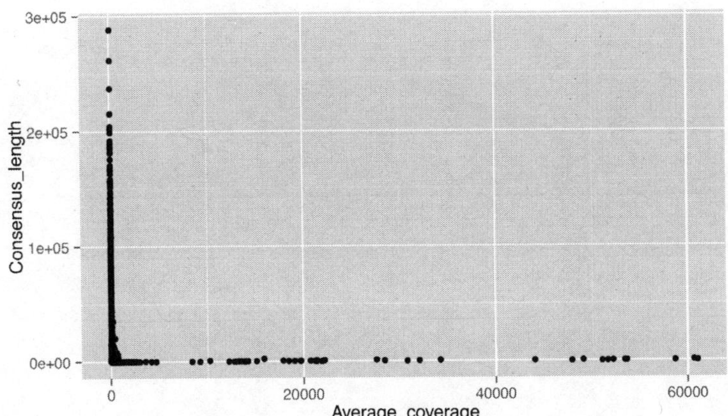

The result isn't very good. There are outliers on both axes, preventing us from seeing any trends in the data. Let's adjust the `x` scale by adding a `scale_x_continuous()`, setting `name =` to fix up the name of the scale, and using `limits =` to set the limits of the scale to `c(0, 1000)`. We'll also use the corresponding `scale_y_continuous()` to set a proper name for the *y* axis. (For discrete scales, such as in the boxplot example above, the corresponding functions are `scale_x_discrete()` and `scale_y_discrete()`.)

```
p <- ggplot() +
     geom_point(data = ctg_stats,
                mapping = aes(x = Average_coverage,
                              y = Consensus_length)) +
     scale_x_continuous(name = "Coverage", limits = c(0, 1000)) +
     scale_y_continuous(name = "Length")
```

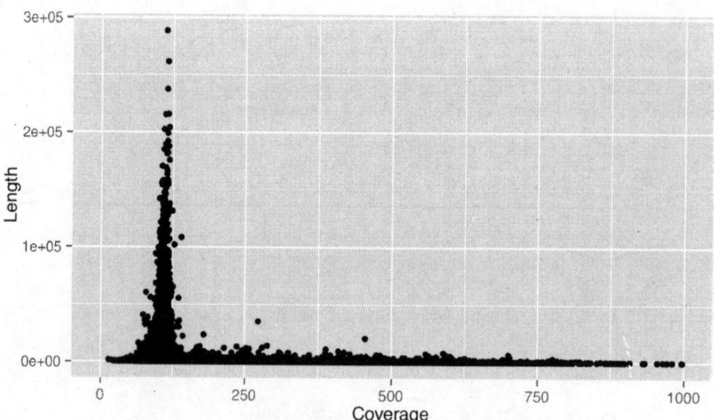

This is much better. It appears that most of the longer contigs have a coverage of about 100X. Sadly, this strategy leaves out many large data points to show the trends in the small majority. Instead, we should either log-adjust the data by plotting `aes(x = log(Average_coverage), y = Consensus_length)`, or log-adjust the scale. If we log-adjusted the data, we'd have to remember that the values are adjusted. It makes better sense to modify the scale itself, and plot the original data on the nonlinear scale. This can be done by supplying a `trans =` parameter to the scale, and specifying the name of a supported transformation function to apply to the scale as a single-element character vector. In fact, let's make both the `x` and `y` scales log-adjusted. While we're at it, we'll specify explicit break marks for the scales, as well as custom labels for those break marks.

```
p <- ggplot() +
     geom_point(data = ctg_stats,
                mapping = aes(x = Average_coverage,
                              y = Consensus_length)) +
     scale_x_continuous(name = "Coverage",
                        trans = "log10",
                        breaks = c(100, 1000, 10000, 100000),
                        labels = c("100", "1K", "10K", "100K")) +
     scale_y_continuous(name = "Length",
                        trans = "log10",
                        breaks = c(100, 1000, 10000, 100000),
                        labels = c("100", "1K", "10K", "100K"))
```

The result is below left, and it looks quite good. For a flourish, we can add an `annotation_logticks(base = 10)` to get logarithmically scaled tick-marks, shown below right.

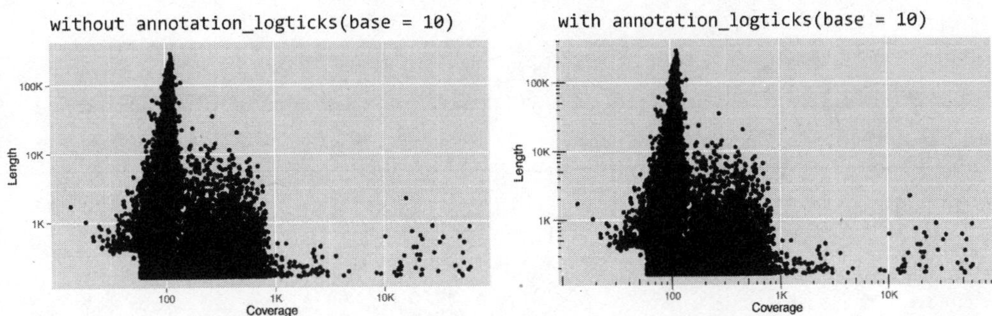

Other adjustment functions we could have used for the `trans =` parameter include `"log2"` or `"sqrt"`, though `"log10"` is a common choice with which most viewers will be familiar.

One of the issues left with our plot is that there are far too many data points to fit into this small space; this plot has an "overplotting" problem. There are a variety of solutions for overplotting, including random sampling, setting transparency of points, or using a different plot type altogether. We'll turn this plot into a type of heat map, where points near each other will be grouped into a single "cell," the color of which will represent some statistic of the points in that cell.[118]

There are two layer functions we can use for such a heat map, `stat_summary_2d()` and `stat_summary_hex()`. The former produces square cells, and the latter hexagons. (The `stat_summary_hex()` layer requires that we have the `"binhex"` package installed via `install.packages("binhex")` in the interactive console.) We'll use `stat_summary_hex()`, as it's a bit more fun. This layer function requires more than the usual number of parameters and aesthetics:

1. The `data =` data frame to plot (as usual).
2. The `mapping =` aesthetics to set via `aes()` (as usual), requiring `x`, the variable to bin by on the `x` axis; `y`, the variable to bin by on the `y` axis; and `z`, the variable to color cells by.
3. A `fun =` parameter, specifying the function to apply to the `z` values for each cell to determine

118. These functions can be used to create heat maps, but generally the rows and columns of a heat map are orderable in different ways. Many packages are available for plotting heat maps of various sorts in R, perhaps one of the more interesting is the `NeatMap` package, which is based on `ggplot2`.

the color.

This might be a bit confusing without an example, so let's replace the dotplot layer with a `stat_summary_hex()` layer plotting `x` and `y` the same way, but coloring cells by the mean `gccontent` of dots within that cell.

```
p <- ggplot() +
     stat_summary_hex(data = ctg_stats,
                      mapping = aes(x = Average_coverage,
                                    y = Consensus_length,
                                    z = gccontent),
                      fun = mean) +
     scale_x_continuous(name = "Coverage",
                        trans = "log10",
                        breaks = c(100, 1000, 10000, 100000),
                        labels = c("100", "1K", "10K", "100K")) +
     scale_y_continuous(name = "Length",
                        trans = "log10",
                        breaks = c(100, 1000, 10000, 100000),
                        labels = c("100", "1K", "10K", "100K")) +
     annotation_logticks(base = 10)
```

The result, below left, has cells colored by the `mean` function applied to all of the `gccontent` values (from the `z` aesthetic), but it doesn't reveal how many points are present in each cell. For this, we can use `fun = length`, which returns the number of elements in a vector rather than the mean, resulting in the plot below right.

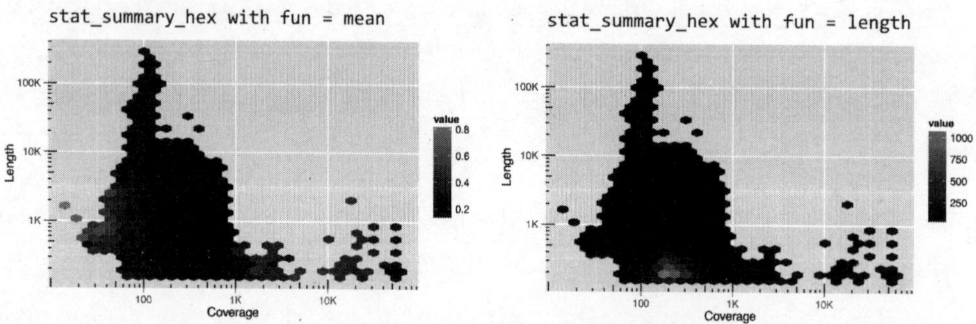

Like the `x` and `y` locations on the scales, the colors of the cells also exist on a scale. In this case, we can control it with the modifier `scale_fill_gradient()`. (Lines and other unfilled colors are controlled with `scale_color_gradient()`, and discrete color scales are controlled with `scale_fill_descrete()` and `scale_color_discrete()`.) This means that color scales can be named, transformed, and be given limits, breaks, and labels. Below, the string `"#BFBCFF"` specifies the light purple color at the top of the scale, based on the RGB color-coding scheme.

```
p <- ggplot() +
     stat_summary_hex(data = ctg_stats,
                      mapping = aes(x = Average_coverage,
                                    y = Consensus_length,
                                    z = gccontent),
                      fun = length) +
     scale_x_continuous(name = "Coverage",
                        trans = "log10",
                        breaks = c(100, 1000, 10000, 100000),
                        labels = c("100", "1K", "10K", "100K")) +
     scale_y_continuous(name = "Length",
                        trans = "log10",
                        breaks = c(100, 1000, 10000, 100000),
                        labels = c("100", "1K", "10K", "100K")) +
     scale_fill_gradient(name = "Count",
                         trans = "log10",
                         breaks = c(1, 10, 100, 1000),
                         low = "black",
                         high = "#BFBCFF") +
     annotation_logticks(base = 10)
```

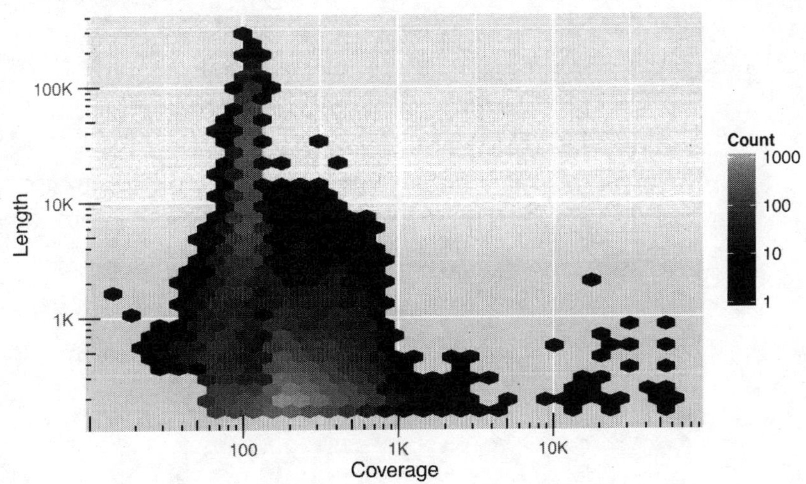

We've also included a `trans = "log10"` adjustment in this color scale, indicating that it can be transformed just as other continuous scales. Using a `log10` adjustment on a color scale may or may not be a good idea. For this data set, it more clearly illustrates the distribution of contigs in the plotted space, but makes comparisons between cells and to the legend more difficult.

The RGB Color System

RBG stands for red, green, blue. Computer monitors and other digital devices display colors by the varying intensities of tiny red, green, and blue light-emitting devices. A triplet of these makes up a single pixel in the display. The RGB scheme encodes a color as `#<RR><GG><BB>`, where `<RR>` encodes the amount of red light, `<GG>` the amount of green, and `<BB>` the amount of blue. These values can range from `00` (off) to `FF` (fully on); these are numbers encoded in *hexadecimal* format, meaning that after `9`, the next digit is `A`, then `B`, and so on, up to `F`. (Counting from `49`, for example, the next numbers are `4A`, `4B`, . . . `4F`, `50`, `51`, etc.) Why red, green, and blue? Most humans have three types of cone cells in their eyes, and each is most responsive to either red, green, or blue light! An RGB scheme can thus represent nearly all colors that humans can see, though in the end we are limited by the gradations (`#<RR><GG><BB>` format can only take about 16.7 million different values) and the quality of the light-emitting devices (many of which can't produce very bright light or be fully turned off in operation). Color blindness is a genetic anomaly caused by having only one or two of the three cone types, and a few rare but lucky individuals possess four cone types along with greater acuity in color vision (tetrachromacy).

Something to consider when devising color scales is that not all of them are created equally—a fair percentage of viewers will have some form of color blindness, and another fair percentage of viewers will prefer to print a plot on a black-and-white printer. The `scale_color_brewer()` function helps the user select good color palettes; it is based on the work found at colorbrewer2.org. Other scale types can be similarly adjusted, including alpha (transparency), and the sizes of points and lines.

Coordinates

In addition to modifying properties of the scales, we can also modify how those scales are interpreted in the overall plot and in relation to each other. Some of the coordinate modifications are less common, but others (like `coord_equal()`, below) are handy. Often, coordinate adjustments are illustrated by considering a dotplot or barplot in polar coordinates.

```
p <- ggplot() +
     geom_point(data = diamonds,
                mapping = aes(x = carat,
                              y = price,
                              color = cut)) +
     coord_polar(theta = "x")
```

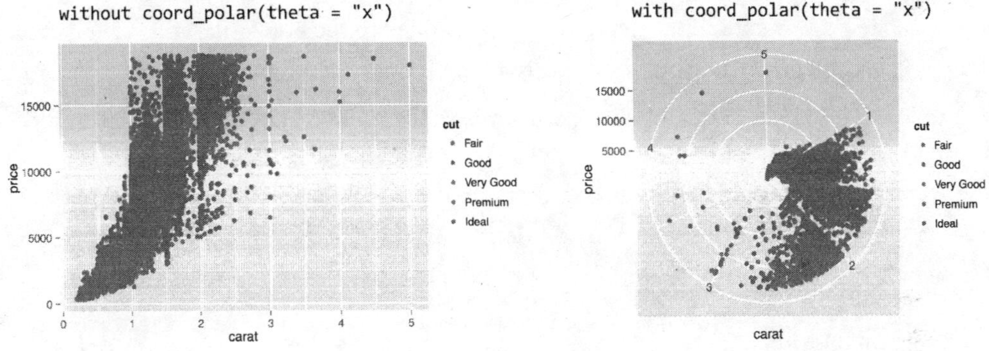

The `coord_polar()` function requires a `theta =` parameter to indicate which axis (`x` or `y`) should be mapped to the rotation angle; the remaining axis is mapped to the radius. This coordinate modification can be used to produce interesting plots, especially when the geom used is `"bar"` or `"line"`.

The `coord_flip()` function can be used to flip the `x` and `y` axes. This feature is especially useful when the desire is to produce a horizontal histogram or boxplot.

```
p <- ggplot() +
     geom_bar(data = diamonds,
              mapping = aes(x = cut, fill = clarity)) +
     coord_flip()
```

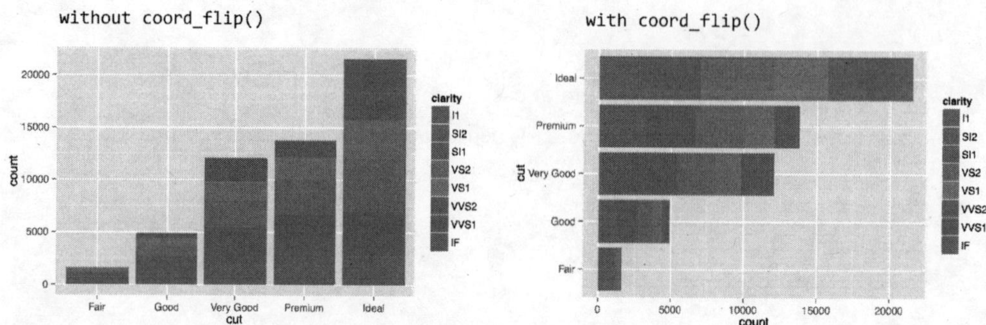

When setting the `fill` aesthetic, the subbars are stacked, which is the default. For a plot with bars presented side by side, one can add the `position = "dodge"` argument to the `geom_bar()` layer call.

For one last coordinate adjustment, the values mapped to the horizontal and vertical axes are sometimes directly comparable, but the range is different. In our random subsample of diamonds `dd`, for example, the `x` and `z` columns are both measured in millimeters (representing the "width" and "height" of the diamond), but the `z` values are generally smaller. For illustration, a single unit on the horizontal axis should be the same size as a single unit on the vertical axis, but `ggplot2` doesn't ensure such sizing by default. We can specify the size by adding a `coord_equal()` coordinate adjustment.

```
p <- ggplot() +
     geom_point(data = dd,
                mapping = aes(x = x, y = z)) +
     coord_equal()
```

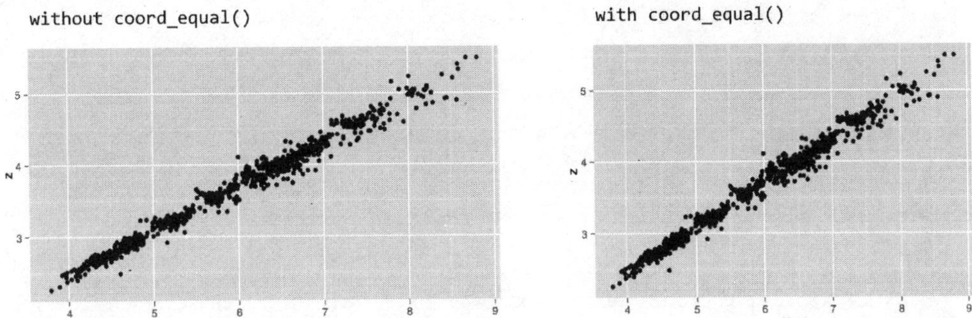

Notice that without `coord_equal()` in the above left, the axes sizes aren't quite comparable, but the grid marks are perfect squares in the above right. The `coord_equal()` adjustment can also be made with log-adjusted plots, and can sometimes produce nicer output even if the two axes aren't in the same units. Here's a comparison with and without `coord_equal()` applied to the final version of the contig length-versus-coverage plot above.

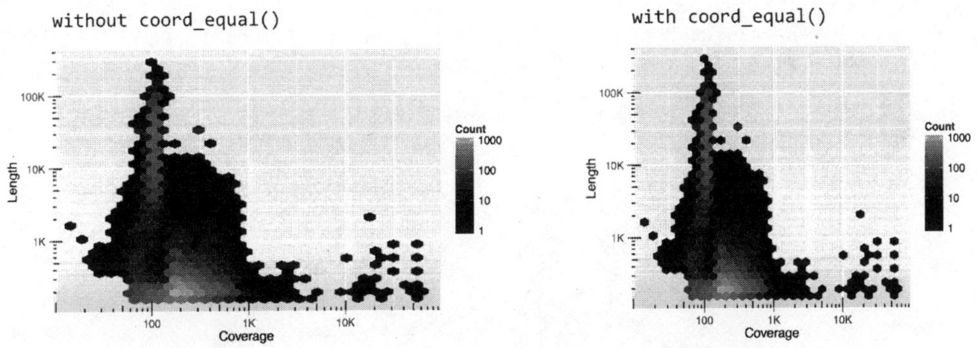

As always, there are a number of other coordinate adjustments that are possible, and we've only covered some of the most useful or interesting. See http://docs.ggplot2.org for the full documentation.

Theming, Annotations, Text, and Jitter

So far, we've covered customizing how data are plotted and axes/coordinates are represented, but we haven't touched on "ancillary" properties of the plot like titles, background colors, and font sizes. These are part of the "theme" of the plot, and many aspects of the theme are adjusted by the `theme()` function. The exception is the addition of a title to a plot, which is accomplished with the `ggtitle()` function.

The text-based parts of a plot are organized hierarchically (see the documentation for `theme()` for the full list). For example, modifying the `text =` parameter will modify all text elements, while modifying `axis.text =` adjusts the tick labels along both axes, and `axis.text.x =` specifies properties of only the *x*-axis tick labels. Other text-theme elements include `legend.text`, `axis.title` (for axis names), `plot.title`, and `strip.text` (for facet labels).

To adjust properties of these elements, we use a call to `element_text()` within the call to `theme()`. We can produce a quick plot counting diamonds by their cut and clarity, for example, setting a plot title and changing the overall text size to `16`, and just the title size to `20`. Shrinking text can be especially helpful for cases when facet labels or theme labels are too large to fit their respective boxes.

```
p <- ggplot() +
     geom_bar(data = dd, mapping = aes(x = clarity, fill = cut)) +
     ggtitle("Diamond Counts by Clarity and Cut") +
     theme(text = element_text(size = 16)) +
     theme(plot.title = element_text(size = 20))
```

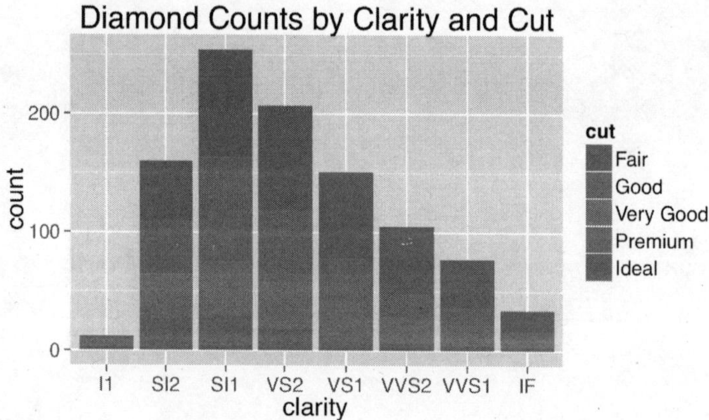

The labels used for breaks are sometimes longer than will comfortably fit on the axes. Aside from changing their size, it sometimes helps to angle them by 30 or 45 degrees. When doing so, it also looks best to set `hjust = 1` to right-justify the labels.

```
p <- ggplot() +
     geom_bar(data = dd, mapping = aes(x = clarity, fill = cut)) +
     ggtitle("Diamond Counts by Clarity and Cut") +
     theme(text = element_text(size = 16)) +
     theme(plot.title = element_text(size = 20)) +
     theme(axis.text.x = element_text(angle = 45, hjust = 1))
```

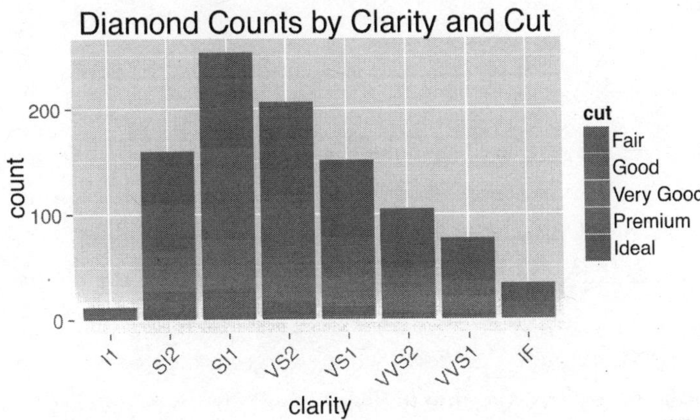

As this example might suggest, theming is itself a complex topic within `ggplot2`, and there are many options and special internal functions modifying nearly all aspects of a plot.

Although `ggsave()` accepts `width =` and `height =` parameters specifying the overall size of the output file, because these parameters include the legend and axis labels, the plotting region has its aspect ratio determined automatically. Indicating that the plotting region should take a specific aspect ratio (defined as the height of the region over the width) also occurs within a `theme()` call.

```
p <- ggplot() +
     geom_bar(data = dd, mapping = aes(x = clarity, fill = cut)) +
     ggtitle("Diamond Counts by Clarity and Cut") +
     theme(text = element_text(size = 16)) +
     theme(plot.title = element_text(size = 20)) +
     theme(axis.text.x = element_text(angle = 45, hjust = 1)) +
     theme(aspect.ratio = 1.5)
```

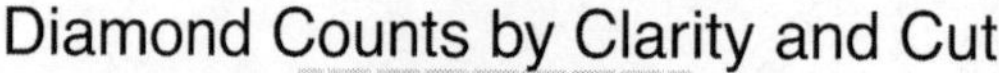

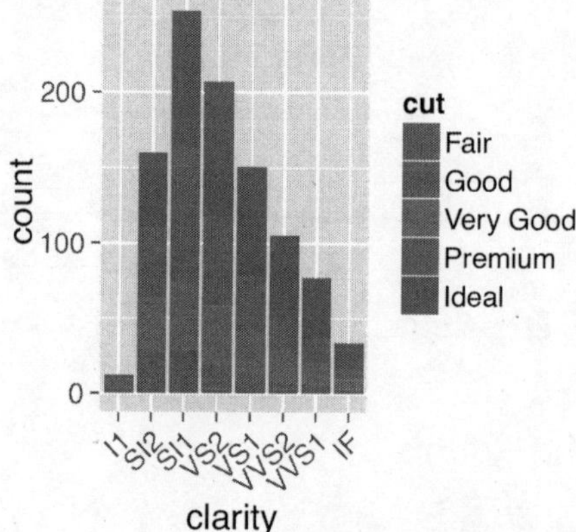

The observant reader might have noticed that, by default, all plotting regions in `ggplot2` use a light gray background. This is intentional: the idea is that a plot with a white background, when embedded into a manuscript, will leave a visual "hole," interrupting the flow of the text. The shade of gray chosen for the default background is meant to blend in with the overall shade of a column of text.

Some users prefer to use a more traditional white background, but doing so requires adjusting multiple elements, including the background itself, grid lines, and so on. So, `ggplot2` includes a number of functions that can change the overall theme, such as `theme_bw()`.

```
p <- ggplot() +
     geom_point(data = dd, mapping = aes(x = price,
                                         y = carat, color = clarity)) +
     facet_wrap(~ cut) +
     theme_bw()
```

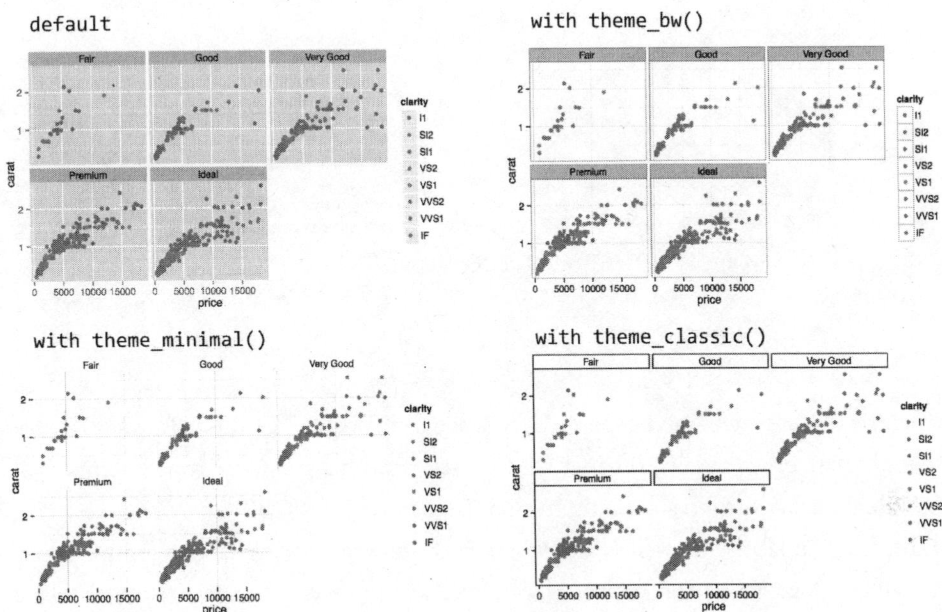

Because calls to `theme_bw()` et al. modify all theme elements, if we wish to also modify individual theme elements with `theme()`, those must be added to the chain after the call to `theme_bw()`.

One feature of `ggplot2` not yet covered is the use of text within plots, which are not theme adjustments but rather special types of plot layers. The `geom_text()` layer function makes it easy to create "dotplots" where each point is represented by a text label rather than a point. Here's an example plotting the first 30 diamonds by carat and price, labeled by their cut.

```
p <- ggplot() +
     geom_text(data = diamonds[1:30, ],
               mapping = aes(x = carat, y = price, label = cut))
```

In the result (below left), it's difficult to see that multiple diamonds are plotted in the same location in some cases. This type of overplotting can happen with points as well; adding a `position = "jitter"` option to the `geom_text()` layer slightly modifies the location of all the geoms so that they stand out (below right).

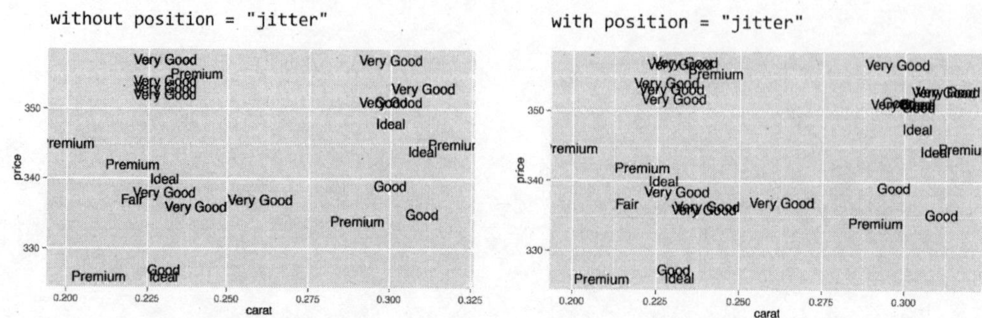

Various aesthetics of the text can be mapped to values of the data, including `size`, `angle`, `color`, and `alpha`. As with other layers, to change the font size (or other property) for all points to a constant value, the instruction should be given outside of the `aes()` call.

Individual text labels—as well as individual line segments, rectangles, points, and other geoms—can be added with an `annotate()` layer. Such a layer takes as its first argument the name of the geom that will be added, and subsequently any aesthetics that should be set for that geom (without a call to `aes()`). Here's an illustration, finishing out the previous length/coverage plot example. (The `hjust = 0` in the text annotation indicates that the text should be left-justified with respect to the reference `x` and `y`.)

```
p <- ggplot() +
  stat_summary_hex(data = ctg_stats,
                   mapping = aes(x = Average_coverage,
                                 y = Consensus_length,
                                 z = gccontent),
                   fun = length) +
  scale_x_continuous(name = "Coverage",
                     trans = "log10",
                     breaks = c(100, 1000, 10000, 100000),
                     labels = c("100", "1K", "10K", "100K")) +
  scale_y_continuous(name = "Length",
                     trans = "log10",
                     breaks = c(100, 1000, 10000, 100000),
                     labels = c("100", "1K", "10K", "100K")) +
  scale_fill_gradient(name = "Count",
                      trans = "log10",
                      breaks = c(1, 10, 100, 1000),
                      low = "black",
                      high = "#BFBCFF") +
  annotation_logticks(base = 10) +
  coord_equal() +
  theme_bw() +
  annotate("rect", xmin = 80, xmax = 150,
                   ymin = 1000, ymax = 400000,
                   alpha = 0.2, fill = "red") +
  annotate("text", x = 200, y = 100000,
                   color = "darkred",
                   label = "Main Assembly Contigs", hjust = 0)
```

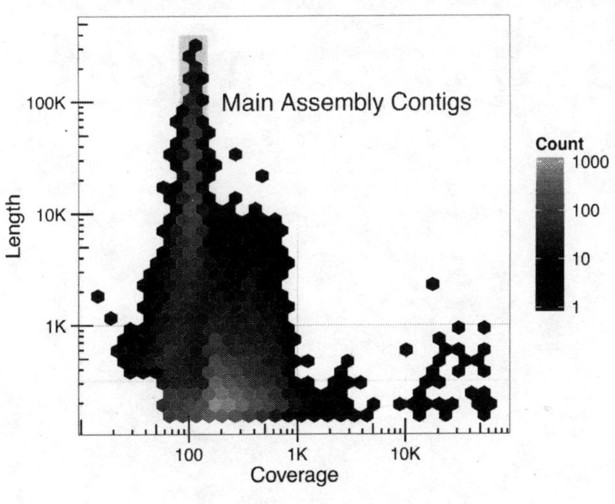

Ideally, we would strive to produce publication-ready graphics with well-documented and easily editable code. In some cases, however, adjustments or annotations are more easily added in graphical editing programs like Adobe Photoshop or Illustrator (or open-source alternatives like Inkscape), so long as the interpretation and meaning of the data are not altered.

Exercises

1. Use `ggplot2` to explore the **`trio.sample.vcf`** we analyzed in previous chapters. Can you effectively visualize the distribution of SNP locations across the chromosomes?
2. Running the `data()` function (with no parameters) in the interactive R console will list the built-in data sets that are available. For example, `USArrests` describes arrest statistics in US states in 1973, with columns for per-capita rates of murder, rape, and assault, and also one for percentage of state residents in urban areas (see `help(USArrests)` for details).

   First, see if you can reproduce this plot, where there is a facet for each state, and a bar for each crime type:

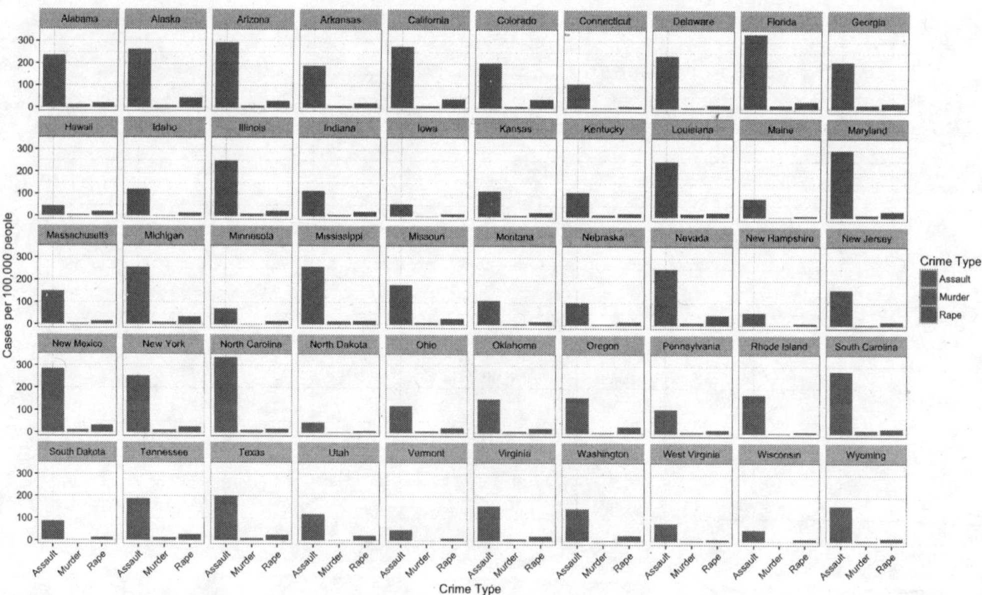

   You might need to first manipulate the data by creating a column for state names (rather than using the row names) and "gathering" some of the columns with `tidyr`.

Next, see if you can reproduce the plot, but in such a way that the panels are ordered by the overall crime rate (Murder + Rape + Assault).

3. Find at least one other way to illustrate the same data. Then, find a way to incorporate each state's percentage of population in urban areas in the visualization.
4. Generate a data set by using observation or experimentation, and visualize it!

Index

Symbols

A

B

G

T

About the Author

Shawn T. O'Neil earned a BS in computer science from Northern Michigan University, and later an MS and PhD in the same subject from the University of Notre Dame. His past and current research focuses on bioinformatics. O'Neil has developed and taught several courses in computational biology at both Notre Dame and Oregon State University, where he currently works at the Center for Genome Research and Biocomputing.